Microbial Enzymes and Biotransformations

METHODS IN BIOTECHNOLOGY™

John M. Walker, SERIES EDITOR

METHODS IN BIOTECHNOLOGY™

Microbial Enzymes and Biotransformations

Edited by

José Luis Barredo

R & D Biology, Antibióticos S. A.,
León, Spain

HUMANA PRESS ✸ TOTOWA, NEW JERSEY

© 2005 Humana Press Inc.
999 Riverview Drive, Suite 208
Totowa, New Jersey 07512

www.humanapress.com

This publication is printed on acid-free paper. ∞
ANSI Z39.48-1984 (American Standards Institute)

Permanence of Paper for Printed Library Materials.

Cover illustration: Figure 6, from Chapter 2, "Enzyme Biosensors," by Steven J. Setford and Jeffrey D. Newman

Cover design by Patricia F. Cleary.

For additional copies, pricing for bulk purchases, and/or information about other Humana titles, contact Humana at the above address or at any of the following numbers: Tel.: 973-256-1699; Fax: 973-256-8341; E-mail: humana@humanapr.com; or visit our website: www.humanapress.com

Printed in the United States of America. 10 9 8 7 6 5 4 3 2 1

Library of Congress Cataloging in Publication Data

Microbial enzymes and biotransformations / edited by José Luis Barredo.
    p. cm.—(Methods in biotechnology)
  Includes bibliographical references and index.
  ISBN 1-58829-253-3 (alk. paper) E-ISBN 1-59259-846-3
  1. Microbial enzymes—Biotechnology—Laboratory manuals. 2. Microbial biotechnology—
Laboratory manuals. 3. Biotransformation (Metabolism)—Laboratory manuals. I. Barredo, José Luis.
II. Series.
  TP248.65.E59M54 2004
  660.6'2–dc22   2004054093

Preface

The two lead chapters of *Microbial Enzymes and Biotransformations* represent overviews of microorganisms as a source of metabolic and enzymatic diversity, and of the fast-developing field of enzyme biosensors. The remaining chapters show comprehensive experimental methods for improving enzyme function by directed evolution, and for manufacturing enzymes used worldwide for human health, nutrition, and environmental protection, including L-glutaminase, D-hydantoinase/D-carbamoylase, fructosyltransferase, food-grade hydrolytic enzymes, phenylalanine dehydrogenase, alkaline enzymes for detergents, conventional and high salinity stable proteases, pectinases, phytases, glucose dehydrogenase, and acetate kinase. Finally, methods for covalently immobilized enzymes, penicillin G acylase, and procedures to microencapsulate enzyme and cells are described.

Microbial Enzymes and Biotransformations has been written by outstanding experts in the field and provides a highly useful reference source for laboratory and industrial professionals, as well as for graduate students in a number of biological disciplines (biotechnology, microbiology, genetics, molecular biology) because of the uncommonly wide applicability of the procedures across the range of areas covered.

I am indebted to the authors who, in spite of their professional activities, agreed to participate in this book, to Dr. John Walker, Series Editor, for his encouragement and advice in reviewing the manuscripts, and to the rest of the staff of The Humana Press for their assistance in assembling this volume and their efforts in keeping this project on schedule. Last but not least, I warmly acknowledge my wife Natalia and our children Diego, José-Luis, Álvaro, and Gonzalo for their patience and support.

José Luis Barredo

Contents

Contributors

OLGA ABIAN • *Departamento de Biocatalysis, Instituto de Catálisis-CSIC, Madrid, Spain*

SIRIN A. I. ADHAM • *Área de Microbiología, Facultad de Ciencias Biológicas y Ambientales, Universidad de León, Spain*

JOSÉ L. ADRIO • *Department of Biotechnology, Puleva Biotech, Granada, Spain*

YASUHISA ASANO • *Biotechnology Research Center, Toyama Prefectural University, Toyama, Japan*

OSKAR BAÑUELOS • *Department of Biotechnology, Puleva Biotech, Granada, Spain*

JOSÉ LUIS BARREDO • *R&D Biology, Antibióticos S. A., León, Spain*

SUSANNE BRAKMANN • *Applied Molecular Evolution, Institute of Biology II, University of Leipzig, Germany*

THOMAS M. S. CHANG • *Artificial Cells and Organs Research Centre, Departments of Physiology, Medicine and Biomedical Engineering, McGill University, Montreal, Canada*

ARNOLD L. DEMAIN • *Charles A. Dana Research Institute for Scientists Emeriti (R.I.S.E.), Drew University, Madison, NJ*

ROBERTO FERNÁNDEZ-LAFUENTE • *Departamento de Biocatalysis, Instituto de Catálisis-CSIC, Madrid, Spain*

GLORIA FERNÁNDEZ-LORENTE • *Departamento de Biocatalysis, Instituto de Catálisis-CSIC, Madrid, Spain*

JAMES G. FERRY • *Department of Biochemistry and Molecular Biology, Center for Microbial Structural Biology, Penn State University, University Park, PA*

MANUEL FUENTES • *Departamento de Biocatalysis, Instituto de Catálisis-CSIC, Madrid, Spain*

JOSÉ A. GIL • *Área de Microbiología, Facultad de Ciencias Biológicas y Ambientales, Universidad de León, Spain*

VALERIA GRAZU • *Departamento de Biocatalysis, Instituto de Catálisis-CSIC, Madrid, Spain*

JOSÉ M. GUISÁN • *Departamento de Biocatalysis, Instituto de Catálisis-CSIC, Madrid, Spain*

YUJI HATADA • *Japan Marine Science and Technology Center, Yokosuka, Japan*

GURINDER S. HOONDAL • *Department of Microbiology, Panjab University, Chandigarh, India*

KOKI HORIKOSHI • *Japan Marine Science and Technology Center, Yokosuka, Japan*

WERNER HUMMEL • *Institute of Molecular Enzyme Technology, Heinrich-Heine-University Düsseldorf, Research Centre Jülich, Germany*

SUSUMU ITO • *Japan Marine Science and Technology Center, Yokosuka, Japan*

PRABHA IYER • *Department of Biochemistry and Molecular Biology, Center for Microbial Structural Biology, Penn State University, University Park, PA*

TAEWAN KIM • *Department of Animal Science, Cornell University, Ithaca, NY*

TOHRU KOBAYASHI • *Tochigi Research Laboratories, Kao Corporation, Tochigi, Japan*

XIN GEN LEI • *Department of Animal Science, Cornell University, Ithaca, NY*

MICHAL LETEK • *Área de Microbiología, Facultad de Ciencias Biológicas y Ambientales, Universidad de León, Spain*

CESAR MATEO • *Departamento de Biocatalysis, Instituto de Catálisis-CSIC, Madrid, Spain*

LUIS M. MATEOS • *Área de Microbiología, Facultad de Ciencias Biológicas y Ambientales, Universidad de León, Spain*

ENCARNACIÓN MELLADO • *Departamento de Microbiología y Parasitología, Facultad de Farmacia, Universidad de Sevilla, Spain*

K. MADHAVAN NAMPOOTHIRI • *Biotechnology Division, Regional Research Laboratory, Trivandrum, India*

HEMRAJ S. NANDANWAR • *Institute of Microbial Technology, Chandigarh, India*

JEFFREY D. NEWMAN • *Institute of BioScience and Technology, Cranfield University, Silsoe, Bedfordshire, UK*

SCOTT J. NOVICK • *BioCatalytics Inc., Pasadena, CA*

JOSÉ M. PALOMO • *Departamento de Biocatalysis, Instituto de Catálisis-CSIC, Madrid, Spain*

ASHOK PANDEY • *Biotechnology Division, Regional Research Laboratory, Trivandrum, India*

ANGELINA RAMOS • *Área de Microbiología, Facultad de Ciencias Biológicas y Ambientales, Universidad de León, Spain*

MARIA DEL CARMEN RONCHEL • *Department of Biotechnology, Puleva Biotech, Granada, Spain*

J. DAVID ROZZELL • *BioCatalytics Inc., Pasadena, CA*

ABDULHAMEED SABU • *Biotechnology Division, Regional Research Laboratory, Trivandrum, India*

CRISTINA SÁNCHEZ-PORRO • *Departamento de Microbiología y Parasitología, Facultad de Farmacia, Universidad de Sevilla, Spain*

CHANDRAN SANDHYA • *Biotechnology Division, Regional Research Laboratory, Trivandrum, India*

STEVEN J. SETFORD • *Institute of BioScience and Technology, Cranfield University, Silsoe, Bedfordshire, UK*

SASCHA NICO STUMPP • *Applied Molecular Evolution, Institute of Biology II, University of Leipzig, Germany*

RUPINDER TEWARI • *Department of Biotechnology, Panjab University, Chandigarh, India*

RAM P. TEWARI • *Department of Microbiology, Panjab University, Chandigarh, India*

NOELIA VALBUENA • *Área de Microbiología, Facultad de Ciencias Biológicas y Ambientales, Universidad de León, Spain*

JAVIER VELASCO • *Department of Biotechnology, Puleva Biotech, Granada, Spain*

ANTONIO VENTOSA • *Departamento de Microbiología y Parasitología, Facultad de Farmacia, Universidad de Sevilla, Spain*

RAKESH M. VOHRA • *Institute of Microbial Technology, Chandigarh, India*

ANDREA WECKBECKER • *Institute of Molecular Enzyme Technology, Heinrich-Heine-University Düsseldorf, Research Centre Jülich, Germany*

1

Microbial Cells and Enzymes

A Century of Progress

José L. Adrio and Arnold L. Demain

Summary

Over the last century, microorganisms have been a great source of metabolic and enzymatic diversity. In recent years, emerging recombinant DNA techniques have facilitated the development of new efficient expression systems, modification of biosynthesis pathways leading to different metabolites by metabolic engineering, and enhancement of catalytic properties of enzymes by directed evolution. More exciting advances are still to come as the complete sequencing of industrially important microbial genomes takes place. Functional genomics and proteomics are already major tools used in the search for new molecules and development of higher-producing strains.

Key Words: Primary metabolites; secondary metabolites; bioconversions; enzymes; hosts; biopharmaceuticals; metabolic engineering; agriculture; polymers.

1. Introduction

Traditional industrial microbiology has existed for thousands of years, preserving milk and vegetables and yielding products such as bread, beer, wine, distilled spirits, vinegar, cheese, pickles, and other fermented materials to satisfy our palates. A second major phase of industrial microbiology began in the early 20th century. This golden era of industrial fermentation featured the first large-scale fermentations devoted to manufacture of solvents, organic acids, vitamins, enzymes, and other products. In mid-century, antibiotic fermentations came onto the scene with the development of processes for the production of penicillin and streptomycin. Investigations on scale-up of these two antibiotic processes, a joint effort between Merck & Co. and Princeton University, as well as work done at Columbia University, gave birth to a new field of biochemical (microbiological) engineering. On the heels of this major breakthrough was the academic development of mutational microbial genetics, which was developed into a new technology of strain improvement.

From: *Methods in Biotechnology, Vol. 17: Microbial Enzymes and Biotransformations*
Edited by: J. L. Barredo © Humana Press Inc., Totowa, NJ

The revolutionary era of recombinant DNA technology, i.e., modern biotechnology, began in the early 1970s after many years of discoveries in basic genetics. The recombinant DNA developments made in 1972 in the laboratories of Stanford University and the University of California at San Francisco propelled the field to new heights and led to the establishment of a new biotechnology industry in the United States and around the world. Biotechnology is having a major effect on health care, diagnostics, agriculture, and production of chemicals, and promises to make inroads in the practices of other industries such as petroleum, mining, foods, and the environment.

2. Traditional Microbial Biotechnology

Microorganisms are important to us for many reasons, but one of the principal ones is that they produce things of value. The main reasons for the use of microorganisms to produce compounds that can otherwise be isolated from plants and animals or synthesized by chemists are: (1) a high ratio of surface area to volume, which facilitates the rapid uptake of nutrients required to support high rates of metabolism and biosynthesis; (2) a wide variety of reactions that microorganisms are capable of carrying out; (3) the facility to adapt to a large array of different environments, allowing a culture to be transplanted from nature to the laboratory flask or the factory fermenter, where it is capable of growing on inexpensive carbon and nitrogen sources and producing valuable compounds; (4) ease of genetic manipulation, both in vivo and in vitro, to increase production up to thousands-fold, to modify structures and activities, and to make entirely new products; (5) simplicity of screening procedures, allowing thousands of cultures to be examined in a reasonably short time; and (6) a wide diversity, in which different species produce somewhat different enzymes catalyzing the same reaction, allowing one flexibility with respect to operating conditions in the reactor.

Microbial products may be very large materials such as proteins, nucleic acids, carbohydrate polymers, or even cells, or they can be smaller molecules, which we usually divide into metabolites essential for vegetative growth and those that are inessential, i.e., primary and secondary metabolites, respectively.

2.1. Primary Metabolites

Primary metabolites are the small molecules of all living cells that are intermediates or end products of the pathways of intermediary metabolism, are building blocks for essential macromolecules, or are converted into coenzymes.

2.1.1. Amino Acids

About 1.6 billion pounds of monosodium glutamate are made annually by fermentation using various species of the genera *Corynebacterium* and

Brevibacterium. Molar yields of glutamate from sugar are 50–60% and broth concentrations reach over 100 g/L. Normally, glutamic acid overproduction would not occur because of feedback regulation. However, due to a modification of the cell membrane, glutamate is pumped out of the cell, thus allowing its biosynthesis to proceed unabated.

Since the bulk of the cereals consumed in the world are deficient in L-lysine, this essential amino acid became an important industrial product. The lysine biosynthetic pathway is controlled very tightly in an organism like *Escherichia coli,* which contains three aspartate kinases, each of which is regulated by a different end product (lysine, threonine, methionine). In addition, after each branch point, the initial enzymes are inhibited by their respective end products. However, in lysine-producing organisms (e.g., mutants of *Corynebacterium glutamicum* and its relatives), there is only a single aspartate kinase, which is regulated via concerted feedback inhibition by threonine plus lysine. By genetic removal of homoserine dehydrogenase, a glutamate-producing wild-type *Corynebacterium* is converted into a lysine-overproducing mutant that cannot grow unless methionine and threonine are added to the medium.

In the nutritional area, the methionine content of the methylotrophic yeast, *Candida boidinii,* has been markedly increased (from 0.5 to 9 mg/g dry cell weight in the intracellular pool and from 6 to 16 mg/g as total cellular methionine) by mutation to ethionine resistance *(1).*

World markets for amino acids amount to $915 million for L-glutamate, $600 million for L-lysine, $198 million for L-phenylalanine, and $43 million for L-aspartate.

2.1.2. Nucleotides

Commercial interest in nucleotide fermentations developed due to the activity of two purine ribonucleoside 5′-monophosphates, namely guanylic acid (GMP) and inosinic acid (IMP), as enhancers of flavor. Techniques similar to those used for amino acid fermentations have yielded IMP titers of 27 g/L. Twenty-five hundred tons of GMP and IMP are produced annually in Japan, with a market of $350 million.

2.1.3. Vitamins

Riboflavin (vitamin B_2) overproducers include two yeastlike molds, *Eremothecium ashbyii* and *Ashbya gossypii,* which synthesize riboflavin in concentrations greater than 20 g/L. New processes using *Candida* sp. have been developed in recent years that produce up to 30 g riboflavin per liter.

Vitamin B_{12} is industrially produced by *Propionibacterium shermanii* or *Pseudomonas denitrificans.* The key to the fermentation is avoidance of feedback repression by vitamin B_{12}. The early stage of the *P. shermanii* fermentation is conducted under anaerobic conditions in the absence of the precursor 5,6-dimethylbenzimidazole. These conditions prevent vitamin B_{12} synthesis and allow for the accumulation of the intermediate, cobinamide. Then the culture is aerated and dimethylbenzimidazole is added, converting cobinamide to the vitamin. In the *P. denitrificans* fermentation, the entire process is carried out under low oxygen. A high level of oxygen results in an oxidizing intracellular environment that represses formation of the early enzymes of the pathway. Production of vitamin B_{12} has reached levels of 150 mg/L and a world market of $71 million.

Carotenoids are a family of yellow to orange-red terpenoid pigments synthesized by photosynthetic organisms and by many bacteria and fungi. They have beneficial health effects and also are desirable commercial products used as additives and colorants in the food industry. Traditionally, carotenoids were obtained by extraction from plants or by direct chemical synthesis. Only a few of the more than 600 identified carotenoids are produced industrially, with β-carotene (a popular additive for butter, ice cream, orange juice, candies, etc.) the most prominent. The green alga *Dunaliella salina* and the fungus *Blakeslea trispora* are the microorganisms of choice for industrial production of β-carotene *(2)*.

2.1.4. Organic Acids

Filamentous fungi have become widely used for the commercial production of organic acids. About 1 billion pounds of citric acid are produced by *Aspergillus niger* per year, with a market of $1.4 billion. The commercial process employs *A. niger* in media deficient in iron and manganese. A high level of citric-acid production is also associated with a high intracellular concentration of fructose 2,6-biphosphate, an activator of glycolysis. Other factors contributing to high citric-acid production are the inhibition of isocitrate dehydrogenase by citric acid, and the low pH optimum (1.7–2.0). In approximately 4–5 d, the major portion (80%) of the sugar provided is converted to citric acid, with titers reaching about 100 g/L. Alternative processes have been developed for the production of citric acid by *Candida* yeasts, especially from hydrocarbons. Such yeasts are able to convert *n*-paraffins to citric and isocitric acids in extremely high yields (150–170% on a weight basis). Titers as high as 225 g/L have been reached.

2.1.5. Ethanol

Ethyl alcohol is produced via fermentation of sugars (or a polysaccharide that can be depolymerized to a fermentable sugar) by *Saccharomyces cerevisiae* in the case of hexoses, and *Kluyveromyces fragilis* or *Candida* species with lactose or a pentose, respectively. Under optimum conditions, approx 10–12% ethanol by volume is obtained within 5 d. Such a high concentration slows down growth and the fermentation ceases. With special yeasts, the fermentation can be continued to alcohol concentrations of 20% by volume, but these concentrations are attained only after months or years of fermentation. At present, all beverage alcohol is made by fermentation.

Fuel ethanol produced from biomass would provide relief from air pollution caused by use of gasoline and would not contribute to the greenhouse effect. The primary difficulty for commercialization of fermentation ethanol as a complete gasoline replacement is its relatively high cost of production. However, several chemical engineering and biological improvements at different steps of this process have been achieved recently. A new countercurrent flow-through process for pretreatment of the biomass yielding up to 82% of hydrolysis of cellulose and near-total depolymerization of xylan has been reported *(3)*. Recombinant DNA technology has been used to convert *E. coli* and its close relatives into efficient producers of ethanol (*see* **Subheading 3.4.1.**). Furthermore, bacteria such as clostridia and *Zymomonas* are being reexamined for ethanol production after years of neglect. *Clostridium thermocellum,* an anaerobic thermophile, can convert waste cellulose (i.e., biomass) and crystalline cellulose directly to ethanol.

2.2. Secondary Metabolites

Microbial secondary metabolites are extremely important to our health and nutrition *(4)*. As a group that includes antibiotics, other medicinals, toxins, pesticides, and animal and plant growth factors, they have tremendous economic importance. Secondary metabolites are usually produced after growth has slowed down, have no function in growth of the producing cultures, are produced by certain restricted taxonomic groups of organisms, and are usually formed as mixtures of closely related members of a chemical family. In nature, secondary metabolites are important for the organisms that produce them, functioning as (1) sex hormones, (2) ionophores, (3) competitive weapons against other bacteria, fungi, amoebae, insects, and plants, (4) agents of symbiosis, and (5) effectors of differentiation *(5)*.

2.2.1. Antibiotics

Antibiotics are the best-known secondary metabolites. They are defined as low-molecular-weight organic natural products made by microorganisms that are active at low concentration against other microorganisms. Of the 12,000 antibiotics known in 1995, 55% were produced by the genus *Streptomyces,* 11% from other actinomycetes, 12% from nonfilamentous bacteria, and 22% by filamentous fungi *(6)*. The search for new antibiotics continues in order to combat evolving pathogens, naturally resistant bacteria and fungi, and previously susceptible microbes that have developed resistance; to improve pharmacological properties; to combat tumors, viruses, and parasites; and to discover safer, more potent, and broader-spectrum compounds. The antibiotic market includes about 160 antibiotics and derivatives such as the β-lactam peptide antibiotics and others *(6)*. The global market for finished antibiotics reached $35 billion in 2000.

2.2.2. Immunosuppressive Agents

Cyclosporin A was originally discovered as a narrow-spectrum antifungal peptide produced by the mold *Tolypocladium nivenum* (previously *Tolypocladium inflatum*). Discovery of immunosuppressive activity led to its use in heart, liver, and kidney transplants and to the overwhelming success of the organ transplant field. Sales of cyclosporin A have reached $1.5 billion. Other important transplant agents include sirolimus (rapamycin) and tacrolimus (FK506), which are produced by actinomycetes. A very old broad-spectrum fungal antibiotic, mycophenolic acid, was never commercialized as an antibiotic, but its 2-morpholinoethylester was approved as a new immunosuppressant for kidney transplantation in 1995 and for heart transplants in 1998. The ester is called mycophenolate mofetil (CellCept®) and is a prodrug that is hydrolyzed to mycophenolic acid in the body.

2.2.3. Hypocholesterolemic Agents

A huge success has been the fungal statins, including lovastatin, pravastatin, and others, which act as inhibitors of 3-hydroxy-3-methylglutaryl-coenzyme A reductase, the regulatory and rate-limiting enzyme of cholesterol biosynthesis in liver. Statins have a very large market of $15 billion. All members of the group are substituted hexahydronaphthalene lactones. Brown et al. *(7)* discovered the first member of this group, compactin (=ML-236B) as an antibiotic product of *Penicillium brevicompactum.* Endo et al. *(8)* discovered compactin in broths of *Penicillium citrinum* and later, Endo *(9)* and Alberts et al. *(10)* independently discovered the more active methylated form of compactin known as lovastatin (monacolin K; mevinolin; Mevacor™) in broths of *Monascus ruber* and *Aspergillus terreus,* respectively. Lovastatin was approved by the FDA in 1987.

2.2.4. Antitumor Agents

Most natural anticancer agents are made by actinomycetes (doxorubicin, daunorubicin, mitomycin, bleomycin) but a successful nonactinomycete molecule is Taxol® (paclitaxel). This compound was originally discovered in plants *(11)* but is also produced by the fungus *Taxomyces andreanae (12)*. It is approved for breast and ovarian cancer and is the only commercial antitumor drug known to act by blocking depolymerization of microtubules. In addition, taxol promotes tubulin polymerization and inhibits rapidly dividing mammalian cancer cells. In 2000, Taxol sales amounted to more than \$1 billion for Bristol Myers-Squibb, representing 10% of the company's pharmaceutical sales and its third-largest-selling product. Currently, the epothilones, which are produced by myxobacteria, have a mode of action similar to Taxol, and, very importantly, are active against Taxol-resistant tumors, are in clinical testing.

2.2.5. Antiparasitic Agents

For years, major therapeutics for nonmicrobial parasitic diseases in animals (e.g., coccidiostats and antihelmintics) came from the screening of synthetic compounds followed by molecular modification. Despite the testing of thousands of such compounds, only a few promising structures were uncovered. Today, microbially produced polyethers such as monensin, lasalocid, and salinomycin dominate the coccidiostat market and are also the chief growth promotants in use for ruminant animals. The avermectins, another group of streptomycete products with a market of more than \$1 billion per year, have high activity against helminths and arthropods.

2.2.6. Bioinsecticides

The insecticidal bacterium, *Bacillus thuringiensis* (BT), owes its activity to its crystal protein produced during sporulation. Crystals plus spores had been applied to plants for years to protect them against lepidopteran insects. BT preparations are highly potent, some 300 times more active on a molar basis than synthetic pyrethroids and 80,000 times more active than organophosphate insecticides. In 1993 BT represented 90% of the biopesticide market and had annual sales of \$125 million.

2.3. Bioconversions

Microorganisms are extremely useful in carrying out biotransformation processes in which a compound is converted into a structurally related product by one or a small number of enzymes contained in the cells. Bioconversion is becoming essential to the fine-chemical industry in that their customers demand single-isomer intermediates. Since the entire 1998 worldwide market for fine chemicals was \$50 billion, the segment for these customers, i.e., \$25 billion for pharmaceuticals and \$10 billion for agrochemicals, is a major portion. Bioconversion processes are characterized by extremely high yields, i.e., 90 to 100%, mild reaction conditions, stereospecificity, and the coupling of reactions using a microorganism containing several enzymes working in series.

One of the most successful examples of biocatalytic production of a commodity chemical is the conversion of acrylonitrile to acrylamide *(13)*. Using *Rhodococcus rhodochrous* J1 or *Pseudomonas chlororaphis* B23, about 20,000 metric tons of acrylamide are produced annually for use as a flocculant, a component of synthetic fibers, a soil conditioner, and a recovery agent in the petroleum industry. The process is the first example of the use of a biological process to compete with a chemical process in the petrochemical industry. The Japanese bioconversion process is carried out at 10°C and the cells can be used many times. The titer is 656 g/L after 10 h, the yield is over 99.9%, and the productivity is higher than 7 kg of acrylamide per gram dry cell weight.

Screening of microorganisms as catalysts for the hydration of adiponitrile led to the isolation of several bacteria with nitrile hydratase activity that regioselectively produced 5-cyanovaleramide (5-CVAM), a starting point for the synthesis of azafenidin, a herbicide, from adiponitrile with a 93% yield and 96% selectivity *(14)*.

Production of nicotinamide (vitamin B$_3$) using a bioconversion process is currently being performed at Lonza's facilities in Guangzhou, China. The process involves immobilized cells of *R. rhodochrous* J1 and reaches 100% conversion, whereas the chemical process produces 4% of a nicotinic acid byproduct.

2.4. Enzymes

The development of industrial enzymes has depended heavily on the use of microbial sources. Microbes are useful because they can be produced economically in short fermentations and inexpensive media. Screening is simple and strain improvement for increased production has been very successful. The industrial enzyme market reached $1.6 billion in 1998 *(15)* divided into the following application areas: food, 45% (of which starch processing represents 11%); detergents, 34%; textiles, 11%; leather, 3%; pulp and paper 1.2%. This does not include diagnostic and therapeutic enzymes. In 2000, the industrial enzyme market reached $2 billion.

Bacterial glucose isomerase (xylose isomerase) in conjunction with fungal α-amylase and glucoamylase is currently used to convert starch to mixtures of glucose and fructose, known as "high-fructose corn syrup," in a $1 billion business. Yeast enzymes useful in the food industry include invertase from *Kluyveromyces fragilis, Saccharomyces carlsbergensis* and *S. cerevisiae* for candy and jam manufacture, β-galactosidase (lactase) from *K. fragilis* or *K. lactis* for hydrolysis of lactose from milk or whey, and α-galactosidase from *S. carlsbergensis* for crystallization of beet sugar.

The major application of proteases in the dairy industry is for the manufacturing of cheese. Calf rennin had been preferred in cheesemaking due to its high specificity, but microbial proteases produced by generally recognized as safe (GRAS) microorganisms such as *Mucor miehei, Bacillus subtilis,* and *Endothia parasitica* are gradually replacing it. The primary function of these enzymes in cheesemaking is to hydrolyze the specific peptide bond (Phe105-Met106) that generates *para-k*-casein and macropeptides *(16)*.

The use of enzymes as detergent additives still represents the largest application of industrial enzymes. Proteases, lipases, amylases, oxidases, peroxidases, and cellulases are added to the detergents, where they catalyze the breakdown of chemical bonds on the addition of water. To be suitable, they must be active under thermophilic (60°C) and alkalophilic (pH 9.0–11.0) conditions as well as in the presence of the various components of washing powders. Proteases in laundry detergents account for approx 25% of the total worldwide sales of enzymes. The first detergent containing a bacterial protease was introduced in 1956 and in 1960, Novo Industry A/S introduced alcalase produced by *Bacillus licheniformis* ("Biotex").

Lipases are commonly used in the production of a variety of products ranging from fruit juices, baked foods, and vegetable fermentations to dairy enrichment. Fats, oils, and related compounds are the main targets of lipases in food technology. Accurate control of lipase concentration, pH, temperature, and emulsion content is required to maximize the production of flavor and fragrance. The lipase mediation of carbohydrate esters of fatty acids offers a potential market for their use as emulsifiers in food, pharmaceuticals, and cosmetics.

In the paper and textile industries, enzymes are increasingly being used to develop cleaner processes and reduce the use of raw materials and production of waste. An alternative enzymatic process in the manufacturing of cotton has recently been developed based on a pectate lyase *(17)*. The process is performed at much lower temperatures and uses less water than the classical method. Another application of increasing importance is the use of lipases in removing pitch (hydrophobic components of wood, mainly triglycerides and waxes). A lipase from *Candida rugosa* is used by Nippon Paper Industries to remove up to 90% of these compounds *(18)*. The use of enzymes as alternatives to chemicals in leather processing has proved successful in improving leather quality and in reducing environmental pollution. Alkaline lipases from *Bacillus* strains, which grow under highly alkaline conditions in combination with other alkaline or neutral proteases, are currently being used in this industry.

Successful application of enzymatic processes in the chemical industry depends mainly on cost-competitiveness with the existing, and well-established, chemical processes, and a significant increase in the use of biocatalysts for chemical synthesis has been observed within recent years *(19)*. Lower energy demand, increased product titer, increased catalyst efficiency, less catalyst waste and byproducts, as well as lower volumes of wastewater streams are the main advantages that biotechnological processes have compared with the well-established chemical processes.

Obtaining enantiomerically pure intermediates and products efficiently and economically is of great importance in the pharmaceutical and chemical industries. Esterases, lipases, and proteases have been widely applied in the preparation of enantiopure compounds from racemic pairs, prochiral or meso compounds, or diastereomeric mixtures *(19)*. The targets of pharmaceutical compounds such as cell-surface receptors or enzymes are chiral biomolecules. Often, just one of the two isomers of a given compound exerts the desired effect, forcing the FDA to require the evaluation of both forms. Some examples are amoxicillin (an antibiotic), captopril (an inhibitor of angiotensin-converting enzyme [ACE]) and erythropoietin (hematopoietic growth factor). The worldwide sales volume for single-enantiomer drugs already exceeds $100 billion.

Application of enzymes in several industrial bioconversions has been broadened by the use of organic solvents replacing water *(20)*, an important development in enzyme engineering. Many chemicals and polymers are insoluble in water and its presence leads to undesirable byproducts and degradation of common organic reagents. Although switching from water to an organic solvent as the reaction medium might suggest that the enzyme would be denatured, many crystalline or lyophilized enzymes are actually stable and retain their activities in such anhydrous environments. Yeast lipases have been used to catalyze butanolysis in anhydrous solvents to obtain enantiopure 2-chloro- and 2-bromo-propionic acids that are used for the synthesis of herbicides and pharmaceuticals *(21)*. Another lipase is used in a stereoselective step, carried out in acetonitrile, for the acetylation of a symmetrical diol during the synthesis of an antifungal agent *(22)*. Enzymatic resolution of racemic amines by esterases or lipases in organic solvents is conducted by BASF in a multiton scale *(23)*. These enantiopure amines may find use as inhibitors of monoamine oxidase in the treatment of diverse neurological disorders such as Parkinson's and Alzheimer's diseases. From a biotechnological perspective, there are many advantages of employing enzymes in organic as opposed to aqueous media *(24)*, including higher substrate solubility, reversal of hydrolytic reactions, and modified enzyme specificity, which result in new enzyme activities. On the other hand, enzymes usually show lower catalytic activities in organic than in aqueous solution.

Enzymes are also used in a wide range of agrobiotechnological processes; the major application is the production of feed supplements to improve feed efficiency. A recent advance in feed enzymes involves the application of phytases to improve plant phosphorus uptake by monogastric animals. New fungal phytases with higher specific activities or improved thermostability have been recently identified *(25)*.

3. Modern Microbial Technology

3.1. Production Hosts

Recombinant DNA technology has been remarkably advanced by the development of efficient and scale-up expression systems to produce enzymes from industrially unknown microorganisms in industrial species such as *E. coli, B. subtilis, S. cerevisiae, Pichia pastoris, Hansenula polymorpha, Aspergillus,* or *Trichoderma.*

3.1.1. Escherichia coli

E. coli has been extensively used as a recombinant host for many reasons, including (1) ease of quickly and precisely modifying the genome, (2) rapid growth to high cell densities, (3) ease of culture in inexpensive media, and (4) ease of reduction of protease activity. *E. coli* can accumulate heterologous proteins up to 50% of its dry cell weight *(26)*. However, there are several drawbacks, such as inability for post-translational modifications, presence of toxic cell wall pyrogens, and formation of inclusion bodies containing inactive, aggregated, and insoluble heterologous protein. In 1993, recombinant processes in *E. coli* were responsible for almost $5 billion worth of products, i.e., insulin, human growth hormone, α, β, γ-interferons, and granulocyte-colony stimulating factor (G-CSF).

3.1.2. Bacillus *Species*

Bacillus is looked upon as a host for recombinant DNA work due to its ability to secrete large amounts of proteins and safety record (US FDA GRAS status). This organism is highly popular in the food and detergent industries, where its enzymes represent about 60% of the global $2 billion industrial-enzyme market. However, many of the species belonging to this genus produce proteases that could destroy the recombinant proteins—e.g., *B. subtilis* has seven known proteases *(27)*, five of which are extracellular. Subtilisin, the major alkaline serine protease, and a neutral protease account for 96–98% of the extracellular protease activity.

3.1.3. Saccharomyces cerevisiae

S. cerevisiae offers certain advantages over bacteria as a cloning host *(28)*. This yeast can be grown rapidly in simple media and to a high cell density, it can secrete heterologous proteins into the extracellular broth, and its genetics are more advanced than those of any other eukaryote. Despite these advantages, *S. cerevisiae* is sometimes regarded as a less-than-optimal host for large scale production of mammalian proteins because of drawbacks such as hyperglycosylation, presence of α-1,3-linked mannose residues that could cause antigenic response in patients, and absence of strong and tightly regulated promoters. Mammalian genes cloned and expressed in *S. cerevisiae* include human interferon, human epidermal growth factor, human hemoglobin, human superoxide dismutase, and interleukin-6. The most commercially important yeast recombinant process has been the production of the genes encoding surface antigens of the hepatitis B virus, resulting in the first safe hepatitis B vaccine.

3.1.4. Pichia pastoris

P. pastoris has become one of the most extensively used expression systems *(29)*. Among the advantages of this methylotrophic yeast over *S. cerevisiae* are (1) it has an efficient and tightly regulated methanol promoter *(AOX1)*, which yields alcohol oxidase at 30% of soluble protein; (2) less extensive glycosylation, due to shorter chain lengths of N-linked high-mannose oligosaccharides, usually up to 20 residues lacking the terminal α-1,3-mannose linkages *(30)*; (3) integration of multicopies of foreign DNA into chromosomal DNA yielding stable transformants, (4) ability to secrete high levels of foreign proteins, and (5) high-density growth and straightforward scale-up *(31)*. Intracellular or extracellular recombinant products made in *P. pastoris* include 6–10 g/L of tumor necrosis factor, 12 g/L of tetanus toxin fragment C, 1.25 g/L of the envelope protein of HIV-1, 4 g/L of intracellular interleukin-2, 4 g/L of secreted human serum albumin, and 0.5 g/L of human insulin-like growth factor. One of the main drawbacks for this excellent expression system is the non-GRAS status of *P. pastoris,* although some products made by this yeast are being evaluated in phase III clinical trials. One example is recombinant hirudin, a thrombin inhibitor from the medicinal leech *Hirudo medicinalis,* where 1.5 g/L of secreted product was obtained *(32)*.

3.1.5. Hansenula polymorpha

Heterologous gene expression in the methylotrophic yeast *H. polymorpha* is similar to that of *P. pastoris.* The promoter of the methanol oxidase gene is used to express foreign genes. As with *AOX1* in *P. pastoris,* the *MOX* gene in *H. polymorpha* is highly expressed and tightly regulated, giving enzyme levels up to 37% of total cell protein *(33)*. One major difference is that expression of the *MOX* gene is significantly derepressed in the absence of glucose or during glucose limitation *(34)* and therefore tight regulation of the *MOX* promoter is lost in the high-glucose conditions usually used for high-biomass fermentations. About 8–9 mg/L of intracellular hepatitis B middle surface antigen, 1 g/L of human serum albumin, and 1.5 g/L of secreted hirudin are levels of expression achieved with this methylotrophic yeast.

3.1.6. Filamentous Fungi

The development of molecular techniques for production of recombinant heterologous proteins in filamentous fungi has been laborious and contrasts markedly with the success achieved in yeasts. The ability to introduce or delete genes remains difficult although some advances in transformation have been recently reported, e.g., restriction enzyme-mediated integration *(35)* or *Agrobacterium tumefaciens*-Ti plasmid-mediated transformation *(36)*. Levels of production of nonfungal proteins are low compared to those of homologous proteins. Different strategies have been developed to overcome these problems, including the construction of protease-deficient strains *(37)*, introduction of a large number of gene copies *(36)*, use of strong fungal promoters, efficient secretion signals *(38,39)*, and gene fusions with a gene that encodes part of or an entire well-secreted protein *(38)*. Little research has been carried out on gly-cosylation in molds, although hyperglycosylation does not seem to occur and low-mannose side chains are formed. Fusion has resulted in levels of secreted proteins of 5 mg/L of human interleukin-6, 2 mg/L of human lysozyme, and 250 mg/L of human lactoferrin. Higher concentrations have been obtained for some of these proteins after mutagenic treatment of high-producing strains, e.g., 5 g/L of human lactoferrin *(40)* and 1 g/L of chymosin *(41)*.

3.2. Biopharmaceuticals

Recombinant therapeutic proteins and biological molecules including cytokines, growth factors, enzymes, antibodies, and steroids form approxi-mately 16% of the $120 billion worldwide pharmaceutical industry *(42)*.

3.2.1. Immunotherapy Products

α-Interferon was initially cleared by the FDA for use against Kaposi's sar-coma, genital warts, and hairy-cell leukemia. Although clinical tests did not support great hopes for the drug as a broad therapeutic agent for cancer, it has become useful for some types of tumors and antiviral therapy. It has been also approved for hepatitis B and C. The market for α-interferon was $1 billion in 1996. γ-Interferon was approved in 1990 for treatment of chronic granuloma-tous disease. β-Interferon (recombinant interferon β-1b) was approved by the FDA for multiple sclerosis in 1993. Interleukin-2 showed activity against renal cell cancer and was approved in 1992. In early 1998, the FDA approved recom-binant interleukin-11 (IL-11) for treatment of cancer chemotherapy-related thrombocytopenia, i.e., low platelet count; IL-11 stimulates platelet production. The market in 1992 was $20 million.

3.2.2. Infectious Disease Combatants

Due to the availability of genomic sequences of bacterial pathogens com-bined with the use of DNA microarrays to measure and compare expression and transcriptional profiles, sensitive proteomic detection and identification tech-niques have radically changed vaccine research *(43)*. The first subunit vaccine on the market was that of hepatitis B virus surface antigen, which is produced in yeast. The market for recombinant hepatitis B vaccine was $650 million in

1994. DNA vaccines are key DNA fragments from a pathogen. Upon injection of the DNA into a human, the human cells produce the encoded protein. The human's class I major histocompatibility (MHC) proteins alert killer cells which launch an immune attack against the protein and thus immunize the human against that pathogen *(44)*. DNA vaccines were in Phase I and II clinical trials in 2000.

3.2.3. Growth Factors for Mammalian Cells

An early effort of the Genentech organization was the production of human growth hormone (hGH), which had immediate application in the treatment of abnormally small children. Sales in Japan in 1998 amounted to $417 million *(45)*; in the United States the market was $355 million and in Europe, $322 million. Although human growth hormone was originally approved only for treating dwarfism, it is now approved as an orphan drug for 11 indications and much of it is sold for burns and aging.

Mammalian growth factors are of great medical interest. Approved in 1991 was Amgen's granulocyte-colony stimulating factor (G-CSF) for chemotherapy-induced white blood cell deficiency (neutropenia). Sales of $260–544 million were achieved in 1992. G-CSF had sales of $1.3 billion in 2001. Immunex's granulocyte macrophage-colony stimulating factor (GM-CSF) was also approved in 1991 for stimulation of white cell growth in autologous bone marrow transplants. The total market for G-CSF and GM-CSF was $870 million in 1994. Other growth factor candidates are epidermal growth factor (EGF), fibroblast growth factor (FGF), platelet-derived growth factor (PDGF), transforming growth factors α and β (TGF-α and TGF-β), and insulin-like growth factor-1 (IGF-1).

There are two other extremely important groups of biopharmaceuticals: blood products (including erythropoietin [EPO], thrombolytic agents [e.g., TPA], coagulant agents [e.g., Factor VIII], and many others) and monoclonal antibodies. These products are not discussed here because they are produced not in microbial cells but chiefly in mammalian cells or transgenic mice. It should be noted, however, that the tremendous development of Chinese hamster cell ovary (CHO) technology and that of other types of mammalian cells was the result of application of microbial technology to such higher cell forms.

3.3. Metabolic Engineering

Metabolic engineering (pathway engineering) has been defined as "the directed improvement of product formation or cellular properties through the modification of specific biochemical reaction(s) or the introduction of new one(s) with the use of recombinant DNA technology" *(46)*. Metabolic engineering is an interactive process that shifts between an analysis part and a synthesis part to keep improving a microorganism for a given purpose. The analysis part includes metabolic flux analysis and metabolic control analysis, which allows the determination of the limiting enzymes and, thus, which genes need to be cloned and overexpressed or, on the other hand, knocked out or decreased in expression *(47)*. Flux analysis can be carried out by a number of techniques including kinetic-based models, control theories, tracer experiments, magnetization transfer, metabolite balancing, enzyme analysis, and genetic analysis.

There are many successful applications *(47–49)* of metabolic engineering including improved production of alcohols, carotenoids, amino acids, organic acids, penicillin, cephalosporins, or the synthesis of complex polyketides in *E. coli.* Some are discussed in the next two sections.

3.4. Primary Metabolites

3.4.1. Ethanol

Development of biomass pretreatment techniques and recombinant strains that are able to ferment all biomass sugars, are resistant to inhibitory byproducts, and are able to produce synergistic combination of cellulases for full cellulose hydrolysis is currently the most active area of this process *(50)*. *E. coli* has been converted into an excellent ethanol producer (43% v/v) by recombinant DNA technology. Alcohol dehydrogenase II and pyruvate decarboxylase genes from *Zymomonas mobilis* were inserted into *E. coli* and became the dominant system for NAD regeneration. Ethanol represents more than 95% of the fermentation products in the genetically engineered strain *(51)*. All these improvements have combined to remarkably reduce the projected cost of bioethanol down to $0.12–0.13 per liter *(50,52)*. These estimations show the potential viability of using bioethanol instead of gasoline without governmental price support.

3.4.2. Amino Acids

Improvements in transformation techniques have been accomplished in *Corynebacterium, Brevibacterium,* and *Serratia* so that recombinant DNA technology is now used to improve these commercial amino acid-producing strains *(53)*. Gene cloning to increase the levels of feedback-resistant aspartate kinase and dihydrodipicolinate synthase has resulted in industrial production yields of 120 g/L and 0.25–0.35 g/L-lysine·HCl per gram of glucose used. Introduction of the proline 4-hydroxylase gene from *Dactylosporangium* sp. into a recombinant strain of *E. coli* producing L-proline at 1.2 g/L led to a new strain producing 25 g/L of hydroxyproline (*trans*-4-hydroxy-L-proline) *(54)*. When proline was added, hydroxyproline reached 41 g/L, with a yield of 87% from proline.

An *E. coli* mutant that had a higher specific growth rate, increased biomass yield, shorter lag phase, less acetate production, and increased stress resistance was further engineered to produce phenylalanine. It made twice as much than its parent engineered in the same manner *(55)*. An engineered strain of *C. glutamicum* producing 50 g/L of L-tryptophan was further modified by cloning in additional copies of its own transketolase gene in order to increase the erythrose-4-phosphate precursor of aromatic biosynthesis *(56)*. A low-copy-number plasmid increased production to 58 g/L, whereas a high-copy-number plasmid decreased production.

3.4.3. Vitamins

Cloning of a biotin operon *(bioABFCD)* on a multicopy plasmid allowed recombinant *E. coli* to produce 10,000 times more biotin than the wild type *(57)*. Strains of *Serratia marcescens* obtained by mutagenesis, selected for resistance to biotin antimetabolites *(58)*, and subjected to molecular cloning produce 600 mg/L in the presence of high concentrations of sulfur and ferrous iron *(59)*.

Vitamin C (ascorbic acid) has traditionally been made in a five-step synthetic process converting glucose to into 2-keto-L-gulonic acid (2-KGA) with a yield of 50%. 2-KGA is then easily converted by acid or base to ascorbic acid. A recombinant strain of *Gluconobacter oxydans* containing genes encoding L-sorbose dehydrogenase and L-sorbosone dehydrogenase from *G. oxydans* T-100 is an improved producer of 2-KGA *(60)*. Cloning of the gene encoding D-sorbitol dehydrogenase from *G. suboxydans* G24 into *Pseudomonas putida* IFO 3738, and introduction of all three genes into *P. putida* and other genetic manipulations, led to production of 2-KGA from D-sorbitol *(61)*. Titer was 16 g/L 2-KGA from 50 g/L D-sorbitol, a yield of 32%.

A recombinant DNA process has been developed for riboflavin in *Corynebacterium ammoniagenes* by cloning and overexpressing the organism's own riboflavin-biosynthesis genes and its own promoter sequences *(62)*. The resulting culture produced 15.3 g/L riboflavin in 3 d. An industrial strain of *B. subtilis* (designed at OmniGene for Roche) was produced by inserting multiple copies of the *rib* operon at two different sites in the chromosome and rounds of mutation *(63)*. The resultant strain produced significantly more than 14 g/L of riboflavin but the final titer was not disclosed.

3.5. Secondary Metabolites

3.5.1. Increasing Productivity by Cloning

Cloning cephamycin path genes from *Streptomyces cattleya* into *Streptomyces lactamgenus* gave a 2.3-fold increase in cephamycin C production *(64)*. An increased dosage of genes encoding β-ketoacyl:acyl carrier protein (ACP) synthase and ACP in *Streptomyces glaucescens* gave a 30% increase in tetracenomycin C formation *(65)*. Two enzymes of cephamycin biosynthesis, δ-(L-α-aminoadipyl)-L-cysteinyl-D-valine synthetase (ACVS) and lysine-ε-aminotransferase (Lat) were found to be limiting in *Streptomyces clavuligerus*. Cloning of *lat* increased cephamycin C by 2- to fivefold (*66*). Replacement of the native promoter of the ACVS-encoding gene in *Aspergillus nidulans* increased penicillin production 30-fold (*67*). Expression of *cefE* from *S. clavuligerus* or *cefEF* from *Acremonium chrysogenum* in *Penicillium chrysogenum* led to recombinant strains able to produce the cephalosporin intermediates adipyl-7-ACA and adipyl-7-ADCA (*68*). Disruption of the *cefEF* gene of *A. chrysogenum* yielded strains accumulating high titers of penicillin N that was subsequently converted to deacetoxycephalosporin C (DAOC) after cloning the *cefE* from *S. clavuligerus* into the high-producing strains (*69*).

3.5.2. Combinatorial Biosynthesis

Combinational biosynthesis is now being used for discovery of new and modified drugs. Recombinant DNA techniques are employed to introduce genes coding for antibiotic synthetases into producers of other antibiotics or into nonproducing strains to obtain modified or hybrid antibiotics. More than 200 new polyketides have been made by combinatorial biosynthesis (*70*).

The isolation of several sugar biosynthesis gene clusters and glycosyltransferases from different antibiotic-producing microorganisms and the increasing knowledge about these biosynthetic pathways have been used to construct macrolides with new sugar moieties (*71*).

Interspecific protoplast fusion between *Streptomyces griseus* and five other species (*Streptomyces cyaneus, S. exfoliatus, S. griseoruber, S. purpureus,* and *S. rochei*) yielded recombinants of which 60% produced no antibiotics and 24% produced antibiotics different from the parent strains (*72*).

3.6. Enzymes

3.6.1. Improving Productivity

Production of calf rennin (chymosin) in recombinant *A. niger* var *awamori* amounted to about 1 g/L after nitrosoguanidine mutagenesis and selection for 2-deoxyglucose resistance (*41*). Further improvement was done by parasexual recombination resulting in a strain producing 1.5 g/L from parents producing 1.2 g/L (*73*). Four recombinant proteases have been approved by the FDA for cheese production (*73*; http://vm.cfsan.fda.gov).

In 1994, Novo Nordisk introduced Lipolase™, the first commercial recombinant lipase for use in a detergent, by cloning the *Humicola lanuginose* lipase into the *Aspergillus oryzae* genome. In 1995, Genencor International introduced two bacterial lipases: Lumafast™, from *Pseudomonas mendocina;* and Lipomax™, from *Pseudomonas alcaligenes.* A new enzyme added only recently to detergents is Mannaway™ (Novozymes), a *Bacillus* mannanase that removes food stains containing guar gum *(75)*. There are three fungal recombinant lipases currently used in the food industry, one from *Rhizomucor miehi,* one from *Thermomyces lanuginosus,* and another from *Fusarium oxysporum,* all being produced in *A. oryzae* (*73;* http://vm.cfsan.fda.gov).

3.6.2. Genetic Means of Improving Properties of Enzymes

The continual expanding list of applications of enzymes for the chemical, pharmaceutical, and food industries is creating a growing demand for biocatalysts that exhibit improved or new properties. New developments in screening methodology are expanding the search for new sources of enzymes (1) in noncultivated organisms *(76;* http://www.diversa.com), (2) in more than 100 sequenced genomes (http://www.tigr.org; http://www.ncbi.nlm.nih.gov), and (3) via the diversity of extremophiles *(77)*. Although enzymes have favorable turnover numbers, there are three major hurdles that must be overcome before a biocatalyst can be considered an industrial enzyme: volumetric productivity, stability, and availability *(78)*. Two key technologies have contributed to expanding biological design capabilities: enzyme engineering *(79)* and protein engineering. The former provides remarkable improvements in nonaqueous media and has been discussed above (*see* **Subheading 2.4.**). The latter provides enzymes with altered structures, functions, and activities and is discussed below.

There are two major ways in which genetic methods can modify enzymes to adapt their functions to applied ends: (1) a rational redesign of existing biocatalysts and (2) a search for the desired functionality in libraries generated at random.

3.6.2.1. RATIONAL REDESIGN

Redesigning nature's catalysts requires detailed information about structures and mechanisms that is not available for many proteins. However, the increasing growth of databases of protein structures and sequences is helping to overcome this lack of information. Comparison of the sequence of a new biocatalyst identified in a screening program with the thousands deposited in the databases can identify related proteins whose functions and/or structures are already known. Because new enzymes evolved in nature by relatively minor modification of active-site structures, the goals of homology-driven experiments include engineering binding sites to fit different substrates as well as construction of new catalytic residues to modify functions and mechanisms *(80)*. Although in many cases results were poor compared to natural enzymes, there have been successful examples *(81,82)*.

Computational protein design starts with the coordinates of a protein main chain and uses a force field to identify sequences and geometries of amino acids that are optimal for stabilizing the backbone geometry *(83)*. Because of the amazing number of possible sequences generated, the combination of predictive force fields and search algorithms is now being applied to functional protein design *(84)*.

3.6.2.2. RANDOM REDESIGN (DIRECTED EVOLUTION)

One of the key factors contributing to the expanding application of biocatalysts in industrial processes is the development of evolutionary design methods using random mutagenesis, gene recombination, and high-throughput screening *(85)*. Directed evolution is a fast and inexpensive way of finding variants of existing enzymes that work better than naturally occurring enzymes under specific conditions.

Diversity is initially created by in vitro mutagenesis of the parent gene using repeated cycles of mutagenic PCR (error-prone PCR) *(86)*, repeated oligonucleotide directed mutagenesis *(87)*, mutator strains *(88)*, or chemical agents *(89)*. A key limitation of these strategies is that they introduce random "noise" mutations into the gene at every cycle and hence improvements are limited to small steps. On the other hand, molecular breeding techniques (DNA shuffling, Molecular Breeding™) come closer to mimicking natural recombination by allowing in vitro homologous recombination *(90)*. The technique of DNA shuffling not only recombines DNA fragments but also introduces point mutations at a very low controlled rate *(91,92)*.

Because the fitness of a gene will increase more rapidly in a breeding population with high genetic variability that is under the influence of selection, an improvement in this breeding technique came from Maxygen's laboratories. DNA FamilyShuffling is based on recombination of several homologous sequences as a starting point to generate diversity. The shuffling of four cephalosporinase genes from diverse species in a single round of shuffling gave a 240- to 540-fold improvement in activity *(93)*. When each gene was shuffled independently, only eightfold improvements were obtained.

Innovations that expand the formats for generating diversity by recombination include formats similar to DNA shuffling and others with few or no requirements for parental gene homology *(94,95)*.

Random redesign techniques are being currently used to generate enzymes with improved properties such as activity and stability at different pH values and temperatures *(96)*, increased or modified enantioselectivity *(97)*, altered substrate specificity *(98)*, stability in organic solvents *(99)*, novel substrate specificity and activity *(100)*, and increased biological activity of protein pharmaceuticals and biological molecules *(94,101)*. Proteins from directed evolution work were already on the market in 2000 *(102)*. These were green fluorescent proteins of Clontech and Novo Nordisk's LipoPrime® lipase.

Recently, a new approach called GenomeShuffling was reported by Codexis to improve pathways rather than individual enzymes. The method combines the advantage of multiparental crossing allowed by DNA shuffling with the recombination of entire genomes. Such recursive genomic recombination has been successfully applied to a population of phenotypically selected bacteria to improve the production of tylosin by *Streptomyces fradiae* *(103)*. Two rounds of shuffling with seven early strains each were sufficient to achieve similar results that had previously required 20 yr of classical strain improvement rounds at Eli Lilly and Company. New shuffled lactobacilli able to grow at pH 3.8 have also been identified using this approach *(104)*. Fermentation at this low pH decreases waste and cost in the purification of the lactic acid produced by these improved bacteria.

3.7. Bioconversions

Recombinant DNA techniques have been useful in developing new bioconversions. Production of 1,3 propanediol from glucose has been achieved with a recombinant *E. coli* strain containing two metabolic pathways, one for conversion of glucose to glycerol and the other for conversion of glycerol to 1,3 propanediol *(105,106)*. The latter is used to produce polymers with very interesting properties.

An oxidative bioconversion of saturated and unsaturated linear aliphatic 12–22 carbon substrates to their terminal dicarboxylic acids was developed by gene disruption and gene amplification *(107)*. Product concentrations reached 200 g/L and problematic side reactions such as unsaturation, hydroxylation, and chain-shortening did not occur.

3-0-acetyl-4″-0-isovaleryltylosin (AIV) is useful in veterinary medicine against tylosin-resistant *Staphylococcus aureus*. It is made by first producing tylosin with *S. fradiae* and then using *Streptomyces thermotolerans* (producer of carbomycin) to bioconvert tylosin into AIV. A new direct-fermentation organism was constructed by transforming *S. fradiae* with *S. thermotolerans* plasmids containing acyl transferase genes *(108)*.

3.8. Agriculture

Plant biotechnology has accomplished remarkable feats. One of the first advances of recombinant DNA technology in agriculture was the development of a strain of *Pseudomonas syringae* that had lost a protein responsible for ice nucleation (and hence frost damage) when the bacterium was present on the plant surface. Since those early experiments, recombinant plants have been produced that are resistant to herbicides, viruses, insects, and microbial pathogens. Microbes have played a major role in these scientific developments by providing the Ti plasmid from *A. tumefaciens* and the *B. thuringensis* (BT) insecticide-encoding gene. In 2000, genetically modified soybeans reached 45% of total US acreage, cotton reached 68%, and corn, 25% *(109)*. The total genetically engineered acreage in the US was 70 million acres, or 41% of the total; worldwide, it amounted to nearly 100 million acres.

By 2000, more than 30 million hectares of transgenic plants had been grown with no human health problems by ingestion of such plants or their products. Furthermore, some transgenic cereal grains rich in iron *(110)* or containing high levels of carotenoids including β-carotene (provitamin A) *(111)* could relieve major problems of mortality in third-world countries.

3.9. Polymers

Ten to twenty thousand tons of microbially produced xanthan are produced annually for use in the food, oil, pharmaceutical, cosmetic, paper, paint, and textile industries. Recombinant DNA manipulation of *Xanthomonas campestris* has increased titers of xanthan by twofold and increased pyruvate content by 45% *(112)*. Cloning genes, which complement xanthan-negative mutants, into wild-type *X. campestris* increased xanthan production by 15%. Xanthan formation by *X. campestris* amounts to 0.6 g/g sucrose utilized *(113)*.

A new environmentally friendly polyester (3G+) has been developed via genetic engineering in a collaborative effort by scientists from Genencor International and DuPont *(114)*. The building block for this polymer is trimethylene glycol (3G). A recombinant organism was constructed that had yeast's ability to convert sugar to glycerol and a bacterium's ability to convert glycerol to 3G. The monomer is chemically converted to the biodegradable polymer textile.

Production of special polymers in organic solvents is another important area. Peroxidase-catalyzed polymerization of phenols leads to the formation of polyphenolic resins that could replace the toxic phenol-formaldehyde resins used as adhesives and photographic developers *(115)*.

Acknowledgments

ALD acknowledges with appreciation the following companies for gifts used for general support of academic activities: ADM, Fujisawa Pharmaceutical Co. Ltd., Kao Corp., Meiji Seika Kaisha Ltd., Pfizer Inc., Schering-Plough Research Institute, Wyeth Research, and Yamasa Corp.

References

1. Tani, Y., Lim, W. J., and Yang, H. C. (1988) Isolation of L-methionine-enriched mutant of a methylotrophic yeast, *Candidi boidinii* No. 2201. *J. Ferm. Technol.* **66**, 153–158.
2. Nelis, H. J. and De Leenheer, A. P. (1991) Microbial sources of carotenoid pigments used in foods and feeds. *J. Appl. Bacteriol.* **70**, 181–191.
3. Torget, R., Kim, J., and Lee, Y. Y. (2000) Fundamental aspects of dilute acid hydrolysis/fractionation kinetics of hardwood carbohydrates. 1. Cellulose hydrolysis. *Ind. Eng. Chem. Res.* **39**, 2817–2825.
4. Demain, A. L. (2000) Microbial biotechnology. *Trends Biotechnol.* **18**, 26–31.

5. Demain, A. L. (1996) Fungal secondary metabolism: regulation and functions, in *A Century of Mycology* (Sutton, B., ed.), Cambridge University Press, Cambridge, MA, pp. 233–254.

6. Strohl, W. R. (1997) Industrial antibiotics: today and the future, in *Biotechnology of Antibiotics,* 2nd ed. (Strohl, W. R., ed.), Marcel Dekker, New York, pp. 1–47.

7. Brown, A. G., Smale, T. C., King, T. J., Hasenkamp, R., and Thompson, R. H. (1976) Crystal and molecular structure of compactin: a new antifungal metabolite from *Penicillium brevicompactum. J. Chem. Soc. Perkin Trans* **I,** 1165–1170.

8. Endo, A., Kuroda, M., and Tsujita, Y. (1976) ML-236B and ML-236C, new inhibitors of cholesterolgenesis produced by *Penicillium citrinin. J. Antibiot.* **29,** 1346–1348.

9. Endo, A. (1979) K Monacolin, a new hypocholesterolemic agent produced by *Monascus* species. *J. Antibiot.* **32,** 852–854.

10. Alberts, A. W., Chen, J., Kuron, G., Hunt, V., Huff, J., Hoffman, C., et al. (1980) Mevinolin: A highly potent competitive inhibitor of hydroxymethylglutaryl-coenzyme A reductase and a cholesterol-lowering agent. *Proc. Natl. Acad. Sci. USA* **77,** 3957–3961.

11. Wall, M. E. and Wani, M. C. (1995) Campothecin and Taxol: discovery to clinic. *Cancer Res.* **55,** 753–760.

12. Stierle, A., Strobel, G., and Stierle, D. (1993) Taxol and taxane production by *Taxomyces andreanae,* an endophytic fungus of Pacific yew. *Science* **260,** 214–216.

13. Yamada, H., Shimizu, S., and Kobayashi, M. (2001) Hydratases involved in nitrile conversion: screening, characterization and application. *Chem. Rec.* **1,** 152–161.

14. Thomas, S. M., DiCosimo, R., and Nagarajan, V. (2002) Biocatalysis: applications and potentials for the chemical industry. *Trends Biotechnol.* **20,** 238–242.

15. Stroh, W. H. (1998) Industrial enzymes market. *Gen. Eng. News* **18,** 11–38.

16. Rao, M. B., Tanksale, A. M., Ghatge, M. S., and Deshpande, V. V. (1998) Molecular and biotechnological aspects of microbial proteases. *Microbiol. Mol. Biol. Rev.* **62,** 597–635.

17. Tzanov, T., Calafell, M., Guebitz, G. M., and Cavaco-Paulo, A. (2001) Bio-preparation of cotton fabrics. *Enzyme Microb. Technol.* **29,** 357–362.

18. Farrell, R. L., Hata, K., and Wall, M. B. (1997) Solving pitch problems in pulp and paper processes by the use of enzymes or fungi. *Adv. Biochem. Eng. Biotechnol.* **57,** 197–212.

19. Koeller, K. M. and Wong, C. H. (2001) Enzymes for chemical synthesis. *Nature* **409,** 232–240.

20. Klibanov, A. (2001) Improving enzymes by using them in organic solvents. *Nature* **409,** 241–246.

21. Kirchner, G., Scollar, M. P., and Klibanov, A. (1995) Resolution of racemic mixtures via lipase catalysis in organic solvents. *J. Am. Chem. Soc.* **107,** 7072–7076.

22. Zaks, A. and Dodds, D. R. (1997) Application of biocatalysis and biotransformations to the synthesis of pharmaceuticals. *Drug Disc. Today* **2,** 513–531.

23. Carrea, G. and Riva, S. (2000) Properties and synthetic applications of enzymes in organic solvents. *Angew. Chem.* **33,** 2226–2254.

24. Lee, M. Y. and Dordick, J. S. (2002) Enzyme activation for nonaqueous media. *Curr. Opin. Biotechnol.* **13,** 376–384.

25. Kirk, O., Borchert, T. V., and Fulgsang, C. C. (2002) Industrial enzyme applications. *Curr. Opin. Biotechnol.* **13,** 345–351.

26. Swartz, J. R. (1996) *Escherichia coli* recombinant DNA technology, in *Escherichia coli and Salmonella: Cellular and Molecular Biology,* 2nd ed. (Neidhardt, F. C., ed.) American Society of Microbiology Press, Washington, DC, pp. 1693–1771.

27. He, X. S., Shyu, Y. T., Nathoo, S., Wong, S. L., and Doi, R. H. (1991) Construction and use of a *Bacillus subtilis* mutant deficient in multiple protease genes for the expression of

eukaryotic genes. *Ann. NY Acad. Sci.* **646,** 69–77.

28. Romanos, M. A., Scorer, C. A., and Clare, J. J. (1992) Foreign gene expression in yeast: a review. *Yeast* **8,** 423–488.

29. Higgins, D. R. and Cregg, J. M. (1998) Introduction to *Pichia pastoris,* in *Pichia Protocols* (Higgins, D. R. and Cregg, J. M., eds.), Humana Press, Totowa, NJ, pp. 1–15.

30. Bretthauer, R. K. and Castellino, F. J. (1999) Glycosylation of *Pichia pastoris*-derived proteins. *Biotechnol. Appl. Biochem.* **30,** 193–200.

31. Romanos, M. A. (1995) Advances in the use of *Pichia pastoris* for high-level expression. *Curr. Opin. Biotechnol.* **6,** 527–533.

32. Sohn, J. H., Kang, H. A., Rao, K. J., Kim, C. H., Choi, E. S., Chung, B. H., and Rhee, S. K. (2001) Current status of the anticoagulant hirudin: its biotechnological production and clinical practice. *Appl. Microbiol. Biotechnol.* **57,** 606–613.

33. Giuseppin, M., van Eijk, H. M., and Bes, B. C. (1988) Molecular regulation of methanol oxidase activity in continuous cultures of *Hansenula polymorpha. Biotechnol. Bioeng.* **32,** 577–583.

34. Egli, T., van Dijken, J. P., Veenhuis, M., Harder, W., and Feichter, A. (1980) Methanol metabolism in yeasts: regulation of the synthesis of catabolite enzymes. *Arch. Microbiol.* **124,** 115–121.

35. Shuster, J. R. and Connelley, M. B. (1999) Promoter-tagged restriction enzyme-mediated insertion mutagenesis in *Aspergillus niger. Mol. Gen. Genet.* **262,** 27–34.

36. Gouka, R. J., Gerk, C., Hooykaas, P. J. J., Bundock, P., Musters, W., Verrips, C. T., and de Groot, M. J. A. (1999) Transformation of *Aspergillus awamori* by *Agrobacterium tumefaciens*-mediated homologous recombination. *Nat. Biotechnol.* **6,** 598–601.

37. van der Hombergh, J. P., van de Vondervoort, P. J., van der Heijden, N. C., and Visser, J. (1997) New protease mutants in *Aspergillus niger* result in strongly reduced in vitro degradation of target proteins; genetical and biochemical characterization of seven complementation groups. *Curr. Genet.* **28,** 299–308.

38. Gouka, R. J., Punt, P. J., and van den Hondel, C. A. M. J. J. (1997) Efficient production of secreted proteins by *Aspergillus:* progress, limitations and prospects. *Appl. Microbiol. Biotechnol.* **47,** 1–11.

39. Moralejo, F. J., Cardoza, R. E., Gutierrez, S., and Martín, J. F. (1999) Thaumatin production in *Aspergillus awamori* by use of expression cassettes with strong fungal promoters and high gene dosage. *Appl. Environ. Microbiol.* **65,** 1168–1174.

40. Ward, P., Cunningham, G. A., and Conneelly, O. M. (1997) Commercial production of lactoferrin, a multifunctional iron-binding glycoprotein. *Biotechnol. Genet. Eng. Rev.* **14,** 303–319.

41. Dunn-Coleman, N. S., Bloebaum, P., Berka, R., Bodie, E., Robinson, N., Armstrong, G., et al. (1991) Commercial levels of chymosin production by *Aspergillus. Bio/Technology* **9,** 976–981.

42. Datamonitor. (2000) *Therapeutic Proteins, Key Markets and Future Strategies.* Reference code DMHC1552, Datamonitor Publications, New York, NY, p. 33.

43. Green, B. A. and Baker, S. M. (2002) Recent advances and novel strategies in vaccine development. *Curr. Opin. Microbiol.* **5,** 483–488.

44. Brown, K. S. (1996) Looking back at Jenner, vaccine developers prepare for 21st century. *The Scientist* **10** (April 1), 14,17.

45. Pramik, M. J. (1999) Recombinant human growth hormone. *Gen. Eng. News* **19**(1), 15,27,32,33.

46. Stephanopoulos, G., Aristodou, A., and Nielsen, J. (eds.) (1998) *Metabolic Engineering.* Academic, San Diego, CA.

47. Ostergaard, S., Olsson, L., and Nielsen, J. (2000) Metabolic engineering of *Saccharomyces cerevisiae*. *Microbiol. Mol. Biol. Rev.* **64,** 34–50.
48. Rohlin, L., Oh, M. K., and Liao, J. C. (2001) Microbial pathway engineering for industrial processes: evolution, combinatorial biosynthesis and rational design. *Curr. Opin. Microbiol.* **4,** 350–355.
49. Bongaerts, J., Kramer, M., Muller, U., Raeven, L., and Wubbolts, M. (2001) Metabolic engineering for microbial production of aromatic amino acids and derived compounds. *Metab. Eng.* **3,** 289–300.
50. Mielenz, J. R. (2001) Ethanol production from biomass: technology and commercialization status. *Curr. Opin. Microbiol.* **4,** 324–329.
51. Ingram, L. O., Conway, T., Clark, D. P., Sewell, G. W., and Preston, J. F. (1987) Genetic engineering of ethanol production in *Escherichia coli*. *Appl. Environ. Microbiol.* **53,** 2420–2425.
52. Lynd, L. (1996) Overview and evaluation of fuel ethanol from cellulosic biomass: technology, economics and policy. *Ann. Rev. Energy Environ.* **21,** 403–465.
53. Eggeling, L. and Sahm, H. (1999) Amino acid production: principles of metabolic engineering, in *Metabolic Engineering* (Lee, S. Y. and Papoutsakis, E. T., eds.), Marcel Dekker, New York, pp. 153–176.
54. Shibasaki, T., Hashimoto, S., Mori, H., and Ozaki, A. (2000) Construction of a novel hydroxyproline-producing recombinant *Escherichia coli* by introducing a proline 4-hydroxylase gene. *J. Biosci. Bioeng.* **90,** 522–525.
55. Weikert, C., Sauer, U., and Bailey, J. E. (1998) Increased phenylalanine production by growing and nongrowing *Escherichia coli* strain CWML2. *Biotechnol. Prog.* **14,** 420–424.
56. Ikeda, M. and Katsumata, R. (1999) Hyperproduction of tryptophan by *Corynebacterium glutamicum* with the modified pentose phosphate pathway. *Appl. Environ. Microbiol.* **65,** 2497–2502.
57. Levy-Schil, S., Debussche, L., Rigault, S., Soubrier, F., Bacchette, F., Lagneaux, D., et al. (1993) Biotin biosyntheric pathway in a recombinant strain of *Escherichia coli* overexpressing *bio* genes: evidence for a limiting step upstream from KAPA. *Appl. Microbiol. Biotechnol.* **38,** 755–762.
58. Sakurai, N., Imai, Y., Masuda, M., Komatsubara, S., and Tosa, T. (1994) Improvement of a d-biotin-hyperproducing recombinant strain of *Serratia marcescens*. *J. Biotechnol.* **36,** 63–73.
59. Masuda, M., Takahashi, K., Sakurai, N., Yanagiya, K., Komatsubara, S., and Tosa, T. (1995) Further improvement of D-biotin production by a recombinant strain of *Serratia marcescens*. *Proc. Biochem.* **30,** 553–562.
60. Saito, Y., Ishii, Y., Hayashi, H., Imao, Y., Akashi, T., Yoshikawa, K., et al. (1997) Cloning of genes coding for L-sorbose and L-sorbosone dehydrogenases from *Gluconobacter oxydans* and microbial production of 2-keto-L-gulonate, a precursor of L-ascorbic acid, in a recombinant *Gluconobacter oxydans* strain. *Appl. Environ. Microbiol.* **63,** 454–460.
61. Shibata, T., Ichikawa, C., Matsuura, M., Takata, Y., Noguchi, Y., Saito, Y., and Yamashita, M. (2000) Cloning of a gene for D-sorbitol dehydrogenase from *Gluconobacter oxydans* G624 and expression of the gene in *Pseudomonas putida* IFO3738. *J. Biosci. Bioeng.* **89,** 463–468.
62. Koizumi, S., Yonetani, Y., Maruyama, A., and Teshiba, S. (2000) Production of riboflavin by metabolically engineered *Corynebacterium ammoniagenes*. *Appl. Microbiol. Biotechnol.* **51,** 674–679.
63. Perkins, J. B., Sloma, A., Hermann, T., Theriault, K., Zachgo, E., Erdenberger, T., et al. (1999) Genetic engineering of *Bacillus subtilis* for the commercial production of riboflavin. *J. Ind. Microbiol. Biotechnol.* **22,** 8–18.

64. Chen, C. W., Lin, H. F., Kuo, C. L., Tsai, H. L., and Tsai, J. F. Y. (1988) Cloning and expression of a DNA sequence conferring cephamycin C production. *Bio/Technology* **6,** 1222–1224.

65. Decker, H., Summers, R. G., and Hutchinson, C. R. (1994) Overproduction of the acyl carrier protein component of a type II polyketide synthase stimulates production of tetracenomycin biosynthetic intermediates in *Streptomyces glaucescens. J. Antibiot.* **47,** 54–63.

66. Malmberg, L.-H., Hu, W.-S., and Sherman, D. H. (1995) Effects of enhanced lysine ε-aminotransferase on cephamycin biosynthesis in *Streptomyces clavuligerus. Appl. Microbiol. Biotechnol.* **44,** 198–205.

67. Kennedy, J. and Turner, G. (1996) δ-L-α-aminoadipyl-L-cysteinyl-D-valine synthetase is a rate limiting enzyme for penicillin production in *Aspergillus nidulans. Mol. Gen. Genet.* **253,** 189–197.

68. Crawford, L., Stepan, A. M., McAda, P. C., Rambosek, J. A., Conder, M. J., Vinci, V. A., and Reeves, C. D. (1995) Production of cephalosporin intermediates by feeding adipic acid to recombinant *Penicillium chrysogenum* strains expressing ring expansion activity. *Bio/Technology* **13,** 58–62.

69. Velasco, J., Adrio, J. L., Moreno, M. A., Diez, B., Soler, G., and Barredo, J. L. (2000) Environmentally safe production of 7-aminodeacetoxycephalosporanic acid (7-ADCA) using recombinant strains of *Acremonium chrysogenum. Nat. Biotechnol.* **18,** 857–861.

70. Rodriguez, E. and McDaniel, R. (2001) Combinatorial biosynthesis of antimicrobials and other natural products. *Curr. Opin. Microbiol.* **4,** 526–534.

71. Mendez, C. and Salas, J. A. (2001) Altering the glycosylation pattern of bioactive compounds. *Trends Biotechnol.* **19,** 449–456.

72. Okanishi, M., Suzuki, N., and Furita, T. (1996) Variety of hybrid characters among recombinants obtained by interspecific protoplast fusion in streptomycetes. *Biosci. Biotech. Biochem.* **6,** 1233–1238.

73. Bodie, E. A., Armstrong, G. L., and Dunn-Coleman, N. S. (1994) Strain improvement of chymosin-producing strains of *Aspergillus niger* var *awamori* using parasexual recombination. *Enzyme Microb. Tech.* **16,** 376–382.

74. Pariza, M. W. and Johnson, E. A. (2001) Evaluating the safety of microbial enzyme preparations used in food processing: update for a new century. *Regul. Toxicol. Pharmacol.* **33,** 173–186.

75. Kirk, O., Borchert, T. V., and Fuglsang, C. C. (2002) Industrial enzyme applications. *Curr. Opin. Biotechnol.* **13,** 345–351.

76. Rondon, M. R., Goodman, R. M., and Handelsman, J. (1999) The earth's bounty: assessing and accessing soil microbial diversity. *Trends Biotechnol.* **17,** 403–409.

77. Schiraldini, C. and De Rosa, M. (2002) The production of biocatalysts and biomolecules from extremophiles. *Trends Biotechnol.* **20,** 515–521.

78. Marrs, B., Delagrave, S., and Murphy, D. (1999) Novel approaches for discovering industrial enzymes. *Curr. Opin. Microbiol.* **2,** 241–245.

79. Schmid, A., Dordick, J. S., Hauer, B., Kiener, A., Wubbolts, M., and Witholt, B. (2001) Industrial biocatalysis today and tomorrow. *Nature* **409,** 258–268.

80. Cedrone, F., Menez, A., and Quemeneur, E. (2000) Tailoring new enzyme functions by rational redesign. *Curr. Opin. Struct. Biol.* **10,** 405–410.

81. Beppu, T. (1990) Modification of milk-clotting aspartic proteinases by recombinant DNA techniques. *Ann. NY Acad. Sci.* **613,** 14–25.

82. Van den Burg, B., de Kreij, A., Van der Veek, P., Mansfeld, J., and Venema, G. (1998) Engineering an enzyme to resist boiling. *Proc. Natl. Acad. Sci. USA* **95,** 2056–2060.

83. Bolon, D. N., Voigt, C. A., and Mayo, S. L. (2002) De novo design of biocatalysts. *Curr. Opin. Struct. Biol.* **6,** 125–129.

84. Shimaoka, M., Shiftman, J. M., Jing, H., Tagaki, J., Mayo, S. L., and Springer, T. A. (2000) Computational design of an integrin I domain stabilized in the open high affinity conformation. *Nat. Struct. Biol.* **7,** 674–678.

85. Arnold, F. H. (2001) Combinatorial and computational challenges for biocatalyst design. *Nature* **409,** 253–257.

86. Leung, D. W., Chen, E., and Goeddel, D. V. (1989) A method for random mutagenesis of a defined DNA segment using a modified polymerase chain reaction. *Technique* **1,** 11–15.

87. Reidhaar-Olson, J., Bowie, J., Breyer, R. M., Hu, J. C., Knight, K. L., Lim, W. A., et al. (1991) Random mutagenesis of protein sequences using oligonucleotide cassettes. *Methods Enzymol.* **208,** 564–586.

88. Bornscheuer, U. T., Altenbuchner, J., and Meyer, H. H. (1998) Directed evolution of an esterase for the stereoselective resolution of a key intermediate in the synthesis of epithilones. *Biotechnol. Bioeng.* **58,** 554–559.

89. Taguchi, S., Ozaki, A., and Momose, H. (1998) Engineering of a cold-adapted protease by sequential random mutagenesis and a screening system. *Appl. Environ. Microbiol.* **64,** 492–495.

90. Ness, J. E., Del Cardayre, S. B., Minshull, J., and Stemmer, W. P. (2000) Molecular breeding: the natural approach to protein design. *Adv. Protein Chem.* **55,** 261–292.

91. Stemmer, W. P. (1994) Rapid evolution of a protein in vitro by DNA shuffling. *Nature* **370,** 389–391.

92. Zhao, H. and Arnold, F. H. (1997) Optimization of DNA shuffling for high fidelity recombination. *Nucleic Acids Res.* **25,** 1307–1308.

93. Crameri, A., Raillard, S. A., Bermudez, E., and Stemmer, W. P. (1998) DNA shuffling of a family of genes from diverse species accelerates directed evolution. *Nature* **391,** 288–291.

94. Kurtzman, A. L., Govindarajan, S., Vahle, K., Jones, J. T., Heinrichs, V., and Patten, P. A. (2001) Advances in directed protein evolution by recursive genetic recombination: applications to therapeutic proteins. *Curr. Opin. Biotechnol.* **12,** 361–370.

95. Lutz, S., Ostermeier, M., Moore, G. L., Maranas, C. D., and Benkovic, S. P. (2001) Creating multiple-crossover DNA libraries independent of sequence identity. *Proc. Natl. Acad. Sci. USA* **98,** 11,248–11,253.

96. Ness, J. E., Welch, M., Giver, L., Bueno, M., Cherry, J. R., Borchert, T. V., et al. (1999) DNA shuffling of subgenomic sequences of subtilisin. *Nat. Biotechnol.* **17,** 893–896.

97. Jaeger, K. E. and Reetz, M. T. (2000) Directed evolution of enantioselective enzymes for organic chemistry. *Curr. Opin. Chem. Biol.* **4,** 68–73.

98. Suenaga, H., Mitsokua, M., Ura, Y., Watanabe, T., and Furukawa, K. (2001) Directed evolution of biphenyl dioxygenase: emergence of enhanced degradation capacity for benzene, toluene, and alkylbenzenes. *J. Bacteriol.* **183,** 5441–5444.

99. Song, J. K. and Rhee, J. S. (2001) Enhancement of stability and activity of phospholipase A(1) in organic solvents by directed evolution. *Biochim. Biophys. Acta* **1547,** 370–378.

100. Raillard, S., Krebber, A., Chen, Y., Ness, J. E., Bermudez, E., Trinidad, R., et al. (2001) Novel enzyme activities and functional plasticity revealed by recombining highly homologous enzymes. *Chem. Biol.* **8,** 891–898.

101. Patten, P. A., Howard, R. J., and Stemmer, W. P. (1997) Applications of DNA shuffling to pharmaceuticals and vaccines. *Curr. Opin. Biotechnol.* **8,** 724–733.

102. Tobin, M. B., Gustafsson, C., and Huisman, G. W. (2000) Directed evolution: the "rational" basis for "irrational" design. *Curr. Opin. Struct. Biol.* **10,** 421–427.

103. Zhang, Y. X., Perry, K., Vinci, V. A., Powell, K., Stemmer, W. P., and del Cardayre, S. B. (2002) Genome shuffling leads to rapid phenotypic improvement in bacteria. *Nature* **415,** 644–646.

104. Patnaik, R., Louie, S., Gavrilovic, V., Perry, K., Stemmer, W. P., Ryan, C. M., and del Cardayre, S. B. (2002) Genome shuffling of *Lactobacillus* for improved acid tolerance. *Nat. Biotechnol.* **20,** 707–712.

105. Tong, I.-T., Liao, J. J., and Cameron, D. C. (1991) 1,3-Propane diol production by *Escherichia coli* expressing genes from the *Klebsiella pneumoniae dha* region. *Appl. Environ. Microbiol.* **57,** 3541–3546.

106. Laffend, L. A., Nagarajan, V., and Nakamura, C. E. (1996) Bioconversion of a fermentable carbon source to 1,3-propanediol by a single microorganism. Patent WO 96/53.796 (E. I. DuPont de Nemours and Genencor International).

107. Picataggio, S., Rohrer, T., Deanda, K., Lanning, D., Reynolds, R., Mielenz, J., and Eirich, L. D. (1992) Metabolic engineering of *Candida tropicalis* for the production of long-chain dicarboxylic acids. *Bio/Technology* **10,** 894–898.

108. Arisawa, A., Kawamura, N., Narita, T., Kojima, I., Okamura, K., Tsunekawa, H., et al. (1996) Direct fermentative production of acyltylosins by genetically-engineered strains of *Streptomyces fradiae. J. Antibiot.* **49,** 349–354.

109. National Academy of Sciences U.S.A. (2000) *Transgenic Plants and World Agriculture,* National Academy Press, Washington, DC.

110. Fox, S. (2000) Golden rice intended for developing world. *Gen. Eng. News* **20**(12), 42,50.

111. Bigelas, R. (1989) Industrial products of biotechnology: Application of gene technology, in *Biotechnology,* vol. 7b (Rehm, H. J. and Reed, G. eds.; Jacobson, G. K. and Jolly, S. O., vol. eds.) VCH, Weinheim, Germany, pp. 229–259.

112. Tseng, Y. H., Ting, W. Y., Chou, H. C., Yang, B. Y., and Chun, C. C. (1992) Increase of xanthan production by cloning *xps* genes into wild-type *Xanthomonas campestris. Lett. Appl. Microbiol.* **14,** 43–46.

113. Letisse, F., Chevallereau, P., Simon, J.-L., and Lindley, N. D. (2001) Kinetic analysis of growth and xanthan gum production with *Xanthomonas campestris* on sucrose, using sequentially consumed nitrogen sources. *Appl. Microbiol. Biotechnol.* **55,** 417–422.

114. Potera, C. (1997) Genencor & DuPont create "green" polyester. *Gen. Eng. News* **17**(11), 17.

115. Akkara, J. A., Ayyagari, M. S., and Bruno, F. F. (1999) Enzymatic synthesis and modification of polymers in nonaqueous solvents. *Trends Biotechnol.* **17,** 67–73.

2

Enzyme Biosensors

Steven J. Setford and Jeffrey D. Newman

Summary

The biosensor field has grown enormously since the first demonstration of the biosensor concept by Leland C. Clark Jr. in 1962. Today's biosensor market is dominated by glucose biosensors, mass-produced enzyme-electrodes for the rapid self-diagnosis of blood glucose levels by diabetes sufferers. Here we take a historical look at the inception, growth, and development of the enzyme biosensor field from a strong commercial viewpoint. The current status of the technology is evaluated and future trends in this dynamic and fast-moving field are also anticipated.

Key Words: Biosensor; enzyme; electrodes; screen-printed; electrochemistry; amperometric; glucose; glucose oxidase; mediator; electrocatalyst; immobilization; membrane; bioreceptor; transducer; biomimic.

1. Introduction

For our purposes, a biosensor can be defined as a compact analytical device incorporating a biological or biologically derived sensing element either integrated within or intimately associated with a physicochemical transducer. The usual aim of such a device is to produce either a discrete or a continuous digital electronic signal that is proportional to a single analyte or a related group of analytes *(1)*.

This definition allows us to identify clearly Professor Leland C. Clark Jr. as the father of the biosensor concept. In 1956, Clark published his definitive paper on the oxygen electrode, a schematic of which is shown in **Fig. 1** *(2)*. Based on this experience and addressing his desire to expand the range of analytes that could be measured in the body, he made a landmark address in 1962 at a New York Academy of Sciences symposium. In this he described "how to make electrochemical sensors (pH, polarographic, potentiometric, or conductometric) more intelligent" by adding "enzyme transducers as membrane enclosed sandwiches." The concept was illustrated by an experiment in which

From: *Methods in Biotechnology, Vol. 17: Microbial Enzymes and Biotransformations*
Edited by: J. L. Barredo © Humana Press Inc., Totowa, NJ

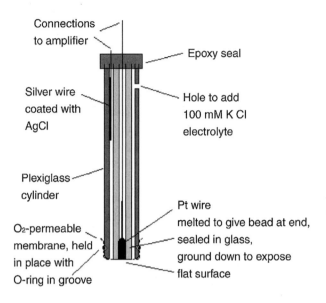

Connections to amplifier

Epoxy seal

Silver wire coated with AgCl

Hole to add 100 mM K Cl electrolyte

Plexiglass cylinder

Pt wire melted to give bead at end, sealed in glass, ground down to expose flat surface

O₂-permeable membrane, held in place with O-ring in groove

Fig. 1. Schematic of Clark oxygen electrode.

glucose oxidase was entrapped at a Clark oxygen electrode using a dialysis membrane. The decrease in measured oxygen concentration was proportional to glucose concentration. In the published paper *(3)*, Clark and Lyons coined the term "enzyme electrode," which many reviewers have mistakenly attributed to Updike and Hicks *(4)*, who expanded on the experimental detail necessary to build functional enzyme electrodes for glucose.

The sensor, essentially invented by Clark, was the basis of numerous variations on the basic design, and many other (oxidase) enzymes were immobilized by various workers as a result. Indeed, Clark's basic design was so successful that many research biosensors and at least one commercial biosensor are still produced using the original concept of oxygen measurement. However, nowadays the preferred alternative is to measure hydrogen peroxide. Most notable of the commercially available biosensors today is probably the range of biosensors sold by the Yellow Springs Instrument Company (YSI; Yellow Springs, Ohio, USA). Their glucose biosensor was successfully launched commercially in 1975 and was based on the amperometric detection of hydrogen peroxide. This was the first of many biosensor-based laboratory analyzers to be built by companies around the world.

Guilbault and Montalvo *(5)* were the first to detail a potentiometric enzyme electrode. They described a urea sensor based on urease immobilized at an ammonium-selective liquid membrane electrode. The use of thermal transducers for biosensors was proposed in 1974 and the new devices were christened "thermal enzyme probes" *(6)* and "enzyme thermistors" *(7)*, respectively. Lubbers and Opitz *(8)* coined the term "optode" in 1975 to describe a fiberoptic sensor with an immobilized indicator to measure carbon dioxide or oxygen. They extended the concept to make an optical biosensor for alcohol by immobilizing alcohol oxidase on the end of a fiberoptic oxygen sensor *(9)*. Commercial optodes are now showing excellent performance for in vivo measurement of pH, pCO_2, and pO_2, but enzyme optodes are not yet widely available.

The biosensor took a further fresh evolutionary route in 1975, when Diviès *(10)* suggested that bacteria could be harnessed as the biological element in "microbial electrodes" for the measurement of alcohol. This paper marked the beginning of a major research effort in Japan and elsewhere into biotechnological and environmental applications of biosensors.

In 1976, Clemens et al. *(11)* incorporated an electrochemical glucose biosensor in a bedside artificial pancreas; this was later marketed by Miles (Elkhart, IN) as the Biostator. Although the Biostator is no longer commercially available, a new semicontinuous catheter-based blood glucose analyzer has recently been introduced by VIA Medical (San Diego, CA). In the same year, La Roche (Basel, Switzerland) introduced the Lactate Analyser LA 640, in which the soluble mediator, hexacyanoferrate, was used to shuttle electrons from lactate dehydrogenase to an electrode. Although this was not a commercial success at the time, it turned out in retrospect to be an important forerunner of a new generation of mediated biosensors and of lactate analyzers for sports and clinical applications. A major advance in the in vivo application of glucose biosensors was reported by Shichiri et al. *(12)*, who described the first needle-type enzyme electrode for subcutaneous implantation in 1982. Companies are still pursuing this possibility, but no device for general use is available as yet.

The idea of building direct "immunosensors" by fixing antibodies to a piezoelectric or potentiometric transducer had been explored since the early 1970s, but it was a paper by Liedberg et al. *(13)* that was to pave the way for commercial success. They described the use of surface plasmon resonance to monitor affinity reactions in real time. The BIAcore (Pharmacia, Uppsala, Sweden) launched in 1990 is based on this technology.

It was during the 1980s, however, that large-scale commercial success was first achieved. YSI had built up a steady and thriving business, but it was not in the same league as the success that had been predicted and, indeed, widely expected. The basic problem lay largely with the cost of producing the biosensors of the time, which made them uncompetitive with the other technologies widely used in the massive rapid testing sector. It was within this sector that the hopes were pinned, since it was (and still is) a huge market.

In 1984, a much-cited paper on the use of ferrocene and its derivatives as an immobilized mediator for use with oxidoreductases was published *(14)*. These were crucial components in the construction of inexpensive enzyme electrodes, and formed the basis for the screen-printed enzyme electrodes launched by MediSense (Cambridge, MA) in 1987 with a pen-sized meter for home blood-glucose monitoring. The electronics have since been redesigned into popular credit-card and computer-mouse style formats, and MediSense's sales showed exponential growth, reaching $175 million per year by 1996, when they were purchased by Abbott (Abbott Park, IL, USA). Boehringer Mannheim (now Roche Diagnostics [Basle, Switzerland]), Bayer (Fernwald, Germany), and LifeScan (Milpitas, CA, USA) now have competing mediated biosensors and the combined sales of the four companies dominate the world market for biosensors and are rapidly displacing conventional reflectance photometry technology for home diagnostics.

Academic journals now contain descriptions of a wide variety of devices exploiting enzymes, nucleic acids, cell receptors, antibodies, and intact cells, in combination with electrochemical, optical, piezoelectric, and thermometric transducers *(15)*. Within each permutation lie myriad alternative transduction strategies and each approach can be applied to numerous analytical problems in health care *(16)*, food and drink *(17)*, the process industries *(18)*, environmental monitoring *(19)*, and defense and security.

A summary of some of the key events in the evolution of biosensors is shown in **Table 1.** As can be seen, the 1980s was a very inventive decade, with commercialization being the theme of the 1990s. There appears to be no sign of the latter theme changing in the early 21st century.

This demonstrates that, in some areas, biosensors have become a mature technology. It must be remembered, however, that this commercial success and maturity are limited to a small number of applications and that they came as a result of a great deal of research and development, and they have really taken place only where market size or share justified significant financial investment.

2. Looking into the Technology

2.1. Back to Basics

There are numerous components to any biosensor configuration. Over the years a great many combinations have been proposed and demonstrated, though far fewer have been commercial successes. A generalized schematic of a biosensor is shown in **Fig. 2.** The basic principle is to convert a biologically induced recognition event into a usable signal. In order to achieve this, a transducer is used to convert the (bio)chemical signal into an electronic one, which can be processed in some way, usually with a microprocessor.

2.2. Assembling the Picture

One of the chief attractions of biosensors is the remarkable specificity that their biological component confers on them. Enzymes are the most commonly

Table 1
Defining Events in the History of Biosensor Development

Date	Event
1916	First report on the immobilization of proteins: adsorption of invertase on activated charcoal
1922	First glass pH electrode
1956	Invention of the oxygen electrode
1962	First description of a biosensor: an amperometric enzyme electrode for glucose
1969	First potentiometric biosensor: urease immobilized on an ammonia electrode to detect urea
1970	Invention of the ion-selective field-effect transistor (ISFET)
1972–75	First commercial biosensor: Yellow Springs Instruments glucose biosensor
1975	First microbe-based biosensor
	First immunosensor: ovalbumin on a platinum wire
	Invention of the pO_2/pCO_2 optode
1976	First bedside artificial pancreas (Miles)
1980	First fiber optic pH sensor for in vivo blood gases
1982	First fiber optic-based biosensor for glucose
1983	First surface plasmon resonance (SPR) immunosensor
1984	First mediated amperometric biosensor: ferrocene used with glucose oxidase for the detection of glucose
1987	Launch of the MediSense ExacTech blood glucose biosensor
1990	Launch of the Pharmacia BIACore SPR-based biosensor system
1992	i-STAT launches hand-held blood analyzer
1996	Glucocard launched
1996	Abbott acquires MediSense for $867 million
1998	Launch of LifeScan FastTake blood glucose biosensor
1998	Merger of Roche and Boehringer Mannheim to form Roche Diagnostics
2001	LifeScan purchases Inverness Medical's glucose testing business for $1.3 billion
2003	i-STAT acquired by Abbott for $392 million
2004	Abbott acquires Therasense for $1.2 billion

used reagents, but many other biologicals and biomimics have also been featured. These include antibodies; whole cells, including microbial, plant, and animal cells; subcellular organelles; tissue slices; lectins; and numerous synthetic molecules with affinity or catalytic properties similar to biologicals, extending to those obtained through parallel synthesis and imprinted polymers.

Since biological components offer such exquisite selectivity (and often sensitivity), why are synthetic molecules so attractive? The answer is frequently that biological reagents are often poorly stable outside of their normal environment.

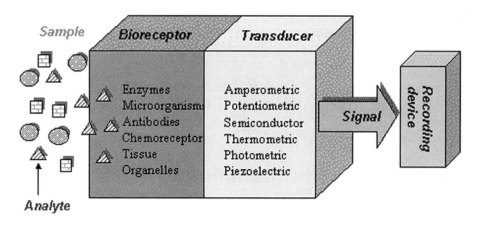

Fig. 2. Generalized schematic of a biosensor.

Table 2
Examples of Transducers Used in Biosensor Construction

Transducer	Examples
Electrochemical	Clark electrode; mediated electrodes; ion-selective electrodes (ISEs); field-effect transistor (FET)-based devices; light addressable potentiometric sensors (LAPS)
Optical	Photodiodes; waveguide systems; integrated optical devices
Piezoelectric	Quartz crystals; surface acoustic wave (SAW) devices
Calorimetric	Thermistor; thermopile
Magnetic	Bead-based devices

Thermal stability is usually poor, resulting in short lifetimes and limited ranges of application. In addition, pH sensitivity is often troublesome, as is the need for cofactors and other reagents in some instances. The aim of the biomimicry approach is to utilize the best features of the biological reagent (sensitivity, specificity, etc.) in a more stable matrix.

There are also numerous transduction mechanisms available. Some of the more popular examples are given in **Table 2.** For further explanation of the transduction principles involved, the interested reader could refer to any of the excellent texts referred to in the **References** at the end of this report. Several shorter descriptions have also been written, including an article by Newman and Turner *(20)*.

2.3. Integration Is Key

Biosensor technologists strive for the simplest possible solution to measurement in complex matrices. While notable success has been achieved with individual sensors, pragmatic solutions to many problems involve the construction of a sensor system in which the carefully optimized performance of the sensor is supported by associated electronics, fluidics, and separation technology.

In process monitoring, for example, the process must remain inviolate while the sensor frequently requires protection from the process and its products. One solution is an integrated system *(21)* consisting of a rotary aseptic sampling system with flow-injection analysis incorporating a reusable, screen-printed electrode. In this instance, the enzyme electrode utilized glucose oxidase immobilized in a hydrophilic gel and detected hydrogen peroxide at a catalytic electrode made of rhodinized carbon. While the enzyme electrode alone exhibited enhanced stability and interference characteristics, a complete solution of the monitoring problem demanded the optimization of the whole system.

There are increasing demands for a systems-oriented approach in other sectors: environmental monitoring places demands on sensor technology that, in many cases, are unlikely to be met by isolated sensors; in clinical monitoring microdialysis offers a useful way forward for measurement in vivo.

The sensor/sampling system biointerface is a key target for further investigation, involving tools such as evanescent wave techniques, and atomic force and electrochemical microscopy to aid further understanding of interactions between biological molecules or their mimics and surfaces. Work on in vivo sensing systems for both glucose and lactate has confirmed the effectiveness of phospholipid copolymers in improving hemocompatability. Immunosensors offer a further general example where microseparations, using, for example, immunochromatographic methods, can be coupled with electrochemical or optical detectors to yield simple dipstick-style devices combining the speed and convenience of sensors with the specificity and sensitivity of immunoassays. The advent of micromachining makes these and other techniques amenable to such a high degree of miniaturization that the distinction between sensor and analytical instrument becomes hazy.

2.4. Sensor Arrays—Looking at the Bigger Picture

The success of single-analyte sensors has been followed by the formulation of arrays of sensors to offer menus relevant to particular locations or situations. The most obvious example is in critical care, where commercially available handheld instruments provide clinicians with information on the concentration of six key analytes in blood samples and benchtop instruments on the ward can measure 16 analytes. These instruments commonly feature biosensors for glucose, lactate, urea, and creatinine.

The dual demands for detection of an increasing range of analytes and miniuarization are driving biosensors toward micro- or even nanoarrays.

Thinking in this area is being stimulated by the demands of the pharmaceutical industry where high-volume, high-throughput drug screening, involving low volumes of analyte, is essential for survival. In the longer term, arrays of 1 million sensors/cm^2 are a realistic target. As can be seen in the later sections, photolithography, microcontact printing, and/or self-assembly techniques offer routes to high-density arrays, but laser desorption is particularly promising and offers the ability to "write" proteins to surfaces with very high resolution.

2.5. Supporting Technologies

The basic components of the biosensor, as outlined above, are only the start of the story. In order to obtain functional devices, which can be manufactured within the necessary performance and cost constraints, numerous other components and technologies are needed.

2.5.1. Membranes and Immobilization

The very first biosensors relied on membranes for their functionality. The Clark oxygen electrode contains a gas-permeable membrane, which allows oxygen to pass while excluding undesirable, interfering species. Many other membranes have been used and these have a wide variety of purposes. They may be used to retain the biological component while allowing the analyte to pass. Another useful function is their ability to extend the (linear) range of a biosensor by acting as a mass transport barrier *(22)*.

Membranes have attracted a great deal of attention in many application areas, including process monitoring *(21)*, food *(23)*, environmental *(24)*, and medical applications *(25)*. However, they have been most prominent, at the forefront of development efforts, in the quest for improved in vivo devices *(26)*. A more detailed description of membrane application in biosensors is provided by Cunningham *(27)*.

The many approaches to immobilization of reagents on biosensors are outside the scope of this report, but the interested reader can obtain details from numerous texts *(1,27,28)*.

2.5.2. Fabrication Techniques

It has long been realized that advanced fabrication techniques are a key to the successful development of commercially viable biosensors in many applications *(29)*. Fortunately, many technologies have been developed for other applications, such as the microelectronics industry, and are therefore available with much greater reliability and at a much lower cost than would otherwise be the case, although they obviously require certain modifications and considerable development.

2.5.3. Screen-Printing

Screen-printing is a thick-film process that has been used for many years in artistic applications and, more recently, for the production of miniature, robust, and cheap electronic circuits. The main developments, from a biosensor viewpoint, have involved ink formulation. There is now a wide array of inks suitable for producing biosensors. Most of the applications to date have involved electrochemical devices, but the technique is applicable to the production of any planar device.

Since the technique has been developed for mass production, it is possible to produce very large numbers of reproducible, inexpensive devices at high speed. The process has been one of the major reasons for the commercial success of many biosensors and is the process by which MediSense (now Abbott) produces more than 1 billion biosensor strips annually.

2.5.4. Liquid-Handling Techniques

The ability to handle small volumes of liquids with high precision is one of the key areas in the development of some of the next generation of biosensors. As devices become smaller and more sophisticated, it becomes increasingly difficult to handle the analytical reagents involved in production. Some of the latest advances in transducer design, for example, make the production of 1 million measurement points on a 1-cm^2 chip a possibility. Currently, the most difficult aspect of the production of these devices is incorporating the biological reagents onto the surface of such arrays.

Ink-jet techniques *(30)* are suitable for depositing droplets of less than 1 nL in volume. This can be achieved at very high speeds (kHz), but the resolution of the droplets is comparatively poor. In addition, although the volume appears very small, the droplet size of 50–100 μm is relatively large compared to the size of the transducer structures that can be produced.

Other liquid-handling techniques include automated syringe-type processes, the best known of which is often referred to as Cavro deposition and usually involves "touching off" a droplet onto a surface. Another method involves picking up reagents on a "pin" that possesses a concave head and depositing it onto the surface of the device, a technique adapted from pharmaceutical applications.

2.5.5. Photolithography

Lithographic techniques are able to produce very small structures with well-defined geometries. Recent advances have enabled submicron structures to be fabricated using oblique evaporation and short-wavelength light and electron beams for fabrication of suitable masks. The lithographic process is very popular, largely due to its applications within microprocessor production.

2.6. Improving Performance

Improvements in the performance of analytical devices are a continuing theme in all areas of their application. Legislators, particularly in environmental applications, change consent levels, often based on what it is possible to detect. Clinicians demand simpler, longer-lasting, less-expensive devices with improved accuracy. The development community itself continuously pushes the boundaries of what is possible.

2.6.1. Sensitivity

Clinicians, food technologists, and environmentalists all have an interest in generally increased sensitivity and limits of detection for a range of analytes. While today's demands for precision may be modest in these respects, few would contest the longer-term benefits of reliable detection of trace amounts of various indicators, additives, or contaminants. With the advent of atomic-force microscopy we can consider single-molecule detection in the research laboratory, but great strides have also been made with conventional sensors. Enzyme electrodes have been designed that preconcentrate the analyte of interest *(31)*.

Advances are not limited to the liquid phase. A gas-phase microbiosensor for phenol, in which polyphenol oxidase was immobilized in a glycerol gel on an interdigitated microelectrode array *(32)*, has been reported, in which phenol vapor partitioned directly into the gel where it was oxidized to quinone. Signal amplification was enhanced by redox recycling of the quinone/catechol couple, resulting in a sensor able to measure 30 ppb phenol. Detection limits of parts per trillion volatile organic carbons are feasible with this approach.

Ultra-low detection limits are achievable with many affinity sensors and electrochemical detection may be readily integrated with chromatographic techniques to yield user-friendly devices *(33)*, an approach that overcomes the need for multiple sample manipulation steps, which was a major drawback of many early sensors of this type. In an alternative approach, double-stranded DNA may be used as a receptor element. "Sandwich"-type biosensors based on liquid-crystalline dispersions formed from DNA-polycation complexes may find application in the determination of a range of compounds and physical factors that affect the ability of a given polycation molecule to maintain intermolecular crosslinks between neighboring DNA molecules *(34)*. In the case of liquid-crystalline dispersions formed from DNA-protamine complexes, the lowest detectable concentration of the hydrolytic enzyme trypsin was 10^{-14} M. Elimination of the cross-links caused an increase in the distance between the DNA molecules that resulted in the appearance of an intense band in the circular dichroism spectrum and a "fingerprint" (cholesteric) texture. Work is in progress to develop mass-producible films and inexpensive instrumentation.

2.6.2. Stability

Arguably the most obvious disadvantage in exploiting the exquisite specificity and sensitivity of complex biological molecules is their inherent instability. Many strategies may be employed to restrain or modify the structure of biological receptors to enhance their longevity. One way, which has been recently demonstrated as a means of stabilization, is to use sol gels as an immobilization matrix. The effectiveness of such materials has been clearly demonstrated in, for example, an optode for glucose detection, using simultaneous fluorescence quenching of two indicators [(2,2'-bipyridyl) ruthenium(II) chloride hexahydrate and 1-hydroxypyrene-3,6,8-trisulfonic acid]. In such a format the excellent optical properties of the gel and enhanced stability of glucose oxidase are highly advantageous *(35)*.

Some desirable activities, however, remain beyond the reach of current technology. Methane monooxygenase is one such case where, despite reports of enhanced stability in the literature, the demands of hydrocarbon detection require stability far beyond that exhibited by the enzyme. In these cases it is valuable to resort to biomimicry to retain the essence of the biocatalytic activity but to house this within a smaller and more robust structure. For example, a simple and rapid method for quantifying a range of toxic organohalides based on their electrocatalytic reaction with a metalloporphyrin catalyst has been demonstrated. This approach can be used to measure lindane and carbon tetrachloride (representative of haloalkane compounds) perchloroethylene (a typical haloalkene) 2,4D and pentachlorophenol (aromatics) and the insecticide DDT *(36)*.

2.6.3. Selectivity

Improvement in the selectivity of biosensors may be sought at two levels: the interface between the transducer and the biological receptor may be made more exclusive, thus reducing interference, and new receptors can be developed with improved or new affinities. The use of mediators as a strategy to improve performance in amperometric biosensors has proved extremely popular and work in this area has continued to explore these possibilities. For example, it has been reported *(37)* that it is possible to use pyrroloquinoline quinone as a "natural" mediator, in combination with glucose oxidase, in an enzyme electrode for the measurement of sugar in drinks.

Alternatively, electrocatalytic detection of the products of enzymatic reactions may be enhanced by the use of chemically modified electrodes such as rhodinized *(22)* or hexacyanoferrate-modified *(38)* carbon. The latter method results in a Prussian blue coating on the electrode, which may then be used for amperometric detection of hydrogen peroxide at both oxidative and reductive potentials in enzyme electrodes for lactate and glucose.

Arguably, a more elegant solution is to seek connection of the redox center of an enzyme to an electrode via a molecular wire. Much has been published about so-called "wired" enzymes, but these papers have generally been concerned with immobilized mediators on various polymer backbones. An alternative approach is to use molecular wires, in their pure sense, for long-distance electron transfer effected via a single molecule with delocalized electrons. Novel heteroarene oligomers, consisting of two pyridinium groups, linked by thiophene units of variable length (thienoviologens) are promising candidates for such conducting molecular wires and may be used in conjunction with self-assembly techniques to produce an insulated electrode that transfers electrons specifically along predetermined molecular paths *(39)*. This design should produce enzyme electrodes free from electrochemical interference.

Advances in computational techniques now allow the modeling of both electron transfer reactions and receptor binding interactions with increasing accuracy. This not only enhances understanding of the receptor/transducer interface, but also allows consideration of the design of new receptors based on biological molecules. To obtain improved binding ligands for use in an optical sensor for glycohaemoglobin (HbA1c), a novel synthetic peptide library composed of 1 million L-amino acid hexapeptides was constructed from 10 amino acids using combinatorial chemistry *(40)*. The hexapeptide library was screened against HbA1c, HbA1b, HbAF, and HbA0 and selected ligands sequenced. Individual ligands or arrays of ligands in conjunction with pattern recognition techniques are being used to design a sensor with improved selectivity.

3. Technology Foresight

It is often difficult to predict the future in any high-technology area, as breakthroughs can happen at any time, are often not the expected, and can occur in fields that are apparently unrelated. There are, however, numerous technologies that have emerged recently, which will have an impact on the future direction of biosensors. There are also many others that will continue to evolve technologically and into new application areas. We will now examine some of these.

3.1. An Insider's Guide to the Technology

Academic research into biosensors continues unabated, and this has been matched by commercial development. The important issues are largely the same as they have always been. End users want devices that last forever, require no calibration, involve no sample extraction or preparation, and, above all, give results that are 100% reliable. We need hardly add that these demands are somewhat excessive. So how far along these lines has the field advanced? In order to answer this, we need to have a close look at the different application areas, as advances have varied.

Table 3
Commercially Available Glucose Biosensor Characteristics

	LifeScan One-Touch Ultra	Roche Diagnostics Accu-Chek Advantage	Bayer Diagnostics Glucometer Elite XL	TheraSense FreeStyle	Medisense Precision Q.I.D.
Alternate site testing	Yes	No	No	Yes	No
Sample size (μL)	1	3–4	2	0.3	3.5
Test time (s)	5	40	30	15	20
Capillary action strip	Yes	Yes	Yes	Yes	No
Temp. range (°C)	5–44	8–39	10–39	10–35	18–30
Test memory	150	100	120	250	10–125
Data downloading	Yes	Yes	Yes	Yes	Yes

3.1.1. Glucose Biosensors

Many of the issues associated with glucose biosensors have very little to do with the device itself. In virtually every case, the basic design of the commercially successful biosensors has not changed significantly for some time. Mediated amperometric designs similar to the original MediSense ExacTech device, which was first launched 15 yr ago, are still the norm.

The active area of the biosensor is universally smaller than it was in the past, enabling smaller samples to be analyzed. As can be seen in **Table 3,** samples need be no larger than a few microliters. Modeling of the response characteristics has also made quicker analyses possible. A result can now be obtained in just 10–15 s.

The way in which a sample is obtained is also an issue undergoing considerable research and discussion. As sample sizes have become smaller, questions have been asked about how representative the sample becomes. This is particularly true in the case of the most minimally invasive of the sampling regimes, which often measure glucose in interstitial fluid.

3.1.2. Sampling Regimes

Most of the current blood-glucose biosensors measure the glucose in a small sample of capillary blood, which is obtained using a lancet (**Fig. 3**). Even though the required samples need be no greater than a few microliters, there has been considerable effort aimed at enabling measurements to be made in other media, thus removing the need to draw blood. The process of drawing blood is inconvenient and uncomfortable, particularly for young diabetics and parents of diabetic children, who find this particularly stressful.

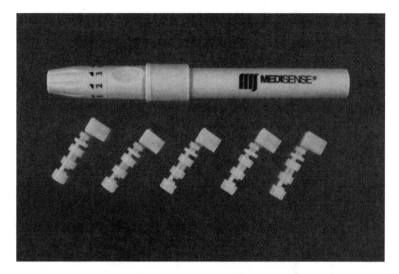

Fig. 3. Lancet for obtaining a sample of capillary blood.

The Bayer Microlet Vaculance® is suitable for non-finger use. The endcap is removed to insert a lancet and the lancet cap removed. After replacing the endcap, the puncture depth is adjusted by rotating and aligning the transparent endcap to one of four settings. The Vaculance is pressed against the puncture site and the lancet fired by completely depressing the plunger. An airtight seal is then formed by slowly releasing the plunger, which creates a vacuum. This causes the skin to bulge into the endcap, dilating the puncture and increasing the flow of blood (**Fig. 4**). The transparent endcap enables the operator to see when sufficient blood has been collected, at which point the vacuum is released by partially depressing the plunger. If this step is omitted, any residual vacuum creates aerosols.

Blood is then applied to a glucose test strip in the usual way. The used lancet is removed, recapped, and discarded. The use of the Vaculance is limited to glucose-measuring systems that aspirate (by capillary draw) small sample volumes of approx 3 μL. Bayer designed it for use with the Glucometer Esprit™ and Elite™.

Several methods have been proposed for obtaining small samples of interstitial fluid (ISF) in a painless manner. The most promising (and those with the most commercial backing) involve laser ablation, ultrasound, or reverse iontophoresis. Laser ablation uses short laser pulses to open up small pores in the epidermis, allowing a small sample of interstitial fluid to be released (**Fig. 5**). The process is painless and causes minimal damage to the skin. Ultrasound can also be used to open pores in the skin. The original application of this technology was envisaged to be opening pores to allow drug administration, but it has

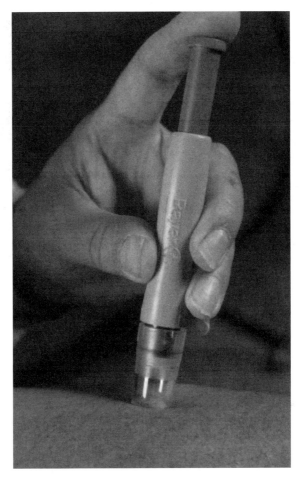

Fig. 4. Bayer Microlet Vaculance®.

been demonstrated for diagnostic applications by Sontra Medical (Cambridge, MA), which has used the technology to open pores to extract a sample. Reverse iontophoresis relies on glucose being carried in an ion flow induced by electrolysis **(Fig. 6)**. The flow of (mainly sodium and potassium) ions carries glucose to the surface of the skin, where it can be measured.

With all the above techniques, relatively conventional measurement technologies can be applied. Other principles are under development, however, that rely on completely different approaches, as discussed in the following sections.

There is, however, some controversy about the validity of measurements made in interstitial fluid. ISF is located in the outermost layers of the skin, above the nerve endings. Hence, withdrawal of ISF from the interstitial space

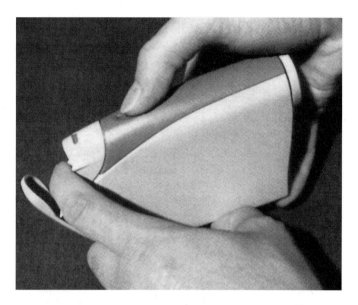

Fig. 5. Cell Robotics Lassette® laser ablation blood sampling device.

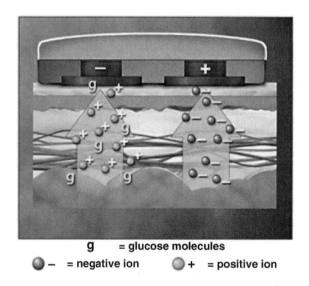

g   = glucose molecules
● − = negative ion   ○ + = positive ion

Fig. 6. Principle of reverse iontophoresis.

above the nerve endings draws no blood and, according to the manufacturers, causes little or no pain. Some research studies conclude that ISF glucose measurements are accurate for diabetes glucose testing, but others say this has not been completely proven, and more research is necessary.

David Klonoff, MD, clinical professor of medicine at the University of California, San Francisco, School of Medicine, is cautious about the use of ISF, stating: "It has not been studied sufficiently for definite answers to be known. Most people agree that interstitial fluid and blood contain essentially identical glucose concentrations, as long as the blood glucose levels are steady and not rising or falling. If the blood levels are rapidly changing, then interstitial fluid glucose levels may trail or possibly precede serum glucose levels." If this lag time is found to be significant, diabetic patients are at an increased risk of becoming hyper- or hypoglycemic without adequate warning. Companies that use ISF measurements claim that clinical studies have shown that this lag time either does not exist, or that it is rarely more than about 5 min or so, and is insignificant.

The first alternative site system to receive FDA approval was Amira's AtLast™ Blood Glucose System. Significantly, it measures blood, rather than ISF glucose. The measurement sites are on the forearm or thigh, which have a lower density of nerve endings than the fingertips. The success of this technology prompted the acquisition of the company by Roche Diagnostics in November 2001. Similar devices have followed from several other major manufacturers.

The fact that ISF-measuring products have been slow to get approval is indicative of the problems that interstitial fluid presents. Nevertheless, the area is being heavily pursued and the recent launch of the Cygnus Glucowatch® Cygnus (Redwood City, CA, USA) is a significant success.

It has also been suggested for some time that it may be possible to measure glucose in saliva, urine, or teardrops. There is some evidence that these are possible alternatives and that correlations with blood glucose levels can be established. However, the issue of a lag time between glucose concentrations in these media and those of blood is even more of a problem in each of these cases.

3.1.3. Noninvasive Glucose Sensors

It has long been suggested that it would be possible to measure blood glucose in a similar way to which blood oxygen can be measured using near-infrared light spectroscopy. This approach was most notably pursued by Futrex (Gaithersburg, MD, USA), for example.

Unfortunately, the glucose spectrum in this light region is diffuse and complex and falls in a region plagued by interference, making analysis difficult. Nevertheless, the approach has attracted a great deal of interest, primarily because of the opportunities presented. The method would allow continuous measurement if necessary and would be completely painless.

The measurement site has attracted a considerable amount of attention. Some of the proposed sites have included the finger, arm, and earlobe. Unfortunately, the problems encountered have been the same for virtually all these sites, as the measurement matrix in each of these is extremely heterogeneous.

One possible way of making an optical measurement, which is less problematic, is to look at the back of the eye, using a reflectance technique. The matrix in this case is much simpler and homogeneous, and the rear of the eye has a blood supply close to its inner surface (www.ieee.org/organizations/pubs/newsletters/leos/apr98/opticalsensor).

3.2. Biomimetics

Despite their numerous advantages, biological materials suffer from several major drawbacks. In some cases, they may be difficult to obtain and hence would be prohibitively expensive. Sometimes there may even be no available biomaterial for a particular task. Those that exist may have limited pH or temperature ranges, which may be inappropriate for the envisioned application. However, it is undoubtedly the poor stability of biological components that has led researchers to look at alternative materials for use in biosensor-type devices. The principle is based on biological inspiration and even biological building blocks (amino acids are a popular example) and chemical synthesis to provide improved materials.

In many ways this approach is not new. One of the earliest ionophores, for use in ion-selective electrodes, is valinomycin (for potassium). This material is biological in nature (it is an antibiotic) and has been in use for more than 30 yr. As far back as the 1960s, chemists have been seeking to imitate materials such as valinomycin with synthetic substitutes, which have improved function, especially stability. This early work focused on relatively simple structures, such as crown ethers, and modeling of the structure featured heavily in these efforts.

More recently, there have been a number of advances in some key technologies, which have greatly increased the efficiency of designing and fabricating potentially useful compounds for diagnostic applications. These advances have largely evolved from the increased complexity of computer-modeling algorithms and some improved manufacturing processes.

Such has been the recent interest in this area that the definition of the biosensor has been changed to incorporate biomimetics and the World Congress on Biosensors, in 2002, included a mini-Congress on biomimetics.

3.2.1. Imprinted Polymers

Work on molecular imprinted polymers (MIPs) began more than 25 yr ago, when Wulff and his coworkers from Dusseldorf University carried out experiments on the preparation of synthetic polymers with receptor properties for sugar derivatives. The process involved the formation of a complex between functional monomer and guest molecules (template) in an appropriate solution and the "freezing" of this complex by polymerization in the presence of a high concentration of cross-linker (**Fig. 7**). The templates were then removed by

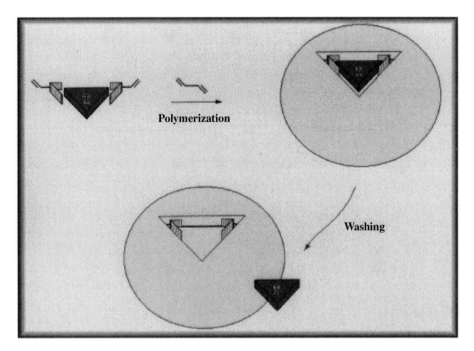

Fig. 7. Formation of imprinted polymers.

simple washing, resulting in the polymer binding sites (imprints), which are structurally specific or complementary to those of template molecules.

Over the years this method, called imprinting or template polymerization, has attracted broad interest from scientists engaged in the development of chromatographic sorbents, sensors, catalysts, enzymes, and receptor mimics. From a sensor point of view, it is interesting to obtain information about the nature of the molecular recognition phenomenon. Primarily, however, interest lies in their considerable advantages in comparison with natural receptors and enzymes, due to their superior stability, low cost, and ease of preparation.

Using such an approach, one can envisage (1) the development of assays and sensors for drugs, toxins, and environmental pollutants using artificial MIP-based receptors and (2) the development of systems for the high-throughput screening of chemical libraries on ligands possessing biological activity. In order to achieve this, work needs to be carried out on the structural analysis of imprinted polymers and computer modeling of the MIP–template interactions.

As was mentioned earlier, it is arguable whether or not devices constructed using these techniques are, indeed, biosensors. It is beyond doubt, though, that they have attracted interest within the biosensor community and, under the banner of biomimetics, they are often included as biosensors.

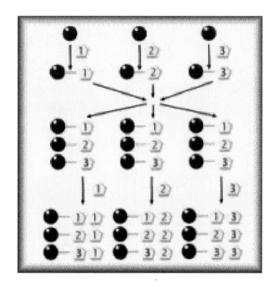

Fig. 8. Schematic of split-and-mix combinatorial synthesis.

3.2.2. Combinatorial Chemistry and Solid-Phase Organic Synthesis

Combinatorial chemistry and solid-phase organic synthesis techniques can be used to prepare libraries for the screening of novel affinity ligands for sensor applications. Current approaches involve the use of peptide libraries using natural and unnatural amino acids as building blocks. Various types of resins, such as Merrifield, Wang, Trityl, and PEGA, are used to prepare libraries using split-and-mix synthesis (**Fig. 8**). The libraries are screened and tethered onto the resin beads, and the positive results can be visualized under a microscope. When libraries are screened in solution, techniques such as surface plasmon resonance (SPR) can be used to measure the binding. Potentially useful ligands can then be structurally analyzed and subsequently resynthesized on a large scale for evaluation and kinetics study.

The use of combinatorial methods for the production of receptors for biosensors is still in its infancy. However, the technique is very widely used in drug discovery by the major pharmaceutical companies. The level of commitment to this technology for pharmaceutical applications means that advances in understanding and equipment will have natural spin-off benefits for analytical applications. However, the investments in terms of time, equipment, and personnel to obtain a ligand with the desired characteristics are not cheap and this may make the technology applicable only to biosensors with considerable market value.

Most importantly, biomimetics of all types have the potential to overcome some of the shortfalls associated with biological components, primarily poor stability and high cost of production. The successful introduction of such materials would enable biosensors to be used in many currently difficult environments.

An alternative way of creating affinity ligands is to design them rationally using a molecular modeling approach. Furthermore, this approach can be used in parallel with combinatorial chemistry or molecularly imprinted polymers to guide the selection of building blocks and to visualize and evaluate the interactions between ligands discovered from combinatorial libraries and the target analytes.

Traditional modeling techniques could involve obtaining protein structural data from a databank and importing it into a modeling program. The next step would involve refining and correcting the protein structure to a reasonable model, followed by visualizing and defining the binding pocket of the protein. From here a designing tool could be used to generate lead ligands for organic synthesis and testing.

3.3. Molecular and Cell Biology, Proteomics, and Genomics

The above approaches are largely synthetic and it could even be argued that biology has very little part in the construction of such "biomolecules." A truly biotechnological approach involves the use of molecular biology to engineer natural receptor molecules and to produce completely new receptors.

Successes in this area have been surprisingly scarce, but it is possible to produce antibodies and antibody fragments for unusual analytes, including mercury, which was considered too small for this approach until recently.

A large amount of information regarding the human genome and its expression will become available over the next few years. This will create new opportunities for biosensors, among other diagnostic devices. These will involve not only gene sequences but also patterns of gene expression and may even report information back from tissues *in situ*. The information, when coupled to computational interpretation, will permit new precision with regard to disease diagnosis and prognosis and we are clearly in for interesting times as we gain an understanding of the molecular basis of life itself.

3.4. Fabrication Issues

The ability to produce devices in large numbers and at a low cost is certainly a major requirement for many biosensors, particularly those aimed at self-testing markets, where one-shot use has many advantages (not least commercially, in terms of continued after-sales). It is also true that the ability to produce better-defined and smaller structures will open up new possibilities. This section looks at the advances already under way and the obstacles that need to be overcome, and then examines some of the opportunities available.

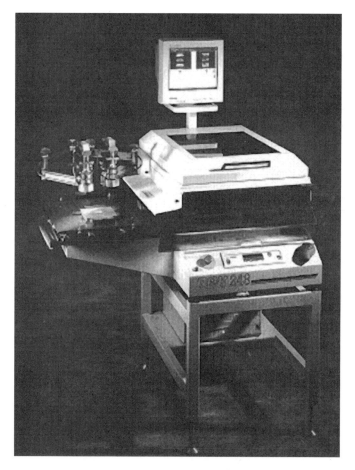

Fig. 9. DEK 248 Screen-Printer.

3.4.1. Screen-Printing Advances

As was discussed earlier, screen-printing has had a huge impact on the commercialization of biosensors. Many of the most successful electrochemically based devices to date have used the technique. A typical screen-printer (for relatively modest production requirements) is shown in **Fig. 9.**

However, the full commercial impact of the technology has been somewhat limited by the range of inks available for the construction of inexpensive devices. The main reason for this is the need for high-temperature curing with many current inks. Whereas carbon inks are widely available for low-temperature applications, those for many metals are not. Temperatures of up to 850°C are often required to cure inks employing gold and many other useful electrode materials and, hence, costly ceramic substrates are often required. It is not impossible to produce relatively inexpensive devices using ceramic bases, but polymeric materials are cheaper and easier to process.

Among others, Gwent Electronic Materials (Pontypool, Gwent, UK) is developing a new range of low-fire materials. Using a new technology, the firing temperature of metallic-based pastes has been reduced from the normal temperatures of 400–450°C to the range of 150–350°C. This type of system means that it is now possible to fire onto polymeric rather than ceramic materials. Additionally, Gwent has found that it is possible to produce a precious-metal coating that is able to connect directly to base-metal systems at a temperature below which oxidation would normally be a problem. These modifications to the metallic film give the ability to render the surface easily solderable. It is expected that these developments will open up new opportunities for inexpensive, precious-metal-based chemical sensors and biosensors.

As a result of the intensive research and expertise acquired in this area, Palintest (Tyne and Wear, UK) has developed screen-printed sensors for the detection of lead, cadmium, and copper. These devices contain no biological components, but they are an interesting spin-off from what was originally a technology developed for biosensors.

3.4.2. Miniaturization

Miniaturization of fabricated structures is reaching the stage where it is possible to imagine devices in the future that have greater capabilities than at present, yet are far smaller. Extrapolation of the recent advances could even predict bioelectrochemical or bio-optical devices that surpass the capabilities of the human brain. Such devices have, of course, featured heavily in science fiction programs, but how close are they to reality?

The limits of what can be constructed, and what can be constructed within realistic timescales and budgets, are currently vastly different. It is possible, for example, to use various technologies to construct monolayers of reagents. Langmuir-Blodgett technology has been around for some time for achieving this. It is also possible to deposit single atoms onto a surface using atomic-force microscopy-based methods. However, in order to produce reasonable numbers of devices, we are limited by methods that are both feasible and economically viable for mass production.

Connecting tracks is one area where extreme miniaturization is important. This applies equally to electrochemical as well as optical biosensors, both of which currently offer the best scopes for miniaturization (compared to the other possible transduction mechanisms). The same problem affects both types of transduction, albeit in different ways. The drawback involves the wavelength of light.

In the case of optical transduction, this means that any confining pathway must have a smallest dimension of the same order of magnitude as the wavelength of the light being transmitted through it. This restricts the size of any structure to being of the order of a micron or so. This limitation also applies to electrochemical devices, although indirectly. In order to manufacture a mask to produce structures using vacuum evaporation, the wavelength of light, once more, comes into play, since the photolithographic technique used to produce such masks is limited by the light used. Once again, structures of many microns are the result.

Recent advances in fabrication have improved matters. It is possible to produce submicron tracks using oblique evaporation, for example. In this case, the tracks are produced using a shadowing effect behind a suitably engineered structure. The obvious method, however, is to use shorter wavelengths of light or electron beams. The microprocessor industry has pioneered advances in this area and at the time of writing, chips with track dimensions of 0.125 μm are in production. These advances are available for biosensor applications when required. However, the cost of adapting the technology, at present, would be considerable. As with all new technologies, one must consider the advantages, which must outweigh the costs. At the moment, it is doubtful whether it is worth pursuing this further, as there are many other problems, which are more important, to be tackled—as we will see later in this report.

3.4.3. Liquid Handling

In order to produce arrays of sensors, reagents need to be handled and deposited onto the correct part of the array. For large arrays, this means that many different liquids need to be deposited at high speed. This is not simple to do in a reliable way. There are many ways of tackling this problem and many companies have become involved in producing deposition equipment. It should be stated that the driving force behind this comes from the high-throughput screening programs within drug discovery businesses. The issues are broadly the same and, indeed, biosensors are one solution to the problems faced within this industry.

The Packard Instrument Company (Meriden, CT), for example, now offers a complete turnkey solution to microarray manufacture and analysis that enables the production and interrogation of DNA chips. The BioChip Arrayer combines PiezoTip™ technology with a robotic arm and stage, with real-time pressure sensing. The PiezoTip, which is licensed from, and was co-developed with, Microdrop GMbH, allows deposition of droplets with a volume of approximately 325 pL. The high resolution X-Y-Z stage allows positioning of the droplets to within a 200-μm spacing (center-to-center), depending on the dried drop size. Another company, Biodot (Huntingdon, UK), produces many different droplet deposition machines based on continuous ink-jet and other related

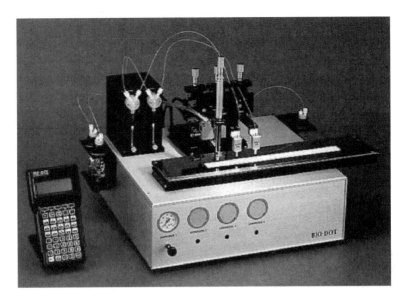

Fig. 10. Biodot Multi-Head Deposition Instrument.

technologies as well as Cavro deposition and "pin" technologies, some of which are illustrated in **Fig. 10.**

A satisfactory solution to the problem of liquid handling will revolutionize the multisensor array field, since production costs and time will be lowered and production volumes raised.

For microsensor arrays, it would also be highly desirable to reduce the volume size still further. It may sound as though 325 pL is a very small volume, but the dimensions of such a droplet are still relatively large, since the sphere will be more than 80 µm in diameter and will spread a little as it is deposited (compared to the submicron structures that can be produced using cutting-edge lithographic processes). The problems of handling such small droplets are significant and complex, but it is predicted that improvements should be possible to the required levels.

3.5. Selective Membranes

Membranes have been a common feature of biosensors since their inception. Clark's original devices were based on an oxygen-permeable membrane, for example. They are incorporated into biosensors for numerous reasons, including exclusion of interferents (size, charge), immobilization of biological components, and extension of (linear) range of sensor.

Advances in conventional membrane design have been ongoing for some years and will probably continue for the foreseeable future. An exciting prospect is the ability to be able to control the membrane porosity or even charge distribution at will. Some of the more promising work in this area suggests that such membranes are a very real possibility and, since they are activated by nothing more complicated than the application of a voltage across the membrane, there is no reason why they could not be manufactured using relatively conventional technology, yet offer the potential of biosensors that can be made to make different responses, depending on environmental conditions.

3.6. Arrays, Computational Intelligence, Noses, and Tongues

The aim of using computational intelligence in conjunction with biosensors (or any other sensors) is to apply computational methods for solving analytical problems, which are complex and problematic using other approaches. It is possible to apply a wide variety of techniques including chemometrics, neural networks, mathematical modeling, expert systems, and data mining to an equally wide range of problems in the fields of data analysis, knowledge discovery, and bioinformatics.

In the field of instrumental data analysis, chemometric techniques such as principal components analysis (PCA), multilinear regression (MLR), time-series analysis, and artificial neural networks (ANNs) are used to extract useful information from analytical responses. These techniques are regularly applied, for example, to various spectra, electrochemical voltammograms, and sensor array responses. For maximum impact, these activities need to be augmented by the development of methods for the effective visualization of multidimensional data produced by such techniques.

An example of a technology that utilizes these methods is the DiagNose instrument (**Fig. 11**) developed by Cranfield University (Bedfordshire, UK). This consists of an electronic "nose," which detects disease states through their characteristic odor. The technology will dramatically reduce the diagnosis time for urinary tract infections (UTIs), which are the initial disease targets. Current tests take between 24 and 48 h and are sent away to a laboratory. DiagNose returns results in a matter of hours and could be installed in most surgeries. The new test can also be done for a fraction of the present cost of UTI diagnosis. The test works by sniffing out characteristic odors of infecting microorganisms from a patient's urine when it is mixed with a specially engineered growth medium. As the microorganisms multiply, they produce odors that give their presence away. Each microorganism produces a different odor, allowing DiagNose to determine the underlying infection so that correct treatment can be given. Furthermore, by detecting the effect of antibiotics added to the medium on the level of bacterial growth, the technology has the potential to provide information on antibiotic resistance and hence determine the most effective therapy.

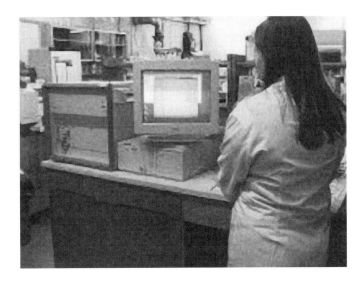

Fig. 11. DiagNose instrument.

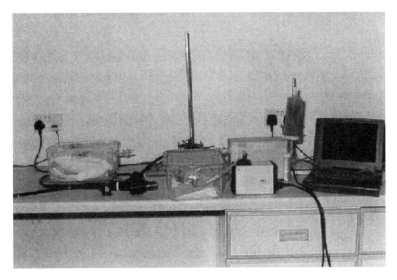

Fig. 12. Sniffing endoscope.

Laboratory trials for the new test have proved very successful, and 80% of UTIs can already be detected. Clinical trials starting shortly will give a full evaluation of DiagNose and diagnostic companies are currently being sought to bring the technology to the market. The Sniffing Endoscope (**Fig. 12**) is a related technology.

Mathematical modeling and analysis of large databases of analytical responses can harness the information present in analytical data to add power to analytical techniques. This can yield a superior understanding of the mechanism of measurement, thereby accelerating the development of new analytical systems, particularly where transducer arrays or complex sample matrices are involved. In addition, new analytical areas open where, previously, complex separation and detection systems were once required. Techniques employed in this area include analysis of variance (ANOVA) and Bayesian belief networks.

All of the above techniques promise advanced sensing possibilities based on "fingerprinting-type" approaches. There is, however, another advantage in using multiple-sensor arrays. It is possible with such technologies to incorporate self-checking capabilities and to have some built-in redundancy. These approaches help to overcome the problems associated with the lability of biological components and concomitant decay in the sensor signal, which is often apparent with such devices.

3.7. Instrumentation Simplification

One of the drawbacks of biosensors compared with other diagnostic techniques is, very often, the need for relatively expensive instrumentation. This is particularly true when they are compared with the various dipstick technologies, which rely on color changes or the appearance of bands, indicating the presence or absence of a particular material. It is often true that such devices are either not quantitative, or only semiquantitative, but the cost differential between them and biosensors can be prohibitive.

There are several technologies that have attempted to overcome this problem. Two examples are shown below.

Holographic biosensors have been developed at the Institute of Biotechnology (University of Cambridge, UK) and are the subject of several patents and patent applications. Academic research to date has centered on the marriage of holography with new polymeric thin films. Data on their performance have been obtained through monitoring the optical properties of the holograms using custom-made laboratory reflection spectrometers. The future of holographic sensors in real applications lies in their use as directly visible image holograms. These "smart holograms" do not need an electronic instrument, as they are a virtual instrument. Holographic sensors are test strips that provide a changing optical image (color, alphanumerics, messages) as the test result. A reflection hologram provides an image when it is illuminated by white light. The image is stored in a thin polymer film using patented photosensitizing technology. The polymer film is also chemically sensitized to react with a

positive   negative

Fig. 13. Holographic sensor display.

substance in, for example, a sample of bodily fluid. During the test, the target substance reacts with the polymer, leading to an alteration in the image displayed by the hologram as shown in **Fig. 13.** In order to take commercial advantage of the opportunities offered by this technique, a spin-off company, Holometrica Limited (Cambridge, UK), has been formed to develop and manufacture sensitive holographic devices for the consumer health care and environment-monitoring markets. The first product is to be aimed at replacing the current alcohol breath test.

In a sensor technology that was developed by researchers at the Center for Fluorescence Spectroscopy at the University of Maryland School of Medicine (Baltimore, MD), only the human eye is needed to detect the sensor response. It consists of a fluorescent probe and an oriented fluorescent film. The probe responds to the analyte while the film does not. Incident light fluoresces the transparent film and is polarized. The same light that fluoresces the film then does the same to the sample. In this case, the fluorescent light is not polarized. At this point the filter removes the incident light and selects only the wavelengths needed for analysis. This light, with the appropriate wavelengths, is sent through a dual polarizer, which polarizes half the light vertically and the other half horizontally. An adjustable polarizer is rotated manually so that both sides of the dual polarizer are of equal brightness. By taking into account the angle of rotation and relating it to a calibration curve, the concentration of the analyte can be measured. Early applications of the technology include glucose, pH, oxygen, and calcium measurement, but the technique is generic.

References

1. Turner, A. P. F., Karube, I., and Wilson, G. S. (eds.) (1987) *Biosensors: Fundamentals and Applications.* Oxford University Press, Oxford, p. 770.
2. Clark, L. C. Jr. (1956) Monitor and control of blood tissue O_2 tensions. *Trans. Am. Soc. Artif. Intern. Organs* **2,** 41–48.
3. Clark, L. C., Jr. and Lyons, C. (1962) Electrode system for continuous monitoring in cardiovascular surgery. *Ann. NY Acad. Sci.* **102,** 29–45.
4. Updike, S. J. and Hicks, J. P. (1967) The enzyme electrode. *Nature* **214,** 986–988.

5. Guilbault, G. G. and Montalvo, J. (1969) A urea specific enzyme electrode. *J. Am. Chem. Soc.* **91,** 2164–2569.

6. Cooney, C. L., Weaver, J. C., Tannebaum, S. R., Faller, S. R., Shields, D. V., and Jahnke, M. (1974) Thermal enzymes probe: a novel approach to chemical analysis, in *Enzyme Engineering 2* (Pye, E. K. and Wingard, L. B. Jr., eds.), Plenum, New York, pp. 411–417.

7. Mosbach, K. and Danielsson, B. (1974) An enzyme thermistor. *Biochim. Biophys. Acta* **364,** 140–145.

8. Lubbers, D. W. and Opitz, N. (1975) Die pCO_2/pO_2 optrode: eine neue pCO_2, pO_2 messonde zur messung der pCO_2 oder pO_2 von gasen und fussigkeiten. *Z. Naturforsch. C: Biosci.* **30c,** 532–533.

9. Voelkl, K. P., Opitz, N., and Lubbers, D. W. (1980) Continuous measurement of concentrations of alcohol using a fluorescence-photometric enzymatic method. *Fres. Z. Anal. Chem.* **301,** 162–163.

10. Diviès, C. (1975) Remarques sur l'oxydation de l'éthanol par une electrode micro-bienne d'acetobacter zylinum. *Annals of Microbiology* **126A,** 175–186.

11. Clemens, A. H., Chang, P. H., and Myers, R. W. (1976) Development of an auto-matic system of insulin infusion controlled by blood sugar, its system for the deter-mination of glucose and control algorithms. *Proc. Journes Ann. de Diabtologie de l'Hôtel-Dieu, Paris,* pp. 269–278.

12. Shichiri, M., Kawamori, R., Yamaski, R., Hakai, Y., and Abe, H. (1982) Wearable artificial endocrine pancreas with needle-type glucose sensor. *Lancet* **2,** 1129–1131.

13. Liedberg, B., Nylander, C., and Lundstrom, I. (1983) Surface plasmon resonance for gas detection and biosensing. *Sensors and Actuators* **4,** 299–304.

14. Cass, A. E. G., Francis, D. G., Hill, H. A. O., Aston, W. J., Higgins, I. J., Plotkin, E. V., et al. (1984) A ferrocene-mediated enzyme electrode for amperometric glu-cose determination. *Anal. Chem.* **56,** 667–671.

15. Turner, A. P. F. (ed.) (1991, 1992, 1993, 1995) *Advances in Biosensors,* I; II; Suppl. I; III. JAI Press, London.

16. Alcock, S. J. and Turner, A. P. F. (1994) Continuous analyte monitoring to aid clin-ical practice. *IEEE Engineering in Medicine and Biology,* June/July, 319–325.

17. Kress-Rogers, E. (ed.) (1996) *Handbook of Biosensors and Electronic Noses: Medicine, Food and the Environment,* CRC Press, Boca Raton, FL.

18. White, S. F. and Turner, A. P. F. (1997) Process monitoring, in *Encyclopedia of Bioprocess Technology: Fermentation, Biocatalysis and Bioseparation* (Flickinger, M. C. and Drew, S. W., eds.), Wiley, New York, pp. 2057–2080.

19. Dennison, M. J. and Turner, A. P. F. (1995) Biosensors for environmental monitor-ing. *Biotechnol. Adv.* **13,** 1–12.

20. Newman, J. D. and Turner, A. P. F. (1992) Biosensors: principles and practice, in *Essays in Biochemistry* (Tipton, K. F., ed.), Portland Press, London, pp. 147–159.

21. White, S. F., Tothill, I. E., Newman, J. D., and Turner, A. P. F. (1996) Development of a mass-producible glucose biosensor and flow-injection analy-

sis system suitable for on-line monitoring during fermentations. *Anal. Chim. Acta.* **321**, 165–172.

22. Newman, J. D., White, S. F., Tothill, I. E., and Turner, A. P. F. (1995) Catalytic materials, membranes and fabrication technologies suitable for the production of amperometric biosensors. *Anal. Chem.* **67**, 4594–4599.

23. Stephens, S. K., Cullen, D. C., and Warner, P. J. (1998) Biosensors for novel rapid assay methods in the food industry. *New Food* **1**, 47–51.

24. Sandstrom, K. J. M. and Turner, A. P. F. (1999) Biosensors in air monitoring. *J. Environ. Monitoring* **1**, 293–298.

25. Adhikari, B. and Majumdar, S. (2004). Polymers in sensor applications. *Progress in Polymer Science* **29**(7), 699–766.

26. Lakard, B., Herlem, G., de Labachelerie, M., et al. (2004). Miniaturized pH biosensors based on electrochemically modified electrodes with biocompatible polymers. *Biosensors and Bioelectronics* **19**(6), 595–606.

27. Cunningham, A. J. (ed.) (1998) *Introduction to Bioanalytical Sensors.* Wiley Interscience, New York.

28. Cass, A. E. G. (ed.) (1998) *Biosensors: A Practical Approach.* IRL Press, Oxford, UK.

29. Newman, J. D. (1998) *Advanced Manufacturing Processes for the Production of Biosensors,* PhD thesis, Cranfield University, UK.

30. Newman, J. D., Turner, A. P. F., and Marrazza, G. (1992) Ink-jet printing for the fabrication of amperometric glucose biosensors. *Anal. Chim. Acta* **262**, 13–17.

31. Saini, S. and Turner, A. P. F. (1995) Multiphase bioelectrochemical sensors. *Trends Anal. Chem.* **14**, 304–310.

32. Dennison, M. J., Hall, J. M., and Turner, A. P. F. (1995) Gas-phase microbiosensor for monitoring phenol vapour at ppb levels. *Anal. Chem.* **67**, 3922–3927.

33. Kim, Y. M., Oh, S. W., Jeong, S. Y., Pyo, D. J. and Choi, E. Y. (2003). Development of an ultrarapid one-step fluorescence immunochromatographic assay system for the quantification of microcystims. *Env. Science & Technol.* **37**(9), 1899–1904.

34. Skuridin, S. G., Yevdokimov, Y. M., Efimov, V. S., Hall, J. M. and Turner, A. P. F. (1996). A new approach for creating double-stranded DNA biosensors. *Biosensors & Bioelectronics* **11**(9), 903–911.

35. Psoma, S. D. and Turner, A. P. F. (1994) The measurement of glucose using simultaneous fluorescence quenching of two indicators, in *Proc. 3rd World Congress on Biosensors* (Turner, A. P. F., ed.), Elsevier Applied Science, Oxford, UK, **2**, 70.

36. Dobson, D. J., Turner, A. P. F., and Saini, S. (1997) Porphyrin-modified electrodes as biomimetic sensors for the determination of organohalide pollutants in aqueous samples. *Anal. Chem.* **69**, 3532–3538.

37. Loughran, M. G., Hall, J. M., and Turner, A. P. F. (1996) Development of a pyrroloquinoline quinone (PQQ) mediated glucose oxidase enzyme electrode for detection of glucose in fruit juice. *Electroanalysis* **8**, 870–875.

38. Jaffari, S. A. and Turner, A. P. F. (1997) Novel hexacyanoferrate(III) modified graphite disc electrodes and their application in enzyme electrodes. *Biosen. Bioelectron.* **12**, 1–9.

39. Albers, W. M., Lekkala, J. O., Jeuken, L., Canters, G. W. and Turner, A. P. F. (1997). Design of novel molecular wires for realizing long-distance electron transfer. *Bioelectrochem. Bioenerg.* **42**(1), 25–33.
40. Chen, B., Bestetti, G., Day, R. M. and Turner, A. P. F. (1998). The synthesis and screening of a combinational peptide library for affinity ligands for glycosylated haemoglobin. *Biosensors & Bioelectronics* **13** (7–8), 779–785.

3

Directed Evolution of Enzymes

How to Improve Nucleic Acid Polymerases for Biocatalysis

Susanne Brakmann and Sascha Nico Stumpp

Summary

Natural selection has created optimal catalysts that exhibit their convincing performance even with a number of sometimes counteracting constraints. Nucleic acid polymerases, for example, provide for the maintenance, transmission, and expression of genetic information and thus play a central role in every living organism. While faithful replication and strict substrate recognition are crucial for long-term survival of species, relaxed fidelity and/or substrate tolerance could be valuable features for biotechnological applications involving the enzymatic synthesis of modified DNA or RNA. Directed evolution that mimics natural evolution on a laboratory time-scale has emerged as a powerful tool for tailoring enzyme functions to specific requirements. In this chapter, we introduce basic techniques necessary for improving enzyme function by directed evolution, and we use the "tuning" of bacteriophage T7 RNA polymerase toward decreased fidelity as an example.

Key Words: RNA polymerase; directed evolution; random mutagenesis; selection; antibiotic resistance.

1. Introduction

Nucleic acids are perhaps the most versatile biological macromolecules. Besides their well-established vital function in storage, transport, and expression of genetic information, these biopolymers can perform a variety of chemical reactions (ribozymes, desoxyribozymes), are the enzymatically active part of the ribosome (RNA), and serve as diagnostic probes (e.g., in mutational analyses) and as effectors because they can bind to other molecules with high affinities.

Often, the biotechnological application of nucleic acids requires their chemical modification, e.g., with fluorescent substituents (for highly sensitive fluorescence detection), with variations at the 2′ position of the ribose moiety (that

From: *Methods in Biotechnology, Vol. 17: Microbial Enzymes and Biotransformations*
Edited by: J. L. Barredo © Humana Press Inc., Totowa, NJ

increases the chemical stability of RNA), or by forced introduction of mutations (to generate mutant libraries for SELEX applications) *(1,2)*. Although single-stranded nucleic acid chains with a maximum of 150 nucleotides (nt) may be synthesized automatically, the enzymatic synthesis of nucleic acids is preferred if longer chains or higher amounts (≥ 1 µmol) are needed. Natural polynu-cleotide polymerases that catalyze the template-instructed synthesis of DNA or RNA, however, evolved together with their natural substrates and thus are "sub-optimal" regarding the incorporation of artificial nucleotides, or regarding nucleic acid library synthesis.

Evolutionary techniques originally developed for the selection of functional nucleic acids (SELEX procedure) are also highly attractive for the functional optimization of enzymes. The principles of "directed evolution" are simple and do not require detailed knowledge of structure, function, or mechanism. Essentially like natural evolution, the methodology involves the iterative imple-mentation of *mutation, selection,* and *amplification* (**Fig. 1**), which are per-formed on a laboratory time-scale using the method repertoire of molecular biology/gene technology. After each round of optimization, the genes of improved variants are deciphered and subsequently serve as parents for anoth-er round of optimization.

During recent years, several successful strategies have been designed for tun-ing polymerase function by directed evolution *(3–8)*. A major goal of these investigations was to elucidate the structural elements that influence poly-merase fidelity and substrate tolerance, two features that are essential for the enzymatic synthesis of modified nucleic acids. We are interested in studying error-prone polymerases that could, for example, be employed for nucleic acid library synthesis, and we started these investigations using bacteriophage T7 RNA polymerase as a model enzyme. Directed evolution of this enzyme was achieved with expression and genetic selection in *Escherichia coli,* and yielded a polymerase variant with a 20-fold higher error rate than exhibited by the wild-type enzyme *(6)*. We will introduce this successful experiment here in order to explain the essential steps that are required for the directed evolution of other enzymes.

1.1. Practical Aspects

The key steps in conducting directed evolution experiments are (1) the gen-eration of a library of mutated genes, (2) their functional expression, and (3) a sensitive assay to identify individuals with the desired properties.

Various techniques exist for the generation of diversity at the nucleotide level *(9)*. Depending on the optimization problem to be solved, the mutagenesis strategies that are applied might range from the insertion of oligonucleotide cassettes *(10)* into those regions of the gene that encode a specific protein motif,

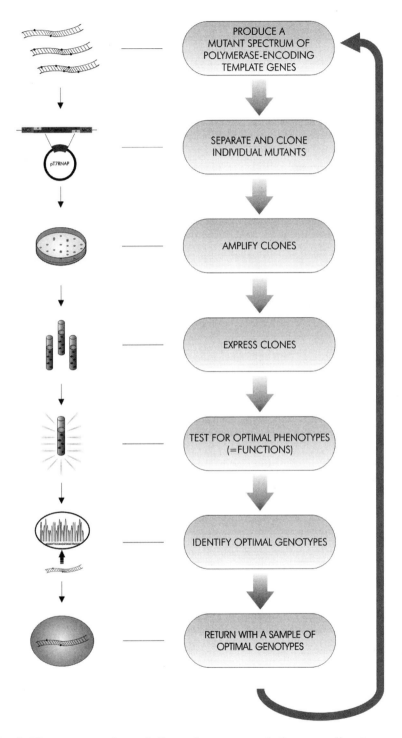

Fig. 1. Flow process chart of directed enzyme evolution according to a control program that was originally conceived by Manfred Eigen and William Gardiner (26).

or error-prone DNA synthesis *(11,12)* that targets whole genes, to diverse methods for the recombination *(shuffling)* of mutant fragments *(13–15)*.

Diversity is created at the DNA level (the *genotype*), but selection or screening acts at the level of encoded protein (the *phenotype*). Therefore, functional expression of DNA libraries is a necessary prerequisite for the detection of improved enzymes. The most common approaches for recombinant protein expression employ the cellular transcription and translation machineries of well-established organisms such as *E. coli, Saccharomyces cerevisiae,* or *Bacillus subtilis.* Cellular expression systems also guarantee the association of a specific protein variant and its encoding gene *(genotype–phenotype coupling)* that is essential for the identification and amplification of desired mutants after selection, or for further rounds of optimization. Alternative strategies exist for the linkage between genotype and phenotype, such as phage display *(16,17)*, as well as for cell-free expression *(18–20)*.

The most challenging step of a directed-evolution experiment is the "sorting" of desired enzyme variants within a library according to their catalytic potential. This step may be accomplished either by applying conditions that allow only variants of interest to appear *(selection)*, or by assaying all members of a library individually *(screening)*. Selection schemes allow for the simultaneous assessment of large populations (approx 10^{10} individual mutants per day) and the enrichment of positives while undesired variants are eliminated. However, the assessment is indirect, and as a consequence, false positives may arise. Screening approaches enable the direct testing of single clones as well as the qualitative and quantitative assay of one or more parameters. The resolution of these analyses is limited only by the detectable minimum change. More detail, however, is gained at the cost of throughput (on average, approx 10^5 individuals can be assayed in few days, if automation is employed).

In this protocol, we illustrate the procedure that is employed for the genetic selection of an error-prone variant of T7 RNA polymerase using *E. coli.* The genetic system that we developed is tailored to the specific selection problem. Nevertheless, our solution may be taken as a guide for designing selection conditions that favor other polymerase (or even other catalytic) activities.

2. Materials

2.1. Generation of a Mutant Library

1. Double-stranded DNA template encoding T7 RNA polymerase (*see* complete sequence of bacteriophage T7 at NCBI [http://www4.ncbi.nlm.nih.gov/PubMed; accession no. NC_001604]).
2. Oligonucleotide primers that bind to the respective gene.
3. Linear "TA cloning vector" pCR2.1 (Invitrogen, Carlsbad, CA).

4. Enzymes: *Taq* DNA polymerase; T4 DNA ligase.
5. Competent *E. coli* strains: XL1-Blue MR for cloning (Stratagene, La Jolla, CA) and XL1-Red for random mutagenesis (Stratagene).
6. Falcon 2059 polypropylene tubes.
7. Luria-Bertani (LB) medium.
8. Ampicillin.

2.2. Plasmid-Based Selection System

1. Plasmid pACYC184 (New England Biolabs, Beverly, MA).
2. 208-bp oligonucleotide 5′-CTA GAT TTC AGT GCA ATT TAT CTC TTC AAA TGT AGC ACC TGA AGT CAG CCC CAT ACG ATA TAA GTT **GCG AAC TTC TGA TAG ACT TCG AAA TTA ATA CGA CTC ACT ATA GGG AGA CCT** TAT CAC AGT TAA ATT GCT AAC GCA GTC AGG CAC CGT GTA TGA AAT CTA ACA ATG CGC TCA TCG TCA TCC TCG GCA CCG TCA CCC TGG ATG C-3′. The sequence in boldface substitutes the original plasmid sequence with the T7 Φ10 promoter (–41 to + 8).
3. 114-bp oligonucleotide 5′-G ATC CTC **CCC** GCC GGA CGC ATC GTG GCC GGC ATC ACC GGC GCC ACA GGT GCG GTT GCT GGC GCC TAT ATC GCC GAC ATC ACC GAT GGG GAA GAT CGG GCT CGC CAC TTC GGG CT-3′. The codon in boldface allows for the substitution of tyrosine 100 (codon TAC) by proline (codon CCC) within the tetracycline resistance gene of pACYC184.
4. Enzymes: Restriction enzymes; T4 DNA ligase.
5. Competent *E. coli* strains: XL1-Blue MR (cloning), and XL1-Blue MR/pT7RNAP (control of selectivity).
6. LB medium.
7. Ampicillin, chloramphenicol, tetracyline.

2.3. Genetic Selection

1. *E. coli* XL1-Blue MR, transformed with selection plasmid pACYC-T7Nontet.
2. Plasmid-encoded mutant library, pPolstar.
3. Low-salt medium: 10 g/L tryptone, 5 g/L yeast extract, and 0.5 g/L NaCl, dissolved in bidistilled water.
4. Bidistilled water.
5. Glycerol, p.A.
6. Liquid nitrogen.
7. SOB medium, pH 6.8–7.0: 20 g/L tryptone, 5 g/L yeast extract, 0.58 g/L NaCl, and 0.1 g/L KCl; to be added after sterilization: 10 mM MgCl$_2$, and 10 mM MgSO$_4$.
8. SOC medium: SOB plus 20 mM glucose.
9. Electroporation cuvets (1 mm) (e.g., Eppendorf, Hamburg, Germany).
10. Electroporator (e.g., Eppendorf).
11. LB medium.
12. Ampicillin, chloramphenicol, tetracycline.
13. Oligonucleotide primers for sequence analysis of the complete T7 RNA polymerase gene.

14. DNA sequencing kit (e.g., FS kit, Applied Biosystems, Foster City, CA) and sequencer.

2.4. Identification and Characterization of Selected Variants

1. T7 RNA polymerase preparation of wild-type, as well as variant, enzyme (according to [*21*]); final enzyme concentration 1–5 μg/μL.
2. 10X transcription buffer: 400 mM Tris-HCl, pH 8.0, 80 mM MgCl$_2$, 100 mM NaCl, 20 mM spermidine, and 300 mM dithiothreitol.
3. Linear template with a T7 promotor, e.g., plasmid pACYC-T7Tet.
4. Oligonucleotide primers for sequence analysis of the transcribed gene.
5. Nucleoside triphosphates (100 mM each).
6. Phenol, equilibrated with TE buffer, pH 7.5–8.0.
7. TE buffer: 10 mM Tris-HCl, pH 8.0, and 1 mM EDTA.
8. Chloroform, p.A.
9. 3.5 M sodium acetate.
10. Ethanol, p.A.
11. Agarose gel electrophoresis equipment.
12. Kit for the elution of nucleic acids from agarose gels, e.g., QIAquick Spin Kit (Qiagen, Valencia, CA).
13. Reverse transcriptase and buffer.
14. *Vent* polymerase, exonuclease-deficient.
15. *Pfu* DNA polymerase.
16. 10X PCR buffer: 100 mM Tris-HCl, pH 8.5, 500 mM KCl, 0.01% gelatin, and 15 mM MgCl$_2$.
17. Deoxynucleoside triphosphates (10 mM each).
18. TA cloning vector, e.g., pCR2.1 (Invitrogen).
19. Competent *E. coli* cloning strain, e.g., XL1-Blue (Stratagene).
20. DNA sequencing kit and sequencer.

3. Methods

The methods described below outline (1) the generation of a mutant library, (2) the construction of a selection plasmid, (3) the selection procedure, and (4) the identification and characterization of selected variants.

3.1. Generation of a Mutant Library

Random mutagenesis at the nucleotide level is a widespread strategy that targets complete genes. The technique is usually applied if knowledge on the functional contribution of defined protein motifs is limited, and if a complex structure-function relationship is assumed. Random mutation is achieved either by passing cloned genes through mutator strains *(22,23)* or by error-prone PCR *(11,12)* (*see* **Note 1**). Our approach includes (a) the construction of an expression vector containing the gene coding for T7 RNA polymerase and (b) random mutagenesis by mutator strain passage.

DNA manipulations (PCR amplification, cloning) were performed according to standard recombinant DNA methodology *(24)* and are not described in detail here. For the sake of simplicity, "TA cloning" was employed: T vectors are linear plasmids with one or a few dT extensions added at the 3′ termini. Apparently, *Taq* DNA polymerase adds (non-template-instructed) dA overhangs to the 3′ termini of PCR products. Since most of the PCR products contain the prolonged termini, they hybridize with the overhangs of the linear T vector and thus can be ligated directly. Vector pCR2.1 contains the ColE1 origin of replication and encodes the ampicillin resistance.

A plasmid (pAR1219) containing the coding sequence for T7 RNA polymerase (T7 gene 1) was kindly provided by A. H. Rosenberg *(25)*. This plasmid served as template for the amplification of T7 gene 1 using PCR. The amplification product was ligated into linear vector pCR2.1 yielding plasmid pT7RNAP **(Fig. 2A)** that was amplified in XL1-Blue MR. This expression construct was verified by restriction digestion as well as DNA sequence analysis. The mutant library was produced by submitting the expression plasmid to 1–5 successive passages in mutator strain XL1-Red by following exactly the instructions given by the supplier (instruction manual for Cat. no. 200129, Stratagene):

1. Thaw competent cells of XL1-Red on ice.
2. Gently mix by hand. Then transfer 100 µL into a prechilled 15–mL Falcon 2059 polypropylene tube.
3. Add 1.7 µL of 1.42 *M* β-mercaptoethanol provided with the kit (final concentration 25 m*M*) and swirl gently.
4. Incubate on ice for 10 min, swirling the tube every 2 min.
5. Add 10–50 ng plasmid pT7RNAP and swirl gently.
6. Incubate on ice for 30 min.
7. Heat pulse in a 42°C water bath for exactly 45 s.
8. Immediately transfer to ice and incubate for 2 min.
9. Add 0.9 mL preheated (42°C) SOC medium, and incubate at 37°C for 1 h with shaking at 225–250 rpm.
10. Plate 100–200-µL aliquots on ampicillin plates. Incubate at 37°C for 24–30 h.
11. Using a sterile inoculation loop, select ≥200 colonies at random from the transformation plates in **step 10** and inoculate 5–10 mL LB broth containing ampicillin. Grow this culture overnight at 37°C. If a higher mutation rate is desired, the cells can be diluted and grown overnight for as many cycles as desired.
12. Perform a miniprep of the overnight culture in order to isolate the randomly mutated plasmid DNA.

The plasmid prepared from mutator strain passage consists of a pool of closely related plasmids that differ in few nucleotide positions (averaged mutation rate after five passages approx 0.1%). The plasmid library is called pPolstar for further experiments.

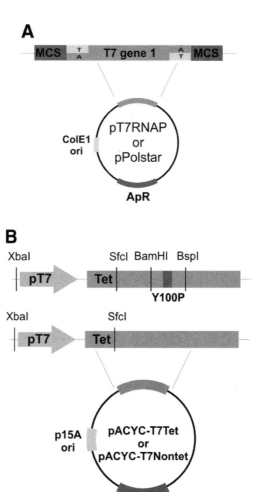

Fig. 2. (**A**) Genetic construction of vectors pT7RNAP, or pPolstar. Origin of replication (ori): ColE1; resistance marker: ampicillin (ApR). (**B**) Genetic construction of pACYC-T7Tet (wildtype), and pACYC-T7Nontet (inactive mutant). Origin of replication: p15A; resistance marker: chloramphenicol (CmR).

3.2. Generation of a Plasmid-Based Selection System

Genetic selection strategies exploit conditions that favor the exclusive survival of desired variants. Most often, this is achieved by *genetic complementation* of a host that is deficient in a certain pathway or activity (*see* **Note 2**). Our approach involves the setup of a selection system consisting of two compatible plasmids that couple mutant polymerase genes in a feedback loop to the essential but inactivated tetracycline resistance gene. Plasmid pACYC184 that represents the basis of our selection system contains a p15A origin of replication (compatible with the ColE1 *ori* of pT7RNAP) and encodes a chloramphenicol as well as a tetracycline resistance. We modified this plasmid in two steps in order to adjust it to our needs by (a) substituting its RNA polymerase promoter with the Φ10 promoter of bacteriophage T7 and (b) inactivating the tetracycline resistance with one mutation (i.e., by substitution of one codon).

First, a 208-bp cassette upstream of the tetracycline *(tet)* resistance gene comprising an *E. coli* RNA polymerase promoter was substituted by a synthetic fragment that introduced the T7 promoter sequence in the correct frame for transcription. Transformation of XL1-Blue MR/pT7RNAP with the resulting plasmid pACYC-T7Tet and selection on ampicillin/tetracycline plates confirmed the correct transcriptional context for the expression of the *tet* resistance as well as the full activity of cloned wild-type T7 RNA polymerase. For inactivation of the tetracycline resistance, a synthetic 114-bp fragment was introduced into pACYC-T7Tet that substituted a fragment of the wild-type *tet* gene and replaced codon 100 (TAC; amino acid: Tyr) with the triplet CCC (amino acid: Pro). Transformation of XL1-Blue MR/pT7RNAP with the resulting plasmid pACYC-T7Nontet **(Fig. 2B)** and selection on ampicillin/chloramphenicol plates as well as replica-plating onto plates with tetracycline (10 µg/mL) revealed that the current construct did not support survival of bacteria over a period of 3 d. Additionally, the construct was analyzed by restriction digestion and DNA sequence analysis.

3.3. Genetic Selection

3.3.1. Production of Electrocompetent Cells of XL1-Blue MR/pACYC-T7Nontet

1. Inoculate 10 mL LB medium containing chloramphenicol (34 µg/mL) with a single, fresh colony of XL1-Blue MR/pACYC-T7Nontet. Incubate overnight at 37°C with shaking (220 rpm).
2. Inoculate 1 L of low-salt medium with 10 mL of the overnight culture (1:100). Incubate at 37°C with shaking at 220 rpm until an optical density (OD_{600}) of 0.6 is reached.
3. Cool the culture on ice for 30 min. During the following steps, keep the cells on ice or in a cooled centrifuge (4°C).
4. Harvest the cells by centrifugation in a cooled centrifuge (4°C, 2000g, 15 min). Decant the supernatant immediately but carefully.
5. Resuspend the cells in 1 L of bidistilled water using a prechilled pipet. Take care not to produce bubbles.
6. Separate the cells by centrifugation (4°C, 2000g, 15 min).
7. Decant the supernatant (*see* **step 4**).

8. Resuspend the cells in 500 mL of bidistilled water.
9. Separate the cells by centrifugation (4°C, 2000g, 15 min).
10. Decant the supernatant (*see* **step 4**).
11. Resuspend the cells in 500 mL bidistilled water containing 10% glycerol.
12. Repeat **steps 9, 10,** and **11.**
13. Resuspend the cells in 5 mL of water containing 10% glycerol.
14. Immediately divide the suspension into 100-µL aliquots and freeze them in liquid nitrogen.
15. Store the electrocompetent cells at –70°C.

3.3.2. Selection Experiment

1. Thaw one aliquot of electrocompetent cells XL1-Blue/pACYC-T7Nontet on ice.
2. Add 1 µL pPolstar (1–20 ng/µL); this amount represents approx 10^8–10^9 molecules. Mix DNA and cells with a pipet tip. Be careful not to produce bubbles.
3. Transfer the cell/DNA mixture to a prechilled electroporation cuvet and place the cuvet in an electroporator.
4. Apply a pulse of appropriate voltage (12–19 kV/cm; e.g., 1800 V).
5. Immediately remove the cuvet from the electroporator and place it on ice.
6. Add 1 mL SOC and transfer the complete suspension to a 1.5-mL tube.
7. Incubate 30 min at 37°C with shaking at 220 rpm.
8. Add 19 mL LB containing ampicillin (100 µg/mL) and chloramphenicol (34 µg/mL) and incubate overnight (37°C; 220 rpm).
9. Dilute the overnight culture into 2 L LB containing ampicillin and chloramphenicol as above (**step 8**) as well as tetracycline (10 µg/mL).
10. Incubate 15–20 h at 37°C with shaking at 220 rpm.

In our selection system, the presence of a T7 promotor upstream of the tetracycline resistance gene *(tet)* resulted in the exclusive dependence of resistance expression on the presence of T7 RNA polymerase. Substitution of a single amino acid of the *tet* fully inactivated the resistance, which is mediated by a hydrophobic, membrane-associated efflux pump. The mutation Y100→P is located within a periplasmic region of the membrane-bound protein that is critical for the performance of resistance. Our selection system is based on the finding that functional interactions exist between the N- and C-terminal domains of the tetracycline efflux pump. As a consequence of this interaction, a deleterious mutation in one domain of the protein might be suppressed by some second-site mutation within the other domain. We attempted to achieve second-site mutations by error-prone transcription by a T7 RNA polymerase variant, followed by normal translation.

Indeed, we observed tetracycline-resistant cells that were grown until the stationary phase after 15 h incubation. We isolated the mixture of both plasmids using a standard plasmid preparation procedure *(24)*. Sequence analysis of the T7 RNA polymerase variant encoded by plasmid pPolstar was achieved by direct sequencing using the plasmid mixture as a template (*see* **Note 3**). We employed Dideoxy sequencing using the *Taq* dye terminator FS kit with resolution of the products on an ABI 373A sequencer.

3.4. Identification and Characterization of Selected Variants

3.4.1. Protein Preparation

While in vivo assays can be performed for characterizing some general polymerase features, only purified enzyme that is assessed in vitro will yield a detailed characterization of activity and substrate selectivity. In the case of T7 RNA polymerase (and its variants), full length protein can be isolated only from *E. coli* strain BL21, which lacks an outer membrane protease *(ompT)*. Strains expressing *ompT* will hydrolyze the 98-kD protein into two fragments (75 kD and 23 kD [*21*]). We purified T7 RNA polymerase, either wild-type or selected variant, following the detailed protocol given by Grodberg and Dunn *(21)*.

3.4.1.1. RUNOFF TRANSCRIPTION

1. Set up two transcription reactions, one with wild-type T7 RNA polymerase and another one with the selected polymerase variant: 10–15 µg linear DNA template (approx 1 pmol), 1 m*M* of each NTP, and 1 pmol T7 RNA polymerase in 250 µL 1X transcription buffer.
2. Incubate both reactions for 3 h at 37°C.
3. Extract once with 250 µL phenol and once with 250 µL chloroform.
4. Precipitate the DNA by adding 1/10 volume of 3.5 *M* sodium acetate and 2 volumes ethanol. Incubate at least 1 h at –20°C.
5. Centrifuge 30 min at 4°C with spinning at 21,700*g*.
6. Decant the supernatant and wash the RNA pellet with 300 µL 70% ethanol. Repeat the centrifugation (**step 5**).
7. Dissolve the dried RNA pellet in 250 µL of water.
8. Purify the the RNA by agarose gel electrophoresis (1.2% agarose), excise the RNA band of correct length using a sharp scalpel, and elute the RNA using a spin kit.

3.4.2. Reverse Transcription and Cloning

1. For reverse transcription, mix approx 0.5–1 pmol RNA with 10 pmol forward primer, heat to 100°C for 3 min, and immediately cool on ice.
2. Add reverse transcriptase buffer (final concentration: 1X), dithiothreitol (to 1 m*M*), dNTPs (to 125 µ*M* each), reverse transcriptase (20 U), and water to a final volume of 20 µL.
3. Incubate 1 h at 42°C; subsequently, inactivate the enzyme during 15 min at 72°C.
4. For second-strand synthesis, add 25 pmol reverse primer, heat to 100°C for 3 min, and immediately cool on ice.
5. Add PCR buffer, dNTPs (to 200 µ*M* each), exonuclease-deficient *Vent* polymerase (1 U), and water to a final volume of 50 µL.

6. Incubate for 30 min at 72°C.
7. Add 25 pmol forward primer, PCR buffer, dNTPs (to 200 μM each), *Pfu* polymerase, and water to a final volume of 100 μL. Perform PCR cycling following standard protocols.
8. Ligate the PCR product with the linear vector pCR2.1 according to the supplier's protocol.
9. Transform *E. coli* with the ligation product.
10. Prepare plasmids from 20–30 individual clones and submit these to a sequence analysis of the region resulting from in vitro transcription.

4. Notes

1. PCR mutagenesis emerged as the most common technique for random mutagenesis, which can result in mutation frequencies as high as 2% per nucleotide position. With some alterations of the PCR conditions, the mutation rate may also be adjusted to lower values. A major drawback is in the ligation and cloning of randomly mutated PCR products: since ligation efficiencies are low in the case of longer DNA fragments (≥ 1000 base pairs), time- and resource-consuming scale-ups of ligation and transformation are necessary in order to yield "functional libraries" of sufficient diversity (that is, a high number of bacterial colonies).
2. Inventing selection schemes for "low-fidelity nucleic acid synthesis" is a complex, nontrivial task: Why, for example, should *E. coli* utilize a plasmid-encoded polymerase that introduces replication errors? Furthermore, it should be noticed that the implementation of genetically defined deficiencies may force the microbial host to circumvent the applied constraint by utilizing a yet unexpected pathway that is not necessarily related to the targeted activity.
3. The sequencing primer combinations used by us were tested in normal PCR experiments prior to the sequencing procedure, and showed high selectivity. If this is not the case, the two plasmids must be separated, e.g., by preparative agarose gel electrophoresis and subsequent elution from the gel.

References

1. Irvine, D., Tuerk, C., and Gold, L. (1991) SELEXION. Systematic evolution of ligands by exponential enrichment with integrated optimization by non-linear analysis. *J. Mol. Biol.* **222,** 739–761.
2. Wilson, D. S. and Szostak, J. W. (1999) In vitro selection of functional nucleic acids. *Ann. Rev. Biochem.* **68,** 611–647.
3. Sweasy, J. B. and Loeb, L. A. (1993) Detection and characterization of mammalian DNA polymerase β mutants by functional complementation in *Escherichia coli.* *Proc. Natl. Acad. Sci. USA* **90,** 4626–4630.
4. Washington, S. L., Yoon, M. S., Chagovetz, A. M., Li, S. X., Clairmont, C. A., Preston, B. D., et al. (1997) A genetic selection system to identify DNA polymerase β mutator mutants. *Proc. Natl. Acad. Sci. USA* **94,** 1321–1326.
5. Kim, B., Hathaway, T. R., and Loeb, L. A. (1996) Human immunodeficiency virus reverse transcriptase. Functional mutants obtained by random mutagenesis coupled with genetic selection in *Escherichia coli.* *J. Biol. Chem.* **271,** 4872–4878.

6. Brakmann, S. and Grzeszik, S. (2001) An error-prone T7 RNA polymerase mutant generated by directed evolution. *ChemBiochem* **2,** 212–219.

7. Ghadessy, F. J., Ong, J. L., and Holliger, P. (2001) Directed evolution of polymerase function by compartmentalized self-replication. *Proc. Natl. Acad. Sci. USA* **98,** 4552–4557.

8. Xia, G., Chen, L., Sera, T., Fa, M., Schultz, P. G., and Romesberg, F. E. (2002) Directed evolution of novel polymerase activities: Mutation of a DNA polymerase into an efficient RNA polymerase. *Proc. Natl. Acad. Sci. USA* **99,** 6567–6602.

9. Brakmann, S. (2001) Discovery of superior enzymes by directed evolution. *ChemBiochem* **2,** 865–871.

10. Horwitz, M. S. and Loeb, L. A. (1986) Promoters selected from random DNA sequences. *Proc. Natl. Acad. Sci. USA* **83,** 7405–7409.

11. Leung, D. W., Chen, E., and Goeddel, D. V. (1989) A method for random mutagenesis of a defined DNA segment using a modified polymerase chain reaction. *Technique* **1,** 11–15.

12. Cadwell, R. C. and Joyce, G. F. (1992) Randomization of genes by PCR mutagenesis. *PCR Meth. Appl.* **2,** 28–33.

13. Stemmer, W. P. C. (1994) DNA shuffling by random fragmentation and reassembly: In vitro recombination for molecular evolution. *Proc. Natl. Acad. Sci. USA* **91,** 10,747–10,751.

14. Zhao, H., Giver, L., Shao, Z., Affholter, J. A., and Arnold, F. H. (1998) Molecular evolution by staggered extension process StEP in vitro recombination. *Nat. Biotechnol.* **16,** 258–261.

15. Shao, Z., Zhao, H., Giver, L., and Arnold, F. H. (1998) Random-priming in vitro recombination: An effective tool for directed evolution. *Nucleic Acids Res.* **26,** 681–683.

16. Smith, G. P. and Petrenko, V. A. (1997) Phage display. *Chem. Rev.* **97,** 391–410.

17. Johnsson, K. and Ge, L. (1999) Phage display of combinatorial peptide and protein libraries and their application in biology and chemistry. *Curr. Top. Microbiol. Immunol.* **243,** 87–105.

18. Hanes, J. and Plückthun, A. (1997) In vitro selection and evolution of functional proteins by using ribosome display. *Proc. Natl. Acad. Sci. USA* **94,** 4937–4942.

19. Roberts, R. W. and Szostak, J. W. (1997) RNA-peptide fusions for the in vitro selection of peptides and proteins. *Proc. Natl. Acad. Sci. USA* **94,** 12,297–12,302.

20. Tawfik, D. S. and Griffith, A. D. (1998) Man-made cell-like compartments for molecular evolution. *Nat. Biotechnol.* **16,** 652–656.

21. Grodberg, J. and Dunn, J. J. (1988) ompT encodes the *Escherichia coli* outer membrane protease that cleaves T7 RNA polymerase during purification. *J. Bacteriol.* **170,** 1245–1253.

22. Cox, E. C. (1976) Bacterial mutator genes and the control of spontaneous mutation. *Ann. Rev. Genet.* **10,** 135–156.

23. Greener, A., Callaghan, M., and Jerpseth, B. (1996) An efficient random mutagenesis technique using an *E. coli* mutator strain. *Meth. Mol. Biol.* **57,** 375–385.

24. Sambrook, J., Fritsch, E. F., and Maniatis, T. (eds.) (1989) *Molecular Cloning: A Laboratory Manual,* Cold Spring Harbor Laboratory Press, Cold Spring Harbor, NY.

25. Davanloo, P., Rosenberg, A. H., Dunn, J. J., and Studier, F. W. (1984) Cloning and expression of the gene for bacteriophage T7 RNA polymerase. *Proc. Natl. Acad. Sci. USA* **81,** 2035–3039.

26. Eigen, M. and Gardiner, W. (1984) Evolutionary molecular engineering based on RNA replication. *Pure Appl. Chem.* **56,** 967–978.

4

L-Glutaminase as a Therapeutic Enzyme of Microbial Origin

Abdulhameed Sabu, K. Madhavan Nampoothiri, and Ashok Pandey

Summary

Enzymes control and regulate all biochemical processes in the body. In a single second, several million enzymes mediate chemical reactions occurring in a living system. Enzymes are highly specific in their function because each enzyme is programmed to carry out one special task. The immense number of enzymes acts as a perfect troop to ensure that enormously complex life mechanisms and processes occur in the right direction. A sufficient amount and optimal function of enzymes present in the human body are essential for life and health. Microbial enzymes play a major role in the diagnosis, curing, biochemical investigation, and monitoring of many dreaded diseases. Microorganisms represent an excellent source of many therapeutic enzymes owing to their broad biochemical diversity and their susceptibility to genetic manipulation. By considering the requirement for large-scale production of such enzymes, a brief outline of the various production techniques, and downstream processing are discussed on the basis of L-glutaminase as a typical microbial therapeutic enzyme.

Key Words: Therapeutic enzyme; L-glutaminase; solid-state fermentation; submerged fermentation; immobilized cells.

1. Introduction

The manufacture or processing of enzymes for use as drugs is an important facet of today's pharmaceutical industry (*1*). Attempts to capitalize on the advantages of enzymes as drugs are now being made at virtually every pharmaceutical research center in the world. Since the later years of the 19th century, crude proteolytic enzymes have been used for gastrointestinal disorders. In fact, other than as digestion aids, enzymes were largely ignored as drugs until Emmerich and his associates observed in 1902 that an extracellular secretion of *Bacillus pyocyaneus* was capable of killing anthrax bacilli. Emmerich deduced that the secretion in question was a nuclease, i.e., it was acting by enzymatically degrading nucleic acids (*2*). This milestone study gradually opened the way for the use

From: *Methods in Biotechnology, Vol. 17: Microbial Enzymes and Biotransformations*
Edited by: J. L. Barredo © Humana Press Inc., Totowa, NJ

Table 1
Some Important Therapeutic Enzymes and Their Applications

Enzyme	Application	Source
L-Glutaminase	Antitumor	*Beauveria bassiana, Vibrio costicola, Zygosaccharomyces rouxii, E. coli.*
L-Asparaginase	Antitumor	*Pseudomonas acidovorans, Acinetobacter* sp.
Superoxide dismutase	Antioxidant, anti-inflammatory	*Mycobacterium* sp., *Nocardia* sp.
Serratio peptidase	Anti-inflammatory	*Serratia marcescens*
Penicillin acylase	Antibiotic production	*E. coli*
Lipase	Lipid digestion	*Candida lipolytica, Candida rugosa, Aspergillus oryzae*
Laccase	Detoxifier	*Trametes versicolor*
Glucosidase	Antitumor	*Aspergillus niger*
β-lactamase	Penicillin allergy	*Citrobacter freundii, Serratia marcescens, Klebsiella pneumoniae*
Protease	Antitumor, digestive aid	*Bacillus polymyxa, Beauveria bassiana*

of parenteral enzymes in the treatment first of infections, then of cancer, and finally of a diverse spectrum of diseases. Enzyme supplements are available in pills, capsules, and powders that often consist of combinations of several different enzymes. John Beard, an English scientist, reported first use of pancreatic enzymes to treat cancer in 1902 *(3)*. He proposed that pancreatic proteolytic enzymes, in addition to their well-known digestive function, represent the body's main defense against cancer. During the first two decades of this century, a number of physicians, both in Europe and in the United States, used injectable pancreatic enzymes to treat advanced human cancer, often with great success.

Therapeutic enzymes have a broad variety of specific uses as oncolytics, as anticoagulants or thrombolytics, as anti-inflammatory agents, and as replacements for metabolic deficiencies. Information on the utilization of microbial enzymes for therapeutic purposes is scarce and the available reports are largely on some anticancer enzymes and on some enzymes active against cystic fibrosis. Development of medical applications for enzymes has been at least as extensive as those for industrial applications, reflecting the magnitude of the potential rewards. A selection of those enzymes that have realized this potential to become important therapeutic agents is shown in **Table 1.**

Therapeutic enzymes are widely distributed in plants, animal tissues, and microorganisms including bacteria, yeast, and fungi. Although microorganisms are a very good potential source of therapeutic enzymes, utilization of microbial enzymes for therapeutic purpose is limited because of their incompatibility with the human body. But there is an increased focus on utilization of microbial enzymes because of their economic feasibility.

1.1. Microbial Enzymes

Microbial enzymes are preferred over plant or animal sources due to their economic production, consistency, ease of process modification, and optimization. They are relatively more stable than corresponding enzymes derived from plants or animals. Further, they provide a greater diversity of catalytic activities. The majority of enzymes currently used in industry are of microbial origin. But once we enter into the therapeutic applications of microbial enzymes, a number of factors severely reduce their potential utility. One of the major problems is due to the large molecular size of the biological catalysts, which prevents their distribution within the somatic cells. Another important problem related to enzyme therapy is the elicitation of immune response in the host cells after injecting the foreign enzyme protein. Through the technique of drug targeting and disguising the enzyme as an apparently nonproteinaceous molecule by covalent modification these problems can be alleviated. L-glutaminase modified by covalent attachment of polyethylene glycol has been shown to retain its antitumor effect while possessing no immunogenicity. Other methods such as entrapment of the enzyme within artificial liposomes, synthetic microspheres and red blood cell ghosts have also been found useful. These inherent problems necessitate the requirement of therapeutic enzymes with a very high degree of purity and specificity *(4)*. In general the favored kinetic properties of these enzymes are low K_m and high V_{max} in order to be maximally efficient even at very low enzyme and substrate concentrations.

1.2. Production Techniques for Microbial Enzymes

There are different methods of fermentation technology, which enables the production of microbial enzymes. As an example, the production techniques and simple downstream processing for L-glutaminase are described in this chapter.

1.2.1. Isolation of Organisms

In an industrial fermentation process, the first step is to isolate potent strains from various sources such as soil, marine sediments, and the like. This will follow screening program to test a large number of isolates. In general a medium containing L-glutamine as the sole carbon and nitrogen source is used for screening glutaminase-producing organisms.

1.2.2. Production Techniques

In general, there are different methods of fermentation technology, which enables the production of most of the microbial enzymes. On a commercial scale, liquid cultures in huge bioreactors are preferred for the bulk production of therapeutic enzymes like glutaminase. Other processes such as solid-state fermentation, fermentation on inert solid supports, or immobilized cells are also widely used for the production of glutaminase. Glutaminase production has been extensively studied by submerged fermentation (SmF) as well as solid-state fermentation (SSF) processes.

1.2.2.1. SUBMERGED FERMENTATION

Submerged fermentation consists of the cultivation of microbial cells in liquid media under controlled conditions, in large vessels called bioreactors, for the production of desirable metabolites. Submerged fermentation offers advantages such as easy online monitoring of process parameters and process automation. Many of the culture medium components are common inexpensive substances containing carbon (e.g., glucose, sucrose, maltose) and nitrogen sources (e.g., ammonium salts, sodium nitrate), vitamins, and minerals and that can be supplied in uniform quantity.

1.2.2.2. SOLID-STATE FERMENTATION

Solid-state fermentation is the culture of microorganisms on moist solid substrates in the absence or near-absence of free water. It is also described as any fermentation process that takes place on solid or semisolid substrate or that occurs on a nutritionally inert solid support, which provides some advantages to the microorganisms with respect to access to nutrients, and the product derived will be of high purity. Recently SSF has generated much interest, because it is a simpler process and the feasibility to use various agroindustrial residues such as wheat bran, oil cakes, and so on. As inert support, materials like polystyrene beads, polyurethane foam, and the like can be used. SSF can be performed in Erlenmeyers flasks, trays, or in glass columns. Nowadays even big bioreactors are also designed for conducting SSF with full automation.

1.2.2.3. IMMOBILIZED CELLS

Immobilization of cells can be defined as the attachment of cells or their inclusion in a distinct solid phase that permits exchange of substrates, products, inhibitors, and so forth but at the same time separates the catalytic cell biomass from the bulk phase containing substrates and products. Cells entrapped in either sodium alginate or agar are widely used.

1.3. Recovery and Purification of L-Glutaminase

Product recovery during any bioprocess is difficult since the diluted and labile products of interest are always mixed with macromolecules of similar properties. Based on the production technique, downstream processes can be designed. It is usually achieved by filtration, centrifugation, and precipitation, and also by various chromatographic procedures such as ion exchange, gel-permeation, and affinity chromatography. To confirm purity of the enzyme and to determine molecular weight, electrophoresis is performed using denaturing polyacrylamide gel electrophoresis (SDS-PAGE).

1.4. Glutaminase Assay

A colorimetric assay using Nessler's reagent to detect the liberated ammonia as a result of the breakdown of the amide substrate (L-glutamine) by the action of enzyme is widely used to determine the glutaminase activity in crude and purified samples.

2. Materials

2.1. Isolation of Glutaminase-Producing Organisms

1. Sample (soil, marine sediment, etc.).
2. Nutrient agar medium: 5.0 g/L peptic digest of animal tissue, 1.5 g/L beef extract, 1.5 g/L yeast extract, 5.0 g/L NaCl, and 15 g/L agar. Final pH at 25°C is 7.4.
3. Minimal agar medium: 0.5 g/L KCl, 0.5 g/L $MgSO_4 \cdot 7H_2O$, 1 g/L KH_2PO_4, 0.1 g/L $FeSO_4 \cdot 7H_2O$, 0.1 g/L $ZnSO_4 \cdot 7 H_2O$, 0.5% L-glutamine, and 0.012 g/L phenol red.

2.2. Production of L-Glutaminase

1. *Pseudomonas* sp. (collected from Cochin, India).
2. ZoBell's marine medium (HiMedia; Mumbai, India): 1.0 g/L yeast extract, 0.1 g/L ferric citrate, 5.0 g/L peptone, 19.45 g/L sodium chloride, 5.9 g/L magnesium chloride, 3.24 g/L sodium sulfate, 1.8 g/L calcium chloride, 0.55 g/L potassium chloride, 0.16 g/L sodium bicarbonate, 0.08 g/L potassium bromide, 0.034 g/L strontium chloride, 0.022 g/L boric acid, 0.004 g/L sodium silicate, 0.0024 g/L sodium fluoride, 0.0016 g/L ammonium nitrate, and 0.008 g/L disodium phosphate. Adjust to pH 7.6 ± 0.2.
3. Mineral salts glutamine (MSG) medium: 1.0 g/L KH_2PO_4, 0.5 g/L $MgSO_4 \cdot 7H_2O$, 0.1 g/L $CaCl_2$, 0.1 g/L $NaNO_3$, 0.1 g/L trisodium citrate, 10.0 g/L NaCl, 5.0 g/L glucose, and 10.0 g/L L-glutamine. Adjust to pH 6.0 *(5)*.
4. Physiological saline: 0.85% NaCl.
5. Bioreactor (e.g., Erlenmeyer flasks, fermenter).
6. Temperature-controlled orbital shaker.
7. Cooling centrifuge.
8. Solid substrate (e.g., agricultural residues such as wheat bran or inert materials like polystyrene beads).
9. Moistening medium: seawater.
10. Temperature regulated incubation chamber.
11. 0.1 *M* phosphate buffer, pH 7.0.
12. 1 *N* NaOH.

13. 1 *N* HCl.
14. *Beauveria bassiana* BTMF S 10 *(6)*.
15. Bennet's agar medium (Hi-Media): 2 g/L casein enzymatic hydrolysate, 1 g/L beef extract, 1 g/L yeast extract, 10 g/L dextrose, and 20 g/L agar. Adjust pH to 7.3 ± 0.2.
16. Production medium: 0.25% L-glutamine, and 0.5% D-glucose in aged seawater. Adjust to pH 9.0. Autoclave at 121°C.
17. 2% sodium alginate.
18. 0.2 *M* CaCl$_2$.
19. Syringe with a pore size of 2 mm.
20. Glass rod.

2.3. Recovery and Purification of L-Glutaminase

1. 0.01 *M* phosphate buffer, pH 8.0.
2. Magnetic stir plate and stir bar.
3. Centrifuge.
4. Dialysis tubing.
5. Glassware.
6. Lyophilizer.
7. Solution A: 1% copper sulfate.
8. Solution B: 2% sodium potassium tartrate.
9. Solution C: 2% sodium carbonate in 0.1 *N* NaOH.
10. Alkaline copper reagent: Mix solutions A, B, and C in ratio 1:1:98.
11. Folin's reagent: 750 mL distilled water, 100 g sodium tungstate, 25 g sodium molybdate, 50 mL of 85% phosphoric acid, 100 mL concentrated HCl, 150 g lithium sulfate, and a few drops of bromine (boiled off).
12. Chromatography column (2.5 × 50 cm).
13. Sephadex G 200 (Pharmacia, Stockholm, Sweden).
14. Fraction collector (Pharmacia).
15. Affinity column filled with L-glutamine activated by cyanogen bromide in 4% beaded agarose (Sigma, St. Louis, MO).
16. Eluent: 2 *M* NaCl in phosphate buffer, pH 8.0.
17. Polyacrylamide gel electrophoresis (SDS-PAGE) equipment.
18. Power pack.
19. 0.002% bromophenol blue.
20. 10% sucrose.
21. Wide-range marker proteins (Sigma).
22. Acrylamide-bisacrylamide (30:0.8): Dissolve 30 g of acrylamide and 0.8 g of bisacrylamide in a total volume of 100 mL of distilled water. Filter the solution through Whatman No. 1 filter paper and store at 4°C in a dark bottle.
23. Whatman No. 1 filter paper (Whatman, Kent, UK).
24. 1.5% (w/v) ammonium persulfate. Prepare fresh just before use.
25. 10% (w/v) sodium dodecyl sulfate (SDS).
26. Stacking gel buffer stock: 0.5 *M* Tris-HCl, pH 6.8. Dissolve 6 g of Tris in 40 mL of distilled water, titrate to pH 6.8 with 1 *M* HCl, and bring to 100 mL final vol-

ume with distilled water. Filter the solution through Whatman No. 1 and store at 4°C.

27. Resolving gel buffer stock: 3.0 *M* Tris-HCl, pH 8.8. Mix 36.3 g of Tris and 48 mL of 1 *M* HCl, and bring to 100 mL with distilled water. Filter through Whatman No. 1 and store at 4°C.

28. Reservoir buffer stock: 0.25 *M* Tris-HCl, 1.92 *M* glycine, and 1% SDS, pH 8.3. Dissolve 30.3 g of Tris-HCl, 144.0 g of glycine, and 10 g of SDS and make up to 1 L with distilled water. Store at 4°C.

29. Slab gels, 1.5-mm thickness. Prepare with either 7.5% or 10% resolving gel for SDS-PAGE.

30. Stacking gel 2.5%.

2.4. Glutaminase Assay

1. Crude enzyme (dialyzed).
2. 0.04 *M* L-glutamine.
3. 0.1 *M* phosphate buffer, pH 8.0.
4. 1.5 *M* Trichloroacetic acid (TCA).
5. Nessler's reagent (HiMedia).
6. Glassware.
7. UV spectrophotometer.

3. Methods

3.1. Isolation of Glutaminase-Producing Organisms

From 0.5 to 1 g of the sample (e.g., soil) was suspended in 5 mL sterile distilled water and vigorously mixed. Aliquots of the clear suspension were plated, after serial dilution, onto minimal agar medium containing 0.5% L-glutamine as the sole carbon and nitrogen source and phenol red as pH indicator. A change from yellow to pink color (*see* **Note 1**) of the agar medium due to the growth of a particular microbial colony is an indication of a change in pH due to the breakdown of amide bond (ammonia is liberated) present in glutamine and it is an indication of the extracellular presence of L-glutaminase produced by the colony.

3.2. Production of L-Glutaminase by Pseudomonas sp.

3.2.1. Maintenance of Culture

The culture is maintained on ZoBell's marine agar slants and subcultured once a month.

3.2.2. Inoculum Preparation

1. Prepare mineral salts glutamine (MSG) medium for the culture of the inoculum (*5*).
2. Transfer aseptically a loopful of cells from 18–24-h-old slant culture to 10 mL of MSG medium in a test tube, and incubate for 24 h on a rotary shaker at 150 rpm and 30°C.

3. Take 1 mL of the culture and inoculate into 100 mL of sterile MSG medium in a 250-mL flask, and incubate for 24 h at 150 rpm at 30°C.
4. After incubation, harvest the cells by centrifugation for 15 min at 8000g and 4°C.
5. Wash twice with sterile physiological saline, and resuspend in 10 mL physiological saline.
6. Transfer aseptically the cell suspension to 1000 mL of sterile MSG medium in a 3-L flask, and incubate at 150 rpm for 24 h at 30°C.
7. Harvest the cells by centrifugation under sterile conditions for 15 min at 8000g and 4°C.
8. Wash twice with sterile physiological saline, resuspend in 100 mL physiological saline, and use as inoculum. 1 mL of the finally prepared cell suspension contains around 10 mg dry weight cells.

3.2.3. Enzyme Production

3.2.3.1. ENZYME PRODUCTION BY SUBMERGED FERMENTATION

1. Prepare mineral salts glutamine (MSG) medium.
2. Inoculate the flasks with 2–5% (v/v) inoculum.
3. Incubate the inoculated flasks at 180 rpm for the desired interval of time.

3.2.3.2. ENZYME PRODUCTION BY SOLID-STATE FERMENTATION

1. Autoclave polystyrene beads of 2–3 mm diameter at 121°C for 15 min (*see* **Note 2**). Use beads of uniform size (1–1.5 mm) for fermentation studies (*see* **Note 3**).
2. Place 10 g of pretreated polystyrene beads in 250-mL Erlenmeyer flasks, and moisten with enzyme production medium (autoclaved and cooled to room temperature before inoculation). If the solid substrate is a nutrient one, such as wheat bran, oil cakes, etc. *(7)*, the substrate doesn't need any pretreatment; a known amount is moistened with salt solution and autoclaved (*see* **Note 4**).
3. Inoculate the sterilized solid substrate media with the inoculum (10 mg dry weight cells/10 g dry solids; arbitrarily selected, need optimization).
4. Mix the contents thoroughly and incubate at 37°C for 24 h under 80% relative humidity (*see* **Note 5**).

3.2.4. Enzyme Recovery

3.2.4.1. ENZYME RECOVERY BY SUBMERGED FERMENTATION

1. Remove the flasks and centrifuge the whole content at 10,000g for 15 min at 4°C to separate the biomass.
2. Use the clear supernatant as the crude enzyme.

3.2.4.2. ENZYME RECOVERY BY SOLID-STATE FERMENTATION

Enzyme from the fermented solid substrates can be extracted after polystyrene solid state fermentation by simple contact method using suitable extractant medium before optimization of the various process parameters (*see* **Note 6**) that influence recovery of enzyme, toward maximum recovery of enzyme.

Table 2
L-glutaminase Production by *Pseudomonas* sp. under Solid-State Fermentation Using Polystyrene as Inert Support

Parameters optimized	Optimum condition
Medium	Polystyrene + seawater/distilled water
Initial moisture	70%
pH	5.0
Substrate (L-glutamine)	1–2%
Concentration of NaCl	1%
Additional carbon source	1% glucose
Inoculum size	3% v/w
Incubation time	18–24 h
Incubation temperature	35°C
Extraction medium	Distilled water + 1% NaCl
Buffer for enzyme recovery	0.1 M phosphate buffer, pH 8
Polystyrene-to-buffer ratio	1:8
Maximum yield obtained	159 U/gds (beads + seawater)

Table 3
L-Glutaminase Production by Saline-Tolerant Yeast *Zygosaccharomyces rouxii* under Solid-State Fermentation Using Wheat Bran (WB)/Sesame Oil Cake (SOC)

Parameters optimized	Optimum condition
Moistening medium	10% NaCl in tap water (WB)/seawater (SOC)
Incubation temperature	30°C
Initial moisture	64%
Inoculum size	2 mL of 48-h-old inoculum /5g substrate
Additional carbon source	Maltose (SOC), glucose (WB), 1% w/w
Incubation time	48 h
Maximum yield	7.57 U/gds (WB), 11.61 U/gds (SOC)

1. Mix the solid fermented substrates with 0.1 M phosphate buffer, pH 7.0, and allow them to remain in contact for 30 min.
2. Press the contents in dampened cheesecloth to recover leachate.
3. Repeat the process twice.
4. Pool the extracts, and centrifuge at 10,000g for 15 min at 4°C in a refrigerated centrifuge.
5. Use the supernatant as the crude enzyme for further assays (*see* **Note 7**).
   Tables 2 and **3** summarize some of the results based on the above experiments.

3.2.5. Parameter Optimization Studies in Solid-State Fermentation

3.2.5.1. INITIAL MOISTURE CONTENT

The effect of initial moisture content of the solid medium for enzyme production is determined by preparing the solid substrates with varying levels of moisture content in the range of 40–90%. This can be achieved by altering the amount of distilled water used in the medium to moisten the substrate.

3.2.5.2. INITIAL pH OF THE MEDIUM

The effect of initial pH of the distilled-water-based medium on enzyme production by *Pseudomonas* sp. during SSF can be studied at a pH range from 4.0 to 11.0, adjusted in the enzyme production medium using 1 N NaOH or 1 N HCl.

3.2.5.3. INCUBATION TEMPERATURE

Optimal temperature required for maximal enzyme production by *Pseudomonas* sp. under SSF can be determined by incubating the inoculated flasks at a temperature range from 20°C to 60°C under 75–80% relative humidity.

3.2.5.4. SUBSTRATE (L-GLUTAMINE) CONCENTRATION

The influence of L-glutamine concentration on glutaminase production by *Pseudomonas* sp. under SSF can be evaluated by supplementing the fermentation medium with different concentrations of glutamine (0.25–3.0% w/v) in the SSF medium.

3.2.5.5. NaCl CONCENTRATION

Optimal NaCl concentration enhancing enzyme production of *Pseudomonas* sp. under SSF can be determined by incorporating NaCl at different concentrations (0.0–10.0% w/v) in the medium.

3.2.5.6. ADDITIONAL NITROGEN SOURCES

Requirement of additional nitrogen sources other than L-glutamine for enhanced enzyme production by *Pseudomonas* sp. can be tested by the addition of different nitrogen sources (beef extract, yeast extract, peptone, glutamic acid, lysine, KNO_3, and $NaNO_3$) at 1% (w/v) in the SSF medium.

3.2.5.7. ADDITIONAL CARBON SOURCES

Requirement of additional carbon sources other than L-glutamine for enhanced enzyme production by *Pseudomonas* sp. can be tested by the addition of different carbon sources (glucose, maltose, starch, lactose, and trisodium citrate) at 1% (w/v) in the SSF medium.

3.2.5.8. Inoculum Concentration

Optimal inoculum concentration for maximal enzyme production under SSF by *Pseudomonas* sp. can be determined at different inoculum concentrations varying from 1.0% to 5.0% (v/w).

3.2.5.9. Incubation Time (Time-Course Experiment)

Optimal incubation time leading to maximal enzyme production by *Pseudomonas* sp. under optimized conditions of SSF can be determined by estimating enzyme yield at regular time intervals.

3.2.6. L-Glutaminase Production by Immobilized Marine
Beauveria bassiana

3.2.6.1. Maintenance of Culture

Beauveria bassiana BTMF S 10 was isolated from marine sediments *(6)*. It is maintained on Bennet's agar slants and subcultured once a month.

3.2.6.2. Preparation of Spore Inoculum

1. Prepare agar slope cultures of *B. bassiana* on Bennet's agar prepared in aged seawater.
2. Add 20 mL sterile physiological saline containing 0.1% Tween-80 to fully sporulated (2 wk old) agar slope cultures.
3. Remove the spores carefully under aseptic conditions by using an inoculation needle.
4. Adjust spore suspension to 12×10^8 spores/mL.
5. Use this suspension as inoculum.

3.2.6.3. Preparation of Beads

1. Mix the spore suspension under sterile conditions with 2% (w/v) sodium alginate solution, at a concentration of 12×10^8 spores /mL and a ratio of 1:2.
2. Mix thoroughly using a sterile glass rod to get spore-alginate slurry.
3. Extrude the slurry dropwise into a solution of 0.2 M $CaCl_2$ from a height of about 10 cm using a syringe with a pore size of 2 mm.
4. Maintain the entrapped calcium alginate beads for 2 h in a solution of 0.2 M $CaCl_2$ for curing.
5. Wash the beads thoroughly 3–4 times with physiological saline and maintain at 4°C until use.

3.2.6.4. Activation of Immobilized Spores

1. Suspend the prepared beads with immobilized fungal spores in a solution of 1% glutamine in aged seawater, pH 9.0, in a conical flask.
2. Incubate on a rotary shaker at 90 rpm and 30°C for 12 h.
3. Remove the activated beads and wash twice with fresh enzyme production medium. The beads are now ready to be used.

Table 4
L-Glutaminase Production by Immobilized Marine
Beauveria bassiana

Parameters optimized	Optimum conditions
Concentration of sodium alginate	3%
Concentration of CaCl$_2$ solution	0.1 M
Initial biomass	12×10^8 spores/mL
Curing time	3 h
Activation time	15 h
Retention time	18 h
pH	9.0
Incubation temperature	27°C
Maximum yield	64.46 U/mL

3.2.6.5. INCUBATION PROCEDURE

1. Weigh aseptically 25 g of beads with immobilized spores.
2. Transfer to 50 mL of enzyme production medium in 250-mL conical flask.
3. Keep on rotary shaker (90 rpm) and incubate for desired interval of time at 30°C.

3.2.6.6. OPTIMIZATION OF IMMOBILIZATION PROCESS CONDITIONS

The important parameters to be considered for optimization of the immobilization process are: (1) alginate concentration, (2) calcium chloride concentration, (3) initial biomass used to prepare the beads, (4) curing time for the beads in CaCl$_2$ solution, (5) activation time required for the beads before transferring to the production medium, and (6) retention time for maximum enzyme production.

Table 4 summarizes some of the results based on the above experiments.

3.3. Recovery and Purification of L-Glutaminase

3.3.1. Ammonium Sulfate Precipitation

1. Slowly add finely powdered enzyme-grade ammonium sulfate into the crude enzyme preparation so as to reach 40% saturation while being stirred continuously at 4°C on a magnetic stirrer.
2. Remove the precipitated protein by centrifugation at 10,000g for 20 min at 4°C.
3. Add fresh ammonium sulfate to the supernatant to increase the concentration to 50% and resuspend the precipitate obtained in a minimal volume of 0.01 M phosphate buffer, pH 8.0.
4. Add fresh ammonium sulfate to the supernatant to increase the concentration to 80% and resuspend the precipitate obtained in a minimal volume of 0.01 M phosphate buffer, pH 8.0 (*see* **Note 8**).

5. Determine the protein content of the fraction by Lowry's method *(8)*: Add 5 mL of alkaline copper reagent to 1 mL of the enzyme. Mix contents thoroughly and leave for 10 min. Add 0.5 mL of Folin's reagent diluted with an equal volume of water to each tube. Measure OD_{750} after 40 min in a UV-visible spectrophotometer. Use bovine serum albumin (BSA) as standard.

3.3.2. Dialysis

1. Dialyze the precipitate obtained after ammonium sulfate fractionation against 0.01 *M* phosphate buffer, pH.8.0, for 24 h at 4°C with continuous stirring.
2. Change the buffer occasionally.

3.3.3. Gel Filtration

1. Concentrate the dialyzate by lyophilization.
2. To prepare a Sephadex G200 column, suspend 7 g of Sephadex G200 in 0.1 *M* phosphate buffer, pH 8.0, and place in a boiling water bath for 5 h.
3. Stir the solution occasionally during the pretreatment procedure.
4. Pour the swollen gel into a chromatographic column (2.5 × 50 cm) and allow settling under gravity while maintaining a slow flow rate through the column. Take care to avoid trapping of air bubbles in the column.
5. Stabilize the column at 4°C and equilibrate by passing above 3 bed volumes of 0.1 *M* phosphate buffer, pH 8.0.
6. Apply the lyophilized dialyzate to the top of the column and elute with buffer at 4°C.
7. Adjust the flow rate to about 50 mL/h and collect fractions of 2.5 mL at 4°C using a fraction collector.
8. Analyze each fraction for glutaminase and protein (*see* **Note 9**).
9. Pool the active fractions and concentrate by lyophilization.

3.3.4. Affinity Chromatography

1. Pool the active fractions, which were collected after gel filtration, and concentrate by lyophilization.
2. To prepare an affinity column, take the column material (L-glutamine activated by cyanogen bromide in 4% beaded agarose) as 10-mL aliquots, and incubate in a beaker for 30 min at 4°C.
3. Pack the gel in a column and allow settling under gravity. Take care to avoid trapping of air bubbles in the column. Stabilize the column and equilibrate by passing above 3 bed volumes of buffer.
4. Apply the lyophilized sample, after Sephadex G200 purification, on the top of the affinity column, and elute at 4°C.
5. Pass samples repeatedly 3 times.
6. Pass the eluent at a flow rate of 25 mL/h at 4°C, and collect fractions of 2.5 mL.
7. Analyze each fraction for glutaminase and protein.
8. Pool the active fractions, dialyze to remove NaCl, and concentrate by lyophilization.
9. Do electrophoretic studies to confirm purity of enzyme.

3.3.5. SDS-PAGE

1. Prepare the sample for SDS-PAGE by mixing 100 μL enzyme solution, 30 μL bromophenol blue (0.002%), 40 μL 10% SDS, 20 μL 10% sucrose, and 10 μL β-mercaptoethanol.
2. Heat the mix in a boiling water bath for 3 min.
3. After cooling to room temperature, remove insoluble materials (if any) by centrifugation.
4. Run SDS-PAGE. For molecular weight determination, use wide-range marker proteins.

3.3.6. L-Glutaminase Assay

L-Glutaminase was assayed by following the methodology of Imada et al. *(9)* with slight modifications, as described below.

1. Mix an aliquot of 0.5 mL of the sample with 0.5 mL of 0.04 *M* L-glutamine solution in the presence of 0.5 mL of distilled water, and 0.5 mL of 0.1 *M* phosphate buffer, pH 8.0.
2. Incubate the mixture at 37°C for 15 min and stop the reaction by the addition of 0.5 mL of 1.5 *M* trichloroacetic acid.
3. Add 3.7 mL of distilled water and 0.2 mL of Nessler's reagent to 0.1 mL of the mixture.
4. Measure optical density at 450 nm using a spectrophotometer (*see* **Note 10**).
5. Plot a standard graph using ammonium chloride as the standard for computation of the concentration of ammonia.
6. Estimate soluble protein content using Lowry's method with bovine serum albumin as the standard (*see* **Subheading 3.3.1.**).

One international unit of glutaminase was defined as the amount of enzyme that liberates 1 μmol of ammonia under optimum conditions. The enzyme yield was expressed as units/gram dry substrate (U/gds).

Table 5 summarizes some of the results based on the above experiments.

4. Notes

1. At acidic or neutral pH, the medium remains yellow and turns to pink at alkaline pH due to the presence of the indicator phenol red. L-glutamine may break down while autoclaving and it can cause a color change in the medium. Hence, filter sterilization is suggested. While designing a medium for screening, the medium must contain its inducers and be devoid of constituents that may repress protease synthesis.
2. After autoclaving, polystyrene beads will shrink and reduced to about one third of their original size.
3. Particle size is an important criterion related to system capacity to interchange with microbial growth and heat and mass transfer during the SSF process. Reduction in particle size provides larger surface area for microbial growth. But mycelium forma-

Table 5
Characterization of L-Glutaminase Produced by *Pseudomonas sp.*

Enzyme characteristics	Optimum level
Activity (pH)	pH 5.0 and 7.0
Enzyme stability (pH)	pH 4.0 and 8.0
Activity (temperature)	35°C
Enzyme stability (temperature)	30–35°C
Substrate specificity	0.02 *M* L-glutamine
K_m value	0.015 *M*
Salt tolerance	7% NaCl
Storage	1 wk without losing activity. Up to 1 yr at –20°C.
Nature and molecular weight	Appears to synthesize isoenzymes, one with 70 kDa and other with 50 kDa. Two active fractions were obtained.

    tion is the main cause for void fraction variation, which provokes canalization during the process. Very tiny particles could produce matrix contraction or compaction, enhanced channeling problems, increasing mass, and heat transfer problems.

4. The use of nutritionally inert materials for SSF facilitates accurate designing of media, monitoring of process parameters, scaling up strategies, and various engineering aspects, which are impossible or difficult with conventional SSF using organic solid substrates such as wheat bran.

5. Care must be taken that no free water is present after inoculation while doing solid-state fermentation.

6. For enzyme extraction, various process parameters that were optimized include suitable extraction medium, buffer system, pH of extraction buffer, solid substrate to extraction buffer ratio, and contact time.

7. The inert materials, on impregnation with a suitable medium, provide a homogeneous aerobic condition throughout the fermenter, and do not contribute impurities to the fermentation product. It facilitates maximal recovery of the leachate with low viscosity and high specificity for the target product.

8. During ammonium sulfate precipitation stirring must be regular and gentle. To ensure maximal precipitation, it is best to start with a protein concentration of at least 1 mg/mL. The salt can be subsequently being removed from the protein preparation by dialysis.

9. Sephadex G-200 has a fractionation range from 5 kDa to 250 kDa, while Sephadex G-100 has a fractionation range from 1 kDa to 100 kDa, and so on. The choice of gel filtration matrix should be based on molecular weight. Matrix flow rate can also be considered when selecting a gel filtration matrix. Faster flow rates will allow a separation to proceed more rapidly, while slower flow rates sometimes offer better resolution of peaks.

10. After the addition of Nessler's reagent, the reading should be taken as early as possible. Otherwise, the liberated ammonia may be lost and, also, sometimes precipitation will occur.

References

1. Cassileth, B. (ed.) (1998) *The Alternative Medicine Handbook.* W. W. Norton & Co., New York.
2. http://www-biol.paisley.ac.uk/Courses/Enzymes/glossary/Therapy1.htm.
3. Gonzalez, N. J. and Isaacs, L. L. (1999) Evaluation of pancreatic proteolytic enzyme treatment of adenocarcinoma of the pancreas, with nutrition and detoxification support. *Nutr. Cancer* **33,** 117–124.
4. Sabu, A., Chandrasekaran, M., and Pandey, A. (2000) Biopotential of microbial glutaminases. *Chem. Today* **11–12,** 21–25.
5. Renu, S. and Chandrasekaran, M. (1992) Extracellular L-glutaminase production by marine bacteria. *Biotech. Letters* **14,** 471–474.
6. Sabu, A. (1999) L-glutaminase production by marine fungi. Ph.D. thesis, Cochin University of Science and Technology, Cochin, India.
7. Kashyap, P., Sabu, A., Pandey, A., Szakacs, G., and Soccol, C. R. (2002) Extracellular L-glutaminase production by *Zygosaccharomyces rouxii* under solid-state fermentation. *Process Biochem.* **38,** 307–312.
8. Lowry, O. H., Rosenbrough, N. N., Farr, A. L., and Randall, R. Y. (1951) Protein measurement with Folin phenol reagent. *J. Biol. Chem.* **193,** 265–275.
9. Imada, A., Igarasi, S., Nakahama, K., and Isono, M. (1973) Asparaginase and glutaminase activities of microorganisms. *J. Gen. Microbiol.* **76,** 85–99.

5

Enzymatic Production of D-Amino Acids

Hemraj S. Nandanwar, Gurinder S. Hoondal, and Rakesh M. Vohra

Summary

Optically pure amino acids are of increasing industrial importance as chiral building blocks for the synthesis of food ingredients, pharmaceuticals, drugs, and drug intermediates. Highly stereoselective enzymatic processes have been developed to obtain either D- or L-amino acids from D,L-mono-substituted hydantoin derivatives. The initial enzymatic reaction step of D,L-mono-substituted hydantoin hydrolysis is catalyzed by a D-hydantoinase after the subsequent racemization of L-isomer to D-isomer and leads to N-carbamoyl-D-amino acid. It is further hydrolyzed to D-amino acid by D-carbamoylase. *Agrobacterium radiobacter* and some other *Agrobacterium* species discussed in this chapter posses both the enzymes viz. D-hydantoinase and D-carbamoylase. In this chapter, D-p-hydroxyphenylglycine production from D,L-p-hydroxyphenyl hydantoin has been studied in more detail. The purpose of this chapter is to provide the comprehensive idea to the readers directly from producing biocatalyst, analytical methods, harvesting the D-amino acids, and its characterization.

Key Words: D-hydantoinase; L-carbamoylase; *Agrobacterium radiobacter;* biocatalysts; D,L-p-hydroxyphenylhydantoin; D-p-hydroxyphenylglycine.

1. Introduction

The preparation of optically active amino acids has been the focus of several efforts both chemical and biological in nature, as a result of commercial demand for them in pharmaceutical industries. Chemical synthesis of α-amino acids generally produces an optically inactive racemic mixture, i.e., both L- and D-isomers *(1)*. Stereospecific synthesis and separation of stereoisomers are still laborious tasks. Enantiomers usually differ greatly in their biological properties. Only one of the isomers fully contributes to the therapeutic action whereas the other often is classified as a medical pollutant *(2)*. Therefore emphasis was given to developing newer methods to produce optically pure enantiomers.

From: *Methods in Biotechnology, Vol. 17: Microbial Enzymes and Biotransformations*
Edited by: J. L. Barredo © Humana Press Inc., Totowa, NJ

The scenario is rapidly changing with the development of stereospecific biocatalytic methods. During the past three decades, applications of biotechnological methods that rely on the help of free cells/enzymes as catalysts and enzymatic production of D-amino acids have replaced chemical methods. Due to a significant revolution and intensive research in the area of biocatalysis, many biological processes have emerged as great breakthrough in the chirality sciences *(3)*.

D-amino acids are utilized in pharmaceuticals, drugs, drug-intermediates, fine chemicals, food additives, and commodity chemicals. D-p-hydroxyphenylglycine and phenylglycine are used as side chains in semisynthetic antibiotics such as amoxicillin, cephadroxyl, cephalexin, and ampicillin. Amoxicillin is one of the most widely used antibiotics because it is a broad-spectrum antibiotic and bacterial resistance toward it is limited *(4)*. Demand for these side-chain D-amino acids is likely to increase in the near future with the introduction of novel semisynthetic antibiotics such as cefbuparzone, aspoxicillin, cefpyramide, and the like *(5)*. D-alanine is used as a conjugate for the novel type of dipeptide artificial sweetner, Alitame®. D-valine and D-serine are used in the synthesis of Fluvalinate® and D-cycloserine®, respectively, which are further used as agrochemicals and pharmaceuticals *(6)*.

Industrial fermentation of microorganisms has been used to produce amino acids for world markets in animal nutrition and food additives for more than 30 yr. The ability to produce optically pure amino acids at high concentration from cheap carbon and nitrogen sources has often offered a significant advantage over alternate enzymatic and chemo-enzymatic methods of large-scale amino acid synthesis, especially L-amino acids such as lysine, threonine, glutamic acid, tryptophan, and phenylalanine *(7)*.

In designing a microbial conversion process many important aspects require careful consideration, including the selection of compound to be synthesized, a survey of available substrates, and the routes of reactions needed. Another important point is to find out microbial enzymes that are suitable for process design and development. The discovery of a new enzyme or a new reaction provides a clue for designing a new microbial transformation process, so novel microbial strains or enzymes are needed. The most preferred and economic route for the production of D-amino acids is the "hydantoinase-carbamoylase" route. Both the enzymes are D-specific, which exclusively react with 5-monosubstituted hydantoins and N-carbamoyl-(D)-amino acids, respectively, to produce D-amino acids *(8–13)*.

It is a two-step enzymatic conversion. The general schematic of the process is shown in **Fig. 1.** The first enzyme, D-hydantoinase, converts 5-monosubstituted hydantoins to N-carbamoyl-(D)-amino acid, and this is further converted to D-amino acid by D-specific carbamoylase, i.e., D-N-carbamoylase.

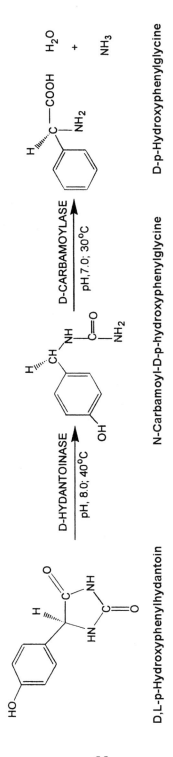

D,L-p-Hydroxyphenylhydantoin

N-Carbamoyl-D-p-hydroxyphenylglycine

D-p-Hydroxyphenylglycine

Fig. 1. Enzymatic route for the production of D-p-hydroxyphenylglycine from D,L-p-hydroxyphenylhydantoin.

In this chapter, we discuss the enzymatic assay methods for both D-hydantoinase and D-carbamoylase. Methods such as preparation of the biocatalyst and its use for biotransformation and product recovery are elaborated. Additionally, analytical methods for qualitative and quantitative analysis of the product and intermediates (TLC, chiral TLC, HPLC, chiral HPLC, and polarimetry) are also discussed.

2. Materials

2.1. Microorganisms with D-Hydantoinase and D-N-Carbamoylase Activity

1. *Agrobacterium radiobacter* NRRL B11291 (Japanese Culture Collection, Osaka, Japan) *(14,15,16)*.
2. *Agrobacterium radiobacter* NCIM (National Collection of Industrial Microorganisms) 2986 (National Chemical Laboratory [NCL], Pune, India).
3. *Agrobacterium radiobacter* NRRL B11291 (Recordati SpA, Milan, Italy) *(17)*.
4. *Agrobacterium* sp. IPI-671 *(18,19)*.
5. *Agrobacterium tumefaciens* NRRL B11291 *(20,21)*.
6. *Agrobacterium tumefaciens* AM10 *(13)*.
7. *Agrobacterium radiobacter* CCRC14924 *(22)*.
8. *Agrobacterium* sp. strain DSM 30147 *(23)*.
9. *Agrobacterium* sp. strain KNK 712 *(24,25)*.

2.2. Chemicals

1. D, L-p-hydroxyphenylhydantoin (*see* **Note 1**).
2. D, L-phenylhydantoin (*see* **Note 2**).
3. Hydantoin (Sigma, St. Louis, MO).
4. D-p-hydroxyphenylglycine (Sigma).
5. D-phenylglycine (Sigma).
6. D- or L-amino acids used as standards or for synthesis of N-carbamoyl derivatives (Sigma).
7. 5-substituted D,L-hydantoins and N-carbamoyl derivatives of amino acids (*see* **Note 3**).
8. p-Dimethylaminobenzaldehyde (p-DMAB).
9. Trichloroacetic acid (TCA).
10. Silica plates for thin-layer chromatography (TLC).
11. Ninhydrin solution: 1% w/v ninhydrin in methanol.

2.3. Medium for Growth

1. Medium A (nutrient broth, NB): 1.0% beef extract, 0.5% sodium chloride, and 1.0% peptone with final pH 7.0.
2. Medium B: 1% glycerol, 0.2% peptone, 0.3%, meat extract, 0.3% yeast extract, 0.03% $MgSO_4 \cdot 7H_2O$, 0.1% K_2HPO_4, and 0.1%. KH_2PO_4. pH of medium is adjusted to 7.0 with 1 *M* NaOH.

2.4. Buffers

1. 0.1 *M* sodium phosphate buffer, pH 8.0.
2. KPB: 0.1 *M* potassium phosphate buffer, pH 7.0.

2.5. Instruments

1. Filtration assembly with 0.22-µm filter paper.
2. Magnetic stirrer with heating arrangement.
3. Refrigerated centrifuge for 10,000*g*.
4. pH meter.
5. UV spectrophotometer.
6. TLC setup.
7. TLC plate silica gel 60F$_{254}$ (Merck, Darmstadt, Germany).
8. Polarimeter.
9. Chiral chromatography setup.
10. Chirapack® WH column for chiral HPLC (Diacel, Tokyo, Japan).
11. Chiroplate plate for chiral TLC (Machery-Nagel, Duren, Germany).
12. HPLC set up with C$_{18}$ (Analytical, ODS, octadecylsilane) column.

3. Methods

3.1. Synthesis of D,L-p-Hydroxyphenylhydantoin, D,L-Phenylhydantoin, and Other 5-Substituted D,L-Hydantoins and N-Carbamoyl Derivatives of Amino Acids

3.1.1. Synthesis of D,L-p-Hydroxyphenylhydantoin

1. Mix 20 g urea, 18.8 g of phenol, 28 mL concentrated HCl, and 7 mL water.
2. Stir the mixture at 90°C for 30 min.
3. Add dropwise, in about 5 h, 16 g glyoxalic acid in 20 mL water.
4. Stir overnight at 90°C.
5. Distill the solvent under vacuum.
6. Separate the crystals by filtration.
7. Recrystallize from hot water.
8. Dry.

3.1.2. Synthesis of D,L-Phenylhydantoin

1. Add 10 g benzaldehyde to a saturated solution of sodium metabisulfite (10 g) and stir.
2. Cool the mixture.
3. Separate adduct and redissolve in a minimum amount of water.
4. Add 50 mL of 13% w/v potassium cyanide to adduct. **Caution:** Perform this step in fume hood.
5. Stir and add alcohol to make the final concentration of 50% and reflux for 3 h at 70–80°C.
6. Cool in an ice bath.

7. Separate the white crystals.
8. Dry.

3.1.3. Synthesis of Other 5-Substituted D,L-Hydantoins and N-Carbamoyl Derivatives of Amino Acids that Are Not Available Commercially

Procedure 1:

1. Dissolve 0.05 moles of amino acid and 4.9 g of potassium cyanate in 30 mL water.
2. Incubate for 30 min at 70°C with stirring.
3. Reflux the reaction mixture for 1 h with 30 mL of 6 N hydrochloric acid.
4. Allow it to react for 30 min at 70°C with stirring.
5. Filter the reaction mixture.
6. Cool overnight in refrigerator.
7. Separate the crystals by filtration.
8. Recrystallize from hot water.

Procedure 2:

1. Dissolve 0.05 moles of amino acid and 4.9 g of potassium cyanate in 30 mL of water.
2. Add 25 mL pyridine to dissolve amino acid.
3. Allow the mixture to react for 1 h at 60°C with stirring.
4. Extract pyridine with 300 mL ethylacetate.
5. Add 30 mL each of 6 N HCl and glacial acetic acid.
6. Reflux aqueous layer for 1 h.
7. Allow the solution to stand overnight in a refrigerator.
8. Filter the crystals.
9. Recrystallize from hot ethanol water.

3.2. Assay for Hydantoinase

1. Harvest 1 mL broth in 1.5-mL microfuge tube.
2. Wash the pellet with 0.1 M sodium phosphate buffer, pH 8.0.
3. Discard the supernatant, and then add 100 µL of substrate containing 50 mM hydantoin in 0.1 M sodium phosphate buffer.
4. Incubate at 50°C for 15 min under shaking condition.
5. Stop the reaction by adding 125 µL of 12 % (w/v) TCA.
6. Add 125 µL of p-dimethyaminobenzaldehyde (10% w/v in 6 N HCl) (*see* **Note 4**).
7. Dilute the above to 1.5 mL by addition of water and centrifuge to remove precipitate.
8. Measure OD_{420} of the supernatant.
9. Follow the same protocol for standard sample of N-carbamoyl-D-amino acid.
10. One unit (U) of enzyme activity is defined as one µmole of N-carbamoyl-D-amino acid formed per minute under the assay condition.

3.3. Assay for Carbamoylase

1. Harvest 1 mL broth, wash with 0.1 M KPB.
2. Add 100 µL of 25 mM of substrate (carbamoyl derivative of amino acid) in 0.1 M KPB. Incubate for 30 min at 30°C under shaking conditions.

3. Stop the reaction with equal volume of absolute alcohol. Dilute to 500 μL with HPLC-grade water (*see* **Note 5**).
4. Analyze the sample by HPLC using C_{18} column at 210 nm, using as mobile phase acetonitrile:water (2.5:97.5), buffered with 10 mM ammonium acetate, pH 5.0, flow rate 1 mL/ min.
5. One unit (U) of enzyme activity is defined as one μmol of D-amino acid formed per minute under the specified assay condition.

3.4. Preparation of Biocatalyst

1. Subculture the *Agrobacterium radiobacter* either from slant or lyophilized ampoule on nutrient agar plate and incubate at 30°C for 24 h.
2. Transfer a loopful of culture from freshly grown plate to a 100 mL Erlenmeyer flask containing 10 mL medium A (NB).
3. Incubate at 30°C and 200 rpm for 10–12 h. This is the "seed culture" for the next step.
4. Transfer the whole seed into 400 mL of either medium A or B in 2.0 L Erlenmeyer flask. Incubate at 30°C and 200 rpm for 18–20 h.
5. Harvest the broth at 10,000g, 4°C for 15 min.
6. Discard the supernatant, wash and resuspend the pellet with 0.1 M sodium phosphate buffer, pH 8.0.
7. Centrifuge again the suspension at 10,000g for 10 min. The resulting cell pellet will be the "biocatalyst" for biotransformation.

3.5. Biotransformation of D,L-p-Hydroxyphenylhydantoin to D-Hydroxyphenylglycine

As explained earlier, biotransformation by *Agrobacterium* is catalyzed by two stereospecific enzymes: D-hydantoinase catalyzes the stereochemical conversion of D,L-p-hydroxyphenylhydantoin (D,L-p-HPH) to D-N-carbamoyl-p-hydroxyphenylglycine (NC-(D)-p-HPG), which is further converted to D-p-hydroxyphenylglycine (D-p-HPG) by D-carbamoylase. Follow the steps given below for the biotransformation experiment.

1. Powder the D,L-p-HPH and make the 100 mL suspension containing 5.0% D,L-p-HPH in 0.1 M sodium phosphate buffer, pH 8.0 (*see* **Note 6**).
2. Add the total pellet accumulated from 400 mL broth on a magnetic stirrer/water bath. Control the temperature to 40°C (*see* **Note 7**).
3. Maintain the pH to 8.0 by 0.5 M HCl. Under this experimental setup, L-hydantoin will simultaneously racemize to D-form and will convert to NC-(D)-p-HPG by D-hydantoinase. NC-(D)-p-HPG is further transformed to D-p-HPG by D-N-carbamoylase (*see* **Note 8**).
4. Incubate for 10 h and analyze by spectrophotometry and by HPLC the amount of NC-(D)-p-HPG and D-p-HPG produced.

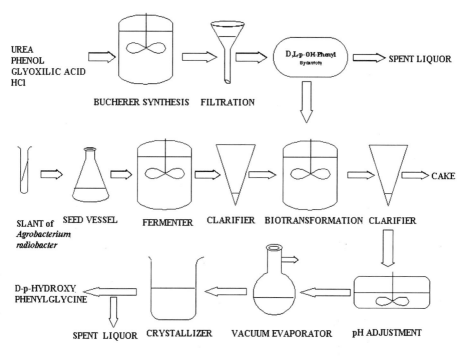

Fig. 2. Flow diagram for enzymatic production of D-p-hydroxyphenylglycine from D,L-p-hydroxyphenylhydantoin.

3.6. Harvesting of Product/Recovery of D-Amino Acid (D-p-HPG)

Product recovery is carried out by a simple crystallization process. The schematic for the production of D-p-HPG from D,L-p-HPH and its recovery from the reaction mixture is shown in **Fig. 2.**

1. Centrifuge the reaction mixture at 10,000g for 15 min.
2. Take the supernatant and adjust to 1.5 with 8 N sulfuric acid.
3. Concentrate the resulting solution by vacuum evaporation and bring its pH to iso-electric point of the corresponding amino acid (*see* **Note 9**).
4. In case of D-p-HPG, bring up the pH of the concentrate to 5 with 4 M sodium hydroxide. D-p-HPG will precipitate out at this point and should be filtered.
5. Dry and crystallize.

3.7. Thin-Layer Chromatography (TLC)

This method is mostly used for qualitative analysis of compounds unless a TLC workstation with scanner and analyzer is not available. The separation of D-p-HPG, NC-(D)-HPG and D,L-p-HPH is shown by TLC in **Fig. 3.**

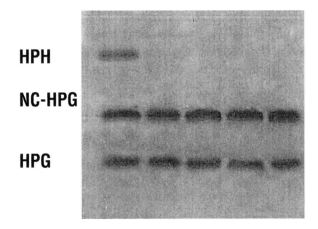

Fig. 3. TLC of D,L-p-hydroxyphenylhydantoin (HPH), NC-(D)-p-hydroxyphenyl-glycine (NC-HPG), and D-p-hydroxyphenylglycine (HPG).

1. Spot the reaction mixture on silica gel $60F_{254}$ TLC plate.
2. Use a mixture of chloroform:methanol:water:formic acid (64:30:4:2) as developing solvent for HPG and other D- or L-amino acid.
3. Dry the plate and spray ninhydrin solution. Heat the plate by hot and dry air. Visualize the spots under a UV illuminator at 254 nm and compare the Rfs and color with standard (*see* **Fig. 3**).

3.8. Chiral Chromatography

This technique is used to see the optical purity of amino acid formed after stereospecific conversion of a racemic mixture of compound. It is done in two ways:

1. Chiral HPLC is ligand-exchange high-performance liquid chromatography. It is performed by using a Chirapak® WH column, 4.6 × 250 mm at 254 nm with 50 mM $CuSO_4$ as the mobile phase at a flow rate of 1 mL/min.
2. Chiral TLC is performed by using a Chiroplate plate. The developing solvent is methanol:water:acetonitrile (1:1:4) with flow rate of 1 mL/min, at 60°C. Activate the plate at 50–60°C for 1 h before spotting the plate for better separation of enantiomers. Amino acids can be visualized by ninhydrin solution.

3.9. Polarimetry

1. Harvest the product.
2. Crystallize it.
3. Dissolve 250 mg of purified product (HPG) in 1 N HCl and analyze in a polarimeter for its optical rotation.

3.10. HPLC and Chiral HPLC

1. Determine the retention time of the product and match it with the retention time of the standard under the specified condition mentioned above **(Subheading 3.3)**.
2. Determine the optical purity of the product by chiral chromatography under the condition described above **(Subheading 3.8)** and calculate its optical/enantiomeric percentage purity.

3.11. Characterization of Product

Analyze the product as described above **(Subheading 3.7–3.10)**. The optical purity of the amino acid, HPG $[\alpha]_D^{25}$ (c = 1%, 1 N HCl) is −153° *(29,30)*. Further characterization can be done by matching IR, liquid chromatography-mass spectrophotometry, and NMR and determining the melting point.

At present, the production of D-amino acids by this route is economically viable. Currently, both enzymes D-hydantoinase and D-carbamoylase have been cloned and overexpressed for making the process even more productive.

4. Notes

1. Hydantoin derivatives can be easily synthesized by various methods. Most important of these is the Bucherer-Berg synthesis, which utilizes respective aldehydes, potassium cyanide, and ammonium carbonate to synthesize D,L-5-substituted hydantoins. The drawback of this method is the use of highly toxic potassium or sodium cyanide. The D,L-5-monosubstituted hydantoins such as D,L-p-hydroxyphenylhydantoin can also be synthesized using glyoxalic acid, phenol urea, and ammonium carbonate *(26)*.
2. D,L-phenylhyadantoin can be synthesized from benzaldehyde and potassium cyanide according to ref. *27*.
3. Other 5-substituted D,L-hydantoins and N-carbamoyl derivatives of amino acids that are not available commercially can be synthesized from corresponding amino acids and potassium cyanate according to ref. *28*.
4. Free amino group from carbamoyl derivative of amino acid reacts with the aldehyde group of p-dimethylaminobenzaldehyde and forms a "chromophore" group under acidic condition, which gives the yellow color. The absorbance at 420 nm of this color complex is proportional to the amount of N-carbamoyl amino acid formed.
5. HPLC is a very sensitive method. So the concentration of components in the reaction mixture has to be brought down to a measurable level for correct analysis. The practice is always to do the analysis of standard and then bring the concentration of components to the level of standard by appropriate dilution.
6. Solubility of (D,L)-p-hydroxyphenylhydantoin is very poor under normal condition. It comes in crystalline form. For better reaction rate, particle size should be as small as possible. So the practice is to pulverize it to a fine powder by mortaring.
7. Solubility of D,L-p-HPH increases with increasing temperature of the reaction condition. So, it is beneficial to maintain the temperature at 40°C or above, as per the physicochemical characteristics of the biocatalyst.

8. The racemization of D,L-p-HPH takes place at pH 8.0–8.5. Since the hydantoinase is D-specific, the racemization is necessary to convert the remaining L-form to D-form.
9. At isoelectric point, amino acids precipitate. So, the practice is to bring down the pH of reaction mixture after biotransformation to isoelectric point to precipitate the amino acid and then use the precipitate for further purification steps.

References

1. Chibata, I. (1980) Development of enzyme-engineering-application of immobilized cell system, in *Food Process Engineering,* Vol. 2 (Linko, P. and Larinkari, J., eds.), Appl. Sci. Publishers, London, pp. 1–39.
2. Ariens, E. J. (1986) Stereochemistry: A source of problem in medicinal chemistry (part 1). *Med. Res. Rev.* **6,** 451–466.
3. Margolin, A. L. (1993) Enzymes in the synthesis of chiral drugs. *Enzyme Microb. Technol.* **15,** 226–280.
4. Louwrier, E. and Knowles, C. J. (1996) The purification and characterization of a novel D(-)-specific carbamoylase enzyme from an *Agrobacterium* sp. *Enzyme Microb. Technol.* **19,** 562–571.
5. Lee, D. C., Lee S. G., and Kim, H. S. (1996) Production of D-p-hydroxyphenylglycine from DL-5-4-hydroxyphenylhydantoin using immobilized thermostable D-hydantoinase from *Bacillus stearothermophillus.* SD1. *Enzyme Microb. Technol.* **18,** 35–40.
6. Boer, L. de and Dijkhnizen, L. (1990) Microbial and enzymatic process for L-phenylalanine, in *Advances in Biochemical Engineering/Biotechnology* (Fiechter, A., ed.), Springer, Berlin, pp. 2–3.
7. Rhem, H. J. and Reed, G. (1996) Amino acids. Technical production and use, in *Biotechnology* (Puhler, A. and Stadler, P., eds.), VCH-Verlag, Weinheim, Germany, 6a, pp. 466–502.
8. Yokozeki, K., Nakamori, S., Eguchi, C., Yamada, K., and Mitsugi, K. (1987) Screening of microorganisms producing D-p-hydroxyphenylglycine from D,L-5-(p-hydroxyphenyl)hydantoin. *Agric. Biol. Chem.* **51,** 355–362.
9. Oliveri, R., Fascetti, E., Angelini, I., and Degen, L. (1981) Microbial transformation of racemic hydantoins to D-amino acids. *Biotechnol. Bioeng.* **23,** 2173–2183.
10. Moller, A., Syldatk, C., Schulze, M., and Wagner, F. (1988) Stereo- and substrate-specificity of a D-hydantoinase and a D-N-carbamoyl amino acid amidohydrolase of *Arthrobacter crystallopoietes* AM2. *Enzyme Microb. Technol.* **10,** 618–625.
11. Runser, S., Chinski, N., and Ohleyer, E. (1990) D-p-hydroxyphenylglycine production from D,L-5-hydroxyphenylhydantoin by *Agrobacterium* sp. *Appl. Microbiol. Biotechnol.* **33,** 382–388.
12. Sharma, R. and Vohra, R. M. (1997) A thermostable D-hydantoinase isolated from a mesophilic *Bacillus* sp. AR9. *Biochem. Biophy. Res. Comm.* **234,** 485–488.
13. Sarin, D., Sharma, R., Nandanwar, H. S., and Vohra, R. M. (2001) Two-step purification of D(-)-specific carbamoylase from *Agrobacterium tumefaciens* AM10. *Protein Expres. Purif.* **21,** 170–175.

14. Chao, Y. P., Juang, T. Y., Chern, J. T., and Lee, C. K. (1999a) Production of D-*p*-hydroxyphenylglycine by *N*-carbamoyl-D-amino acid amidohydrolase-overproducing *Escherichia coli* strains. *Biotechnol. Prog.* **15,** 603–607.

15. Deepa, S., Sivasankar, B., Jayararman, K., Prabhakaran, K., George, S., Palani, P., et al. (1993) Enzymatic production and isolation of D-amino acids from the corresponding 5-substituted hydantoins. *Process Biochem.* **28,** 447–452.

16. Chao, Y. P., Fu, H., Lo, T. E., Chen, P. T., and Wang, J. J. (1999b) One-step production of D-*P*-hydroxyphenylglycine by recombinant *Escherichia coli* strains. *Biotechnol. Prog.* **15,** 1039–1045.

17. Buson, A., Negro, A., Grassato, L., Tagliaro, M., Basaglia, M., Grandi, C., et al. (1996) Identification, sequencing and mutagenesis of the gene for a D-carbamoylase from *Agrobacterium radiobacter.* *FEMS Microbiol. Lett.* **145,** 55–62.

18. Runser, S., Chinski, N., and Ohleyer, E. (1990) D-*p*-hydroxyphenylglycine production from D,L-*p*-hydroxyphenylhydantoin by *Agrobacterium species.* *Appl. Microbiol. Biotechnol.* **33,** 382–388.

19. Meyer, P. and Runser, S. (1993) Efficient production of the industrial biocatalysts hydantoinase and *N*-carbamoylamino acid amidohydrolase: novel non-metabolizable inducers. *FEMS Microbiol. Lett.* **109,** 67–74.

20. Park, H. H., Kim, G. J., and Kim, H. S. (2000) Production of D-amino acids using whole cells of recombinant *Escherichia coli* with separately and coexpressed D-hydantoinase and *N*-carbamoylase. *Biotechnol. Prog.* **16,** 564–570.

21. Ley, C. J., Kirchmann, S., Burton, S. G., and Dorrington, R. (1998) Production of D-amino acids from D,L-5-substituted hydantoins by an *Agrobacterium tumefaciens* strain and isolation of a mutant with inducer-independent expression of hydantoin-hydrolysing activity. *Biotechnol. Lett.* **20,** 707–711.

22. Hsu, W. H., Chien, F. T., Hsu, C. L., Wang, T. C., Yuan, H. S., and Wang, W. C. (1999) Expression, crystallization, and preliminary X-ray diffraction studies of N-carbamoyl-D-amino-acid amidohydrolase from *Agrobacterium radiobacter. Acta Crystallogr. D* **D55,** 694–695.

23. Morin, A., Hummel, W., and Kula, M. R. (1986) Rapid detection of microbial hydantoinase on solid medium. *Biotechnol Lett.* **8,** 573–576.

24. Nakai, T., Hasegawa, T., Yamashita, E., Yamamoto, M., Kumasaka, T., Ueki, T., et al. (2000) Crystal structure of N-carbamoyl-D-amino acid amidohydrolase with a novel catalytic framework common to amidohydrolases. *Structure Fold, Des.* **15,** 729–737.

25. Nanba, H., Ikenaka, Y., Yamada, Y., Yajima, K., Takano, M., and Takahashi, S. (1998) Isolation of *Agrobacterium* sp. KNK712 that produces *N*-carbamoyl-D-amino acid amidohydrolase, cloning of the gene for this enzyme, and properties of the enzyme. *Biosci. Biotech. Biochem.* **62,** 875–881.

26. Ohasi, T., Takahashi, S., Nagamachi, T., and Yoneda, H. (1981) A new method for 5-(4-hydroxyphenyl)-hydantoin synthesis. *Agric. Biol. Chem.* **45,** 831–838.

27. Henze, H. R. and Speer, R. J. (1942) Identification of carbamoyl compounds through conversion into hydantoin. *J. Am. Chem. Soc.* **64,** 522–523.

28. Suzuki, T., Igarashi, K., Hase, K., and Tuzimura, K. (1973): Optical rotatory dispersion and circular dichorism of amino acid hydantoin. *Agric. Biol. Chem.* **37,** 411–416.

29. Yamada, S., Hongo, C., and Chibata, I. (1978) Preparation of D-p-hydroxyphenyl-glycine: optical resolution of D,L-p-hydroxyphenylglycine by preferential crystallization procedure. *Agric. Biol. Chem.* **42,** 1521–1526.

30. Takahashi, S., Ohashi, T., Kii, Y., Kumagai, H., and Yamada, H. (1979) Microbial transformation of hydantoins to N-carbamoyl-D-amino acids. *J. Ferment. Technol.* **57,** 328–332.

6

Screening of Microorganisms for Enzymatic Biosynthesis of Nondigestible Oligosaccharides

Oskar Bañuelos, Maria del Carmen Ronchel, José Luis Adrio, and Javier Velasco

Summary

Nondigestible oligosaccharides (NDOs), namely prebiotic oligosaccharides, are functional food ingredients that possess properties that are beneficial to the health of consumers. These include noncariogenicity, a low calorific value, and the ability to stimulate the growth of beneficial bacteria in the colon. Fructooligosaccharides (FOS) represent one of the major classes of oligosaccharides in terms of their production volume. They are manufactured by two different processes, which result in slightly different end products. First, FOS are produced by controlled hydrolysis of inulin. Secondly, FOS can be produced from sucrose by using the transfructosylating activity of the enzyme β-fructofuranosidase (EC 3.2.1.26) at high concentrations of the starting material. This chapter summarizes the methods used for the detection and characterization of enzymes of fungal origin with fructosyltransferase activity and the analytical methods utilized to identify the oligosaccharides produced.

Key Words: Fructooligosaccharides; prebiotics; fructosyltransferase; *Aspergillus*.

1. Introduction

Prebiotics are defined as nondigestible food ingredients that beneficially affect the host by selectively stimulating the growth and/or activity of one or a limited number of bacteria in the colon *(1)*. Nondigestible oligosaccharides (NDOs) are used in a variety of foods not only for their prebiotic effect but also because they are noncariogenic, they represent less-sweet sweeteners, and they enhance the physical properties of foods *(2)*.

The most abundantly supplied and utilized group of NDOs as food ingredients are fructooligosaccharides (FOS), which are generally produced by enzymatic transglycosylation because of adequate supply of the raw materials and the high efficiency of the reaction. Industrial production of FOS is carried out by the transglycosylation activity of β-fructofuranosidases (EC 3.2.1.26) at

From: *Methods in Biotechnology, Vol. 17: Microbial Enzymes and Biotransformations*
Edited by: J. L. Barredo © Humana Press Inc., Totowa, NJ

high sucrose concentrations *(3)*. These enzymes have also hydrolytic activity in reaction mixtures containing low concentrations of sucrose. Thus, hydrolysis of sucrose and formation of FOS occurred by the mechanism of transfructosylation. Sucrose and FOS may act as fructose donors and simultaneously, sucrose, FOS, and water act as acceptors for fructose. The sources of enzymes can be either microorganisms (*Aspergillus* sp., *Aureobasidium* sp., *Penicillium* sp., *Fusarium* sp., etc.) or plants *(4)*. The production yield of FOS using enzymes from plants is low and mass production of enzyme is limited by seasonal conditions. On the other hand, fructosyltransferases derived from fungi provide high yields of FOS and mass production is straightforward. It has been reported that yeast invertases (β-fructofuranosidases) have more hydrolytic activity, whereas fungal enzymes have more fructosyltransferase activity *(5)*.

In this chapter, the screening of fungal strains able to produce enzymes with fructosyltransferase activity will be used to illustrate the approach applied to detect and quantify this activity and the analytical methods utilized to identify the oligosaccharides produced.

2. Materials

2.1. Culture of Fungal Strains

1. YPS medium: 1% yeast extract, 2% peptone, and 5% sucrose.
2. Nylon filters, 30 μm pore (SST Thal, Thal, Switzerland).

2.2. Fructosyltransferase Assay

1. Substrate solution: 1 M sucrose in McIlvain buffer, pH 5.0.
2. McIlvain buffer: 100 mM citric acid and 200 mM Na_2HPO_4, pH 5.0.
3. Glucose-detection stock solution 1: 80 mM NADP (nicotine adenine nucleotide phosphate). Store at –20°C.
4. Glucose-detection stock solution 2: 20 mM MTT (methylthiazoletetrazolium). Store at –20°C.
5. Glucose-detection stock solution 3: 60 mM PMS (phenazine methosulfate). Store at –20°C.
6. Glucose-detection stock solution 4: 2 M imidazole-HCl, pH 6.8.
7. Glucose-detection stock solution 5: 100 mM $MgCl_2$.
8. Glucose-detection stock solution 6: 20 g/L BSA.
9. ATP (adenosine triphosphate).
10. Hexokinase (EC 2.7.1.1).
11. Glucose-6-phosphate dehydrogenase (EC 1.1.1.49).
12. Phosphoglucose isomerase (EC 5.3.1.9).
13. Spectrophotometer.

2.3. Fructosyltransferase Activity Staining in Semi-Native PAGE

1. SDS-PAGE (sodium dodecyl sulfate-polyacrylamide gel electrophoresis) equipment.
2. Washing buffer: 50 mM sodium acetate, pH 5.6, and 0.5 % (v/v) Triton X-100.

3. Substrate solution: 1 M sucrose in 100 mM McIlvain buffer, pH 5.0.
4. Staining solution: 1% (w/v) 2,3,5-triphenyl tetrazolium chloride (TTC) in 250 mM NaOH at 100°C.
5. Stop solution: 5% (v/v) acetic acid.
6. Coomassie solution: 0.1% Coomassie blue, 10% v/v ethanol, and 7.5% v/v acetic acid.

2.4. Partial Purification of the Enzyme and Enzymatic Synthesis of Fructooligosaccharides

1. Ultrafiltration device with 100 kDa cut-off membranes (Millipore Bedford, MA).
2. Whatman filter paper #1 (Whatman, Kent, UK).

2.5. Fructooligosaccharides Analysis

2.5.1. Thin-Layer Chromatography (TLC)

1. Silica gel on aluminium TLC plates, 200 μm thickness, 2–25 μm particle size, and developing container.
2. Fume extraction cabinet.
3. Mobile phase: butanol-ethanol-water (4:5:1, v/v/v).
4. Detection solution: 5% ceric sulphate in 15% H_2SO_4.
5. FOS Actilght® 950P (Beghin-Meiji Industries, Neully Sur Seine, France).

2.6. High-Performance Anion Exchange Chromatography (HPAEC)

1. Millex filters (Millipore).
2. Bio-LC system (Dionex, Sunnyvale, CA).
3. Chromatographic column Carbopac PA100, 4 × 250 mm column, with matching guard column.
4. Mobile phase: 500 mM sodium acetate, and 100 mM NaOH.

3. Methods

The methods described below summarize (1) the culture of fungal strains in optimal conditions for the production of fructosyltransferases, (2) the quantification of the activity fructosyltransferase, (3) the detection of enzymes with fructosyltransferase activity after seminative polyacrylamide gel electrophoresis, (4) the enzymatic synthesis of oligosaccharides, and (5) the characterization of the oligosaccharides formed.

3.1. Culture of Fungal Strains

The criteria for the selection of fungal strains for fructosyltransferase production were (1) to belong to genera with generally recognized as safe (GRAS) status, (2) the capacity to secrete the enzyme, and (3) to show a high fructosyltransferase activity. Twenty-two microorganisms from our own culture collection morphologically identified as filamentous fungi of the genera *Aspergillus, Aureobasidium,* or *Penicillium* were tested.

1. Culture the fungal strains by adding 500 µL of a concentrated suspension of spores (*see* **Note 1**) to 250-mL Erlenmeyer flasks containing 50 mL of YPS.
2. Incubate the flasks for 120 hours at 28°C on a shaker with an agitation rate of 250 rpm.
3. Remove the mycelium from the culture by filtration through a 30-µm pore size Nytal filter (*see* **Note 2**).

3.2. Fructosyltransferase Assay

Filamentous fungi can produce enzymes that carry out both the hydrolysis of sucrose (S + H_2O → G + F) and the synthesis of FOS by transfructosylation (2 S → G + GF_2). The extracellular fructosyltransferase and hydrolytic activities produced by the fungal strains were determined by measuring the amounts of glucose and fructose release, respectively. This miniaturized method is based on the colorimetric method described by Cairns *(6)*. One unit of fructosyltransferase was defined as the amount of enzyme required to release 1 µM of glucose per minute under the reaction conditions. One unit of hydrolytic activity is defined as the amount of enzyme that catalyzes the formation of 1 µM of fructose under these conditions.

The transferase/hydrolase (T/H) ratio is a good indicator of the biosynthetic capability of the enzymes present in the culture broth.

1. Mix 900 µL of substrate solution and 100 µL of culture broth, and incubate the reaction mix at 45°C for 15 min. Boil at 100°C for 5 min to stop the reaction.
2. Just before use, mix 1 mL of each glucose detection stock solution (1 to 6) with 22 mg ATP, and bring the mixture to a final volume of 19 mL by adding 13 mL of water. To this solution add 4 units of hexokinase (EC 2.7.1.1.) and 2 units of glucose-6-phosphate dehydrogenase (EC 1.1.1.49).
3. Take 950 µL of the above mix and add 50 µL of the stopped reaction (**step 1**). Incubate at 37°C for 20 min in the dark and measure OD_{595} in a spectrophotometer. The amount of glucose release (G_1) is determined by interpolation with a standard curve of pure glucose (1 to 10 mM) processed in the same way (*see* **Note 3**).
4. Add 0.1 U of phosphoglucose isomerase (EC 5.3.1.9.) and, after incubating the reaction again at 37°C for 20 min in the dark, measure OD_{595} again and determine the concentration of glucose (G_2).
5. The concentration of fructose in the sample is determined as the difference between G_2 and G_1. The transfructosylation ratio is calculated as $G_1/(G_2–G_1)$.

3.3. Fructosyltransferase Activity Staining in Semi-Native PAGE

Using the screening method described in **Subheading 3.2.**, *Aspergillus* BT18 was selected as the best producer showing a fructosyltransferase activity of 11 U/mL and a T/H ratio of 5. In order to determine the time of maximum production and a rough estimate of the molecular weight of the enzyme, a time-course experiment was carried out by culture of *Aspergillus* BT18 as described

kDa

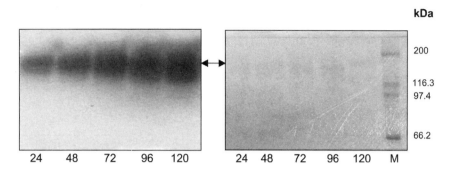

Fig. 1. Semi-native PAGE of culture samples taken at 24, 48, 72, 96, and 120 h from a culture of *Aspergillus* BT18 rich in fructosyltransferase activity. **Right:** gel stained with Coomassie brilliant blue. **Left:** activity stain with 1, 2, 3 triphenyltetrazolium chloride. M: molecular weight marker. The band with fructosyltransferase activity is marked by an arrow.

in **Subheading 3.1.** One-mL samples of the culture broth were taken every 24 h and the enzyme present in the supernatant was analysed by semi-native PAGE and fructosyltransferase activity staining **(Fig. 1)** according with the method described by Heyer and Wendenburg *(7)*.

1. Prepare a 6% semi-native PAGE gel *(8)* containing 0.1% SDS.
2. Prepare samples of 10 μL of culture supernatant in 20 μL loading buffer containing the same amount of SDS but omitting β-mercaptoethanol and the heating step.
3. After the electrophoresis, rinse the gel extensively with washing buffer for 15 min to remove SDS.
4. Incubate the gel in substrate solution for 30 min at room temperature.
5. Wash the gel repeatedly with water to remove sucrose.
6. To visualize glucose release, pour onto the gel 10 mL of staining solution at 100°C. A red insoluble formazan will form in the areas where a fructosyltransferase is present.
7. Once the red area is visible, discard the substrate solution, and stop the staining by adding 10 mL of stop solution (*see* **Note 4**).

3.4. Partial Purification of the Enzyme and Enzymatic Synthesis of Fructooligosaccharides

The results obtained in the semi-native PAGE confirm those previously described *(9–11)* suggesting a high molecular weight (>100 kDa) for the enzyme with fructosyltransferase activity. Thus, in order to concentrate and partially purify the enzyme, an ultrafiltration step (cut-off 100 kDa) was performed after the scale-up of the fermentation of the selected strain.

1. Grow *Aspergillus* BT18 in 5 Erlenmeyer flasks of 500 mL containing 100 mL of YPS medium by inoculation of 1 mL of a spore suspension (10^7 spores/mL) per flask. Incubate the flasks at 28°C for 5 d with agitation (250 rpm).
2. Collect the supernatant solution (400 mL containing 4400 U of fructosyltransferase activity) by filtration through 30-μm pore Nytal filters.
3. Concentrate the proteins by ultrafiltration (cut-off 100 kDa) to a final volume of 20 mL. To this solution add 40 mL of cold ethanol and let stand for 1 h at –20°C.
4. Filter the mixture through Whatman filter paper no. 1 and wash the precipitated protein with cold ethanol.
5. Dry the filtrated pellet in a extraction cabinet. The resulting dried powder is stable at 4°C for several months.
6. Determine the fructosyltransferase activity of the powder as described in **Subheading 3.3,** and prepare a solution of enzyme 10 U/mL in 10 m*M* McIlvain buffer.
7. For oligosaccharides biosynthesis add 50 μL of enzyme solution to 950 μL of a 40% (w/v) sucrose solution in 10 m*M* McIlvain buffer and incubate for 1 h at 45°C with gentle agitation (*see* **Note 5**).
8. Stop the reaction by heating the test tube for 5 min at 100°C.

3.5. Fructooligosaccharides Analysis

3.5.1. Thin-Layer Chromatography

In order to qualitatively determine the production of fructooligosaccharides, the reaction products were analyzed by thin-layer chromatography (TLC) **(Fig. 2)**.

1. Dilute the stopped reaction mixture with water (1:1 v/v) and "spot" 1 μL of the dilution in TLC plates.
2. Start the run the chromatography by placing the plate in a developing container previously saturated with mobile phase, cover the developing container, and leave it undisturbed on the benchtop. Run until the solvent is about 0.5 cm below the top of the plate. Allow the mobile phase to evaporate in a fume extraction cabinet and repeat the run two more times.
3. Visualize the sugars by spraying with detection solution (*see* **Subheading 2.5.1.**) and heating for 10 min at 120°C.

3.5.2. High-Performance Anion-Exchange Chromatography

The oligosaccharide composition of the reaction mixture is established using high-performance anion-exchange chromatography with pulsed amperometric detection (HPAEC-PAD). For this purpose a Bio-LC system that includes a quaternary gradient pump, an eluent gas (He) module, and a 4 × 250 mm Carbopac PA100 column with matching guard column (pulse amperometric detection mode) was used **(Fig. 3)**.

1. Dilute 1 μL of the reaction samples with 999 μL of MilliQ water.
2. To clean the sample, filter through Millex filters.

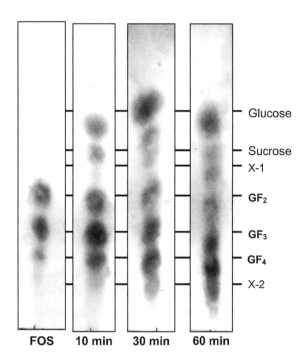

Fig. 2. TLC analysis of the oligosaccharides produced with the partially purified enzyme produced by *Aspergillus* BT18 at 10, 30, and 60 min of reaction. FOS: Commertial FOS Actilight® 950P, GF_2: 1-kestose; GF_3: nystose; and GF_4: 1-fructofura-nosyl nystose; X-1 and X-2: unidentified compounds.

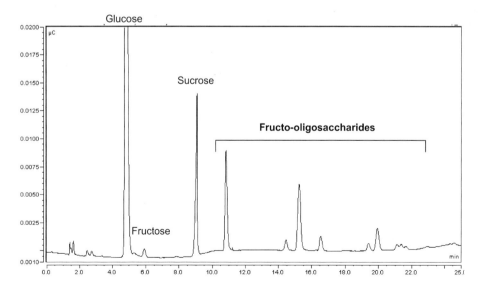

Fig. 3. HPAEC chromatogram of oligosaccharides produced from sucrose by the fructosyltransferase partially purified from *Aspergillus* BT18.

3. Inject 20 µL of the sample into the system using an autosampler.
4. Elute the sample by using a linear gradient of 0 to 100 mM sodium acetate in 100 mM NaOH for 0 to 30 min with a flow rate of 1 mL/min.
5. Monitor the effluent using a pulsed electrochemical detector containing a gold electrode with an Ag-Ag reference electrode.
6. Wash the column for 5 min with 1 M sodium acetate and equilibrate again for 15 min with 100 mM NaOH before the next run.

4. Notes

1. The morphological features of the filamentous fungi have a significant effect on the rheological properties of the cultivation medium, reflected directly in the production and excretion of enzymes. In submerged cultivation involving filamentous fungi, the morphology can vary from discrete compact pellets of hyphae to homogeneous suspension of dispersed mycelia. The morphological feature of fungi not only affects the enzyme production and excretion but also has a significant effect on the reactivity. In our experience pellet formation, which gives the better results for fructosyltransferase production, is induced by inoculation of a large number of spores per milliliter.
2. Filtration is recommended over centrifugation because it is faster and more efficient. Centrifugation of cultures of filamentous fungi usually gives rise to loose pellets that are difficult to manipulate.
3. The determination of glucose and fructose can be easily miniaturized to be performed in 96-well flat-bottom plates. In this case the volumes are reduced by a factor of 5 (reaction mix 190 µL, sample 10 µL) and the absorbance is measured in a microplate reader.
4. Molecular-weight markers and other proteins present in the culture broth can be visualized by running a parallel gel containing the same samples but stained with Coomassie solution for 3 h and destained with 5% v/v methanol/ 7.5% acetic acid.
5. Sucrose concentrations up to 80% can be used as substrate but preparation of the solution is problematical and mixing is impaired by the high viscosity of the syrup formed. In the conditions described (40% sucrose), yields of 60% (g of FOS per g of sucrose) can be typically obtained.

Acknowledgments

The authors thank A. Carceles and C. Membrilla for the excellent technical assistance. This work was supported by a contract with Azucarera Ebro S.A. and by PROFIT and CDTI grants (Spanish Ministry of Science and Technology).

References

1. Gibson, G. R. and Roberfroid, M. B. (1995) Dietary modulation of the human colonic microbiota: Introducing the concept of prebiotics. *J. Nutr.* **125,** 1401–1412.
2. van Loo, J., Cummings, J., Delzenne, N., Englyst, H., Franck, A., Hopkins, M., et al. (1999) Functional food properties of non-digestible oligosaccharide: a consen-

sus report from the ENDO project (DGXII AIRII-CT94-1095). *British J. Nutr.* **81**, 121–132.

3. Hidaka, H, Hirayama, M., and Yamada, K. (1991) Fructooligosaccharides: enzymatic preparation and biofunctions. *J. Carbohydrate Chem.* **10**, 509–522.
4. Yun, J. W. (1996) Fructooligosaccharides: Occurrence, preparation and application. *Enzyme Microb. Technol.* **19**, 107–117.
5. Hidaka, H., Hirayama, M., and Sumi, N. A. (1988) A fructooligosaccharide-producing enzyme from *Aspergillus niger* ATCC20611. *Agric. Biol. Chem.* **52**, 1181–1187.
6. Cairns, A. J. (1987) Colorimetric microtiter plate assay of glucose and fructose by enzyme linked formazan production: applicability of the measurement of fructosyl transferase activity in higher plants. *Anal. Biochem.* **167**, 270–278.
7. Heyer, A. G. and Wendenburg, R. (2001) Gene cloning and functional characterization by heterologous expression of the fructosyltransferase of *Aspergillus sydowi* IAM 2544. *Appl. Environ. Microbiol.* **67**, 363–370.
8. Laemmli, U. K. (1970) Cleavage of structural proteins during the assembly of the head of bacteriophage T4. *Nature* **227**, 680–685.
9. Boddy, L. M., Berges, T., Barreau, C., Vainstein, M. H., Dobson, M. J., Balance, D. J., and Peberdy, J. F. (1993) Purification and characterization of an *Aspergillus niger* invertase and its DNA sequence. *Curr. Genet.* **24**, 60–66.
10. Somiari, R. I., Brzeski, H., Tate, R., Bielecki, S., and Polak, J. (1997) Cloning and sequencing of an *Aspergillus niger* gene coding for β-fructofuranosidase. *Biotechnol. Lett.* **19**, 1243–1247.
11. Yanai, K., Nakene, A., Kawate, A., and Hirayama, M. (2001) Molecular cloning and characterization of the fructooligosaccharide-producing β-fructofuranosidase gene from *Aspergillus niger* ATCC 20611. *Biosci. Biotechnol. Biochem.* **65**, 766–773.

7

Food-Grade Corynebacteria for Enzyme Production

José A. Gil, Angelina Ramos, Sirin A. I. Adham, Noelia Valbuena, Michal Letek, and Luis M. Mateos

Summary

The expression of genes coding for heterologous extracellular enzymes or proteins in corynebacteria has provided new capacities to these industrially important microorganisms, such as the use of the culture media as sources of essential amino acids and hydrolytic enzymes that can be used as complements in animal food or for the production of enzymes with industrial, clinical, or pharmaceutical applications. Using genetic manipulation techniques, several corynebacteria strains expressing genes coding for hydrolytic enzymes or proteins have been constructed in different laboratories. Such strains carry antibiotic resistance genes and consequently they cannot be used in the food industry due to the stringent regulations on genetically manipulated microorganisms. To solve this problem, here we describe a general method for the construction of engineered corynebacteria bearing a single copy of a gene coding for a hydrolytic enzyme or a desired protein in its chromosome where it is stably maintained with no selective pressure and lacking any antibiotic resistance gene.

Key Words: Amino acids; extracellular enzymes; amylase; cellulase; xylanase; cell division genes.

1. Introduction

Escherichia coli has been the choice microorganism for the expression of recombinant proteins or enzymes, but it does not secrete proteins and proteins are often accumulated as inclusion bodies. Therefore, downstream processes to recover the recombinant or heterologous enzyme are expensive. On the other hand, *E. coli* is not a food-grade bacterium.

Bacillus and *Streptomyces* are Gram-positive microorganisms able to produce not only antibiotics, but also a variety of extracellular enzymes. They can be used for the production of heterologous enzymes, although they have high proteolytic activity, which will degrade them.

From: *Methods in Biotechnology, Vol. 17: Microbial Enzymes and Biotransformations*
Edited by: J. L. Barredo © Humana Press Inc., Totowa, NJ

Over the past two decades, considerable advances have been made in the understanding of the molecular biology of *Corynebacterium glutamicum/ Brevibacterium lactofermentum* through the development of effective transformation systems and plasmid vectors that have been used for the cloning of many genes involved in amino acid biosynthesis. However, little attention has been dedicated to the possible use of corynebacteria for the production of extracellular enzymes, probably owing to the failure to detect extracellular hydrolytic enzymes (cellulases, glucanases, proteases, xylanases, etc.) in these microorganisms.

C. glutamicum is a Gram-positive, nonsporulating bacterium used for the industrial production of amino acids that have been used in human food, animal feed, and pharmaceutical products for several decades. It is a nonpathogenic bacterium, produces no hazardous toxins, and there is much accumulated experience with it in the fermentation industry. As a result, *C. glutamicum* is considered a food-grade bacterium that can be used for the secretion of heterologous extracellular enzymes. Owing to the lack of protease activity and the presence of the Sec and Tat secretion pathways, we propose the use of coryneform bacteria for the production and secretion of hydrolytic enzymes or proteins of high added value.

1.1. Plasmid Vectors for the Introduction of Exogenous DNA

Most plasmid vectors used for the introduction of exogenous DNA in corynebacteria derive from the cryptic endogenous plasmids pBL1 *(1)* and pHM1519 *(2)*. A complete review of the variety of endogenous plasmids from corynebacteria can be found in a recent review *(3)*. Antibiotic resistance genes from *Streptomyces, Bacillus, E. coli,* or from transposon Tn*5* have been cloned into those cryptic plasmids and a variety of monofunctional vectors are currently available. In our hands, the best antibiotic resistance marker for corynebacteria is the *kan* gene from Tn*5,* initially used by Santamaría et al. *(1),* followed by the hygromycin resistance gene *(hyg)* from *Streptomyces hygroscopicus.* Examples of these monofunctional vectors are plasmid pULRS6 and pULRS8 *(4).*

Shuttle *E. coli*-corynebacteria vectors have also been constructed in different laboratories *(5,6),* and these facilitate genetic manipulations because all plasmid constructions can be performed in *E. coli,* and once the desired construction has been obtained it can be transferred to corynebacteria by either transformation or electroporation. One example from our laboratory is pUL609M, a hybrid plasmid constructed from the corynebacterial plasmid pULRS6 (a pBL1 derivative) and the *E. coli* vector pUC13. pUL609M contains kanamycin and hygromycin resistance markers that are expressed in corynebacteria as well as in *E. coli,* and the ampicillin resistance marker from pUC13 for the selection of transformants in *E. coli.* Transformation of protoplasts and electroporation of *C. glutamicum* with pUL609M is obtained with high efficiency (10^6 transformants/μg DNA).

Shuttle *E. coli*-corynebacteria mobilizable vectors based on RP4 were initially described by Schafer et al. *(7)*. Plasmid pECM1 (***E****scherichia*-**c**orynebacteria-**m**obilizable) resulted from the fusion of pSUP102 (mobilizable from *E. coli*) with *C. glutamicum* plasmid pHM1519 *(2)*. Later, Qian et al. *(8)* described the construction of the mobilizable shuttle plasmids pXZ911, pBZ51, pBZ52, which carry the Mob site (RP4) and replication origins of *E. coli* and corynebacterial plasmids. These shuttle mobilizable vectors can be transferred from *E. coli* to corynebacteria by conjugation.

Many of the suicide mobilizable vectors used in Gram-positive bacteria are also based on the broad-host-range plasmid RP4 bearing the *cis*-acting DNA recognition site for conjugative DNA transfer between bacterial cells (Mob site) and the kanamycin resistance gene from Tn5 *(9)*. Priefer et al. *(10)* developed the pSUP series, broad-host-range vectors that are based on conventional *E. coli* vectors (pBR325 and pACYC184) modified to include the mobilization and broad-host-range replication functions of the IncQ plasmid RSF1010. pSUP vectors are efficiently mobilized by RP4 and are thus of particular interest for bacteria refractory to transformation. Small mobilizable vectors based on the *E. coli* plasmids pK18 and pK19 have been described by Schafer et al. *(11);* these vectors combine the useful properties of the pK plasmids (multiple cloning site, *lacZ* alpha fragment, kanamycin resistance) with the broad-host-range transfer machinery of plasmid RP4 and have been named pK18/19mob. Vectors pK18/19mob can be transferred by RP4-mediated conjugation into a wide range of Gram-negative and Gram-positive bacteria, including corynebacteria, and they behave as suicide vectors.

The literature contains only a few references to the construction of promoter probe vectors for corynebacteria and the work described to date is based on the promoterless *kan* gene from Tn5 *(12)*, the *cat* gene from Tn9 *(13,14)*, the β-glucuronidase gene from *E. coli (15)*, the α-amylase gene from *Bacillus subtilis (16)* or from *Streptomyces griseus (17)*, the β-galactosidase and permease genes from *E. coli (18)*, and the *melC* operon from *Streptomyces glaucescens (19)*. Some of these promoter-probe plasmids are monofunctional plasmids, whereas others are bifunctional *E. coli*-corynebacteria plasmids.

1.2. Introduction of Exogenous DNA into Corynebacteria

All these vectors are monofunctional or bifunctional plasmids that must be introduced into *C. glutamicum* protoplasts or electroporated into treated *C. glutamicum* cells. Although protoplast transformation *(20)* and electroporation *(21,22)* have been reported to be efficient methods for transforming coryneform bacteria with plasmid DNA, conjugation is also a good method for introducing plasmids into *C. glutamicum (23,24)*.

1.2.1. Transformation of Protoplasts

All the published procedures for corynebacteria protoplast transformation are based on a process that can be split up into three stages: (1) treatment of cells to partial or complete removal of the cell wall (spheroplast or protoplast generation); (2) polyethylene-glycol (PEG)-assisted introduction of naked DNA; (3) regeneration of an intact cell wall followed by scoring of transformants. Obtaining and regenerating protoplasts from corynebacteria requires weakening of the cell walls by growing the cells in subinhibitory concentrations of penicillin *(1)* or in the presence of glycine *(25)*. Efficiencies of up to 10^6 transformants per μg of DNA have been obtained by optimizing the age of the cultures used to prepare protoplasts and the DNA and PEG type and concentrations (PEG 6000, 30%) *(20,26)*.

1.2.2. Electroporation

Introduction of DNA into bacterial cells via brief high-voltage electric discharges has become the preferred method for transformation of recalcitrant species, because it is simple, fast, and reproducible. When cells are placed in an electric field of a critical strength, a reversible local disorganization and transient breakdown occurs, allowing molecular flux (influx and efflux). During pore formation, transforming DNA can diffuse into the bacterial cells. Electroporation has the advantage that it does not require the preparation and regeneration of protoplasts, which is a limiting step for PEG-mediated transformation. Although the reported efficiencies of electroporation with plasmid DNA are relatively high (up to 10^7 transformants per μg of DNA) *(21,22)*, routine results are less satisfactory (10^4 to 10^5 transformants per μg of DNA).

1.2.3. Conjugal Plasmid Transfer from E. coli to Corynebacteria

Conjugal transfer by broad-host-range IncP-type resistance plasmids from Gram-negative to Gram-positive bacteria is well known *(27)*, and Schafer et al. *(7)* applied this system to corynebacteria. For mating experiments, plasmids were introduced into the mobilizing strain *E. coli* S17-1 carrying an RP4 derivative integrated into the chromosome, which provides the transfer functions necessary for mobilization. In our hands, conjugation has been a very efficient method to introduce plasmid DNA, reaching up to 10^{-4} transconjugants/donor *E. coli* cell.

The protocol for conjugation consisted in growing separately to late exponential phase *E. coli* S17-1 containing the suicide or the shuttle conjugative plasmid and *C. glutamicum;* an equivalent number of donor and recipient cells were joined and spread onto a 0.45-μm pore size cellulose acetate filter (for mating) placed on a TSA medium plate. After 20 h of incubation at 30°C, cells were washed from the filter and transconjugants were selected by plating serial dilutions on TSA medium with nalidixic acid (a bacteriostatic agent for *E. coli*) and suitable pressure (second antibiotic) for transconjugant selection.

1.3. Genes Coding for Interesting Enzymes/Proteins Expressed in Corynebacteria

As a rule, heterologous genes are poorly expressed in *C. glutamicum/B. lactofermentum* from their own promoters *(28,29)*. Therefore, to express het-erologous genes in corynebacteria these should be cloned downstream from corynebacterial promoters or from promoters recognized by corynebacterial RNA polymerase. If the desired protein is to be secreted, its leader peptide can be processed by the *C. glutamicum/B. lactofermentum* signal peptidase, which recognizes the same cleavage amino acid sequence as *Streptomyces, Bacillus, Staphylococcus,* and *Cellulomonas.* To date, the only foreign signal peptides shown to be functional in *C. glutamicum* have come from the Gram-positive bacteria indicated above. Analysis of the genome of *C. glutamicum* revealed the presence of three putative signal peptidases (Cgl1586, Cgl1987, and Cgl2088) that might be responsible for the cleavage and removal of the leader peptide. The heterologous genes expressed in corynebacteria described below are not present in the genome of *C. glutamicum,* and therefore their expression will afford this strain new biochemical capabilities.

1.3.1. Amylase

Genes coding for α-amylases, either from *Bacillus* or *Streptomyces,* have been expressed in *C. glutamicum/B. lactofermentum.*

The *amy* gene from *Bacillus amyloliquefaciens* was cloned onto a plasmid able to replicate in *B. lactofermentum* and transformants expressed amylase activity at low levels but produced material that was cross-reactive to amylase antibody *(30)*. The reason for the low production is that the *amy* gene has been expressed from its own promoter.

The gene coding for an α-amylase *(amy)* from *S. griseus* IMRU3570 *(31)* was cloned in the monofunctional *B. lactofermentum* plasmid pULRS8 *(4)* under the control of Pkan, affording pULMJ91. Culture supernatants of *B. lactofermentum* [pULMJ91] displayed two protein bands (57 and 50 KDa) with amylase activity, representing about 90% of the total extracellular protein. Those proteins had the same NH_2-end and exactly matched the mature amylase obtained in *S. lividans (28)*.

1.3.2. Xylanase

The whole *xysA* (catalytic domain + cellulose binding domain) gene from *Streptomyces halstedii* JM8 *(32)* and the catalytic domain with a tag of 6 His at the C-terminus (Δ*xysA-His6*) were expressed in *B. lactofermentum* only when cloned under the control of Pkan in the bifunctional plasmid pUL880M *(29)*. More than 95% of the xylanase activity was found in the culture supernatant and the NH_2-end of the xylanase secreted by *B. lactofermentum* was sequenced, with the finding that it was identical to the amino-end of the xylanase produced by *Streptomyces*.

1.3.3. Cellulase

There are two examples in the scientific literature of the expression of cellulase genes in corynebacteria: (1) The *cenA* gene from *Cellulomonas fimi* was subcloned into the *E. coli/B. lactofermentum* shuttle vector pBK10 and expressed in *B. lactofermentum* from its own promoter. The level of expression was not very high and most of the CenA was extracellular *(33)*. (2) The promoterless *celA1* gene from *S. halstedii (34)* was cloned under Pkan and flanked by two terminators (T1 and its own terminator Tcel) in the bifunctional plasmid pUL880M, giving rise to plasmid p880C2A *(29)*. *E. coli* and *B. lactofermentum* cells transformed with this plasmid exhibited cellulase activity in carboxy-methyl-cellulose plates after staining with Congo red and after growth in liquid medium; most of the cellulase activity was present in the *B. lactofermentum* supernatants *(29)*.

1.3.4. β-Galactosidase

Two different β-galactosidase-producing *C. glutamicum/B. lactofermentum* strains were constructed in two different laboratories by cloning genes coding for β-galactosidase *(lacZ)* from *E. coli (35)* and from *S. lividans (28)*. In both cases, efficient expression of the *lacZ* gene required the presence of a mutagenically altered *lacZ* promoter *(35)* or a plasmid pBL1 promoter *(28)* for efficient transcription in *C. glutamicum*. Transformants expressed β-galactosidase activity after 1 or 3 d; this was observed as a blue color around the transformant colonies growing in medium containing X-Gal (40 μg/mL). The galactose moiety of lactose was not utilized but accumulated in the culture broth. *C. glutamicum* strains carrying only the *lacZ* (β-galactosidase) gene, but not *lacY* (lactose permease), were not able to grow in lactose minimal medium *(35)*.

1.3.5. Protease

Two serin proteases of the subtilisin family have been expressed in *C. glutamicum:* subtilisin I168 from *Bacillus subtilis* I168 and a basic protease from *Dichelobacter nodosus* 198 (formerly *Bacteroides nodosus*, the causative agent of footrot in sheep). The gene encoding subtilisin I168 *(aprE) (36)* was expressed and secreted in *C. glutamicum* using its own leader peptide and promoter, whereas the gene encoding the basic protease (**bprV**) *(37)* was not expressed with its natural promoter and secretion signal *(38)*. However, *bprV* was expressed and secreted when it was located downstream from the promoter and leader peptide of *aprE*.

1.3.6. Nuclease

The gene for an extracellular nuclease of *Staphylococcus aureus (39)* was expressed using its own promoter and signal peptide in *C. glutamicum,* exported, and correctly processed *(40)*. It was also possible to achieve IPTG-inducible overexpression and secretion of the extracellular nuclease in *C. glutamicum* by using a plasmid containing *lacI*q and the gene encoding nuclease under the control of the *tac* promoter *(40)*.

1.3.7. Transglutaminase

Transglutaminases (TG) have been used in the food industry for the modification of proteins *(41)* and the catalysis of acyl transfer reactions between a γ-carboxyamide group of a glutamine residue in a peptide chain and a γ-amino group of a lysine residue, resulting in the formation of an ε-(γ-glutamyl) lysine cross-link. A calcium-independent transglutaminase from *Streptoverticillium mobaraense (42)* was efficiently secreted by *C. glutamicum* as pro-mature TG when it was coupled to the promoter and signal peptide from *cspB (43)* but not using its own signal peptide *(44)*. Amino-terminal sequencing of pro-mature TG secreted by *C. glutamicum* revealed that the signal peptide was correctly processed. The pro-mature TG was converted into active-form TG when processed by a subtilisin-like protease from *Streptomyces albogriseolus (45)* co-secreted by *C. glutamicum (44)*.

1.3.8. Ag85

The gene encoding the fibronectin-binding protein 85A (Ag85) from *Mycobacterium tuberculosis (46)* was weakly expressed in *C. glutamicum* under the control of the *tac* promoter but efficiently so under the control of the *cspB* promoter *(47)*. The recombinant 85A protein was correctly processed in *C. glutamicum;* it was biologically active, retaining its full B- and T-cell immunogenicity *(48)*, and it was released into the extracellular culture medium *(47)*.

1.3.9. Ovine Interferon

The gene encoding ovine γ-interferon *(49)* was fused with the gene for glutathione S-transferase and expressed in *C. glutamicum* as a fusion protein *(50)*. Expression of recombinant protein in *C. glutamicum* from the *tac* promoter and *lacI*q was unstable, but it was stabilized using *lacI*[18K], a *lacI* expressed from the promoter, RBS, and ATG initiation codon from a mycobacterial gene *(51)*.

Under those conditions, *C. glutamicum* intracellularly expressed ovine γ-interferon at high levels (in inclusion bodies); it was biologically active and was recognized by anti-γ-interferon monoclonal antibodies *(50)*.

1.4. Construction of Food-Grade Corynebacteria Producing a Desired Heterologous Protein

All the above reports concerning the construction of genetically manipulated corynebacteria are a marked improvement for all industrial strains because of the lower cost of the substrates (xylane, cellulose, and whey) used for growth and for the production of extracellular enzymes or proteins of high added value (γ-interferon). However, the previously mentioned strains carried plasmids and antibiotic resistance genes and thus might be not useful in the near future because of the stringent regulations on genetically manipulated microorganisms, especially when the fermentation product is to be used in human or animal food (food-grade microorganisms) *(52,53)*.

To overcome this problem, we have developed a method that can be used to integrate any gene in the genome of *B. lactofermentum* by double recombination *(23)*. There are several descriptions of plasmids that can be used to integrate genes in the genome of corynebacteria based on the presence of a chromosomal gene in the plasmid *(54,55)* or very short homologous DNA segments (8 to 12 bp) in the vector and in the host DNA *(56)*. However, all of them require the presence of antibiotics for selection. The generation of a drug-resistant recombinant strain can both reduce the *in vivo* applicability of the strain and exclude the use of recombinant vectors carrying the same drug resistance marker.

Here we describe how to construct food-grade (lacking any antibiotic resistance gene) corynebacteria strains able to secrete xylanase or cellulase, or both enzymes, from *Streptomyces halstedii* JM8, although the methodology can also be used for any of the heterologous proteins already expressed in corynebacteria.

2. Materials
1. *E. coli* strain S17-1 *(11)*.
2. *B. lactofermentum* R31 *(20)*.
3. Antibiotics: kanamycin and nalidixic acid.
4. Tryptone soy broth (TSB): 17 g/L casein peptone, 3 g/L soy peptone, 5 g/L NaCl, and 2.5 g/L glucose.
5. Tryptone soy agar (TSA): 17 g/L casein peptone, 3 g/L soy peptone, 5 g/L NaCl, 2.5 g/L glucose, and 15 g/L agar.
6. Luria-Bertani (LB) medium: 10 g/L bacto-tryptone, 5 g/L bacto-yeast extract, 10 g/L NaCl, and 15 g/L agar.
7. pK18mob *(11)*.
8. 3.3 kb *Bam*HI fragment of *B. lactofermentum* DNA containing 3 ORFs located downstream from *ftsZ* *(57)* and not required for cell growth and viability *(58)*.

9. pK18-3 *(23)*.
10. pXHis-Npro *(29)*.
11. pXC6 (Ramón I. Santamaría, Univ. Salamanca, Spain).
12. 0.45-μm-pore-size cellulose acetate filter (Millipore, Bedford, MA).
13. Remazol brilliant blue R-D-xylan (RBB-xylan, Sigma Chemical, St. Louis, MO).
14. Carboxy-methyl cellulose (CMC, Sigma).
15. Congo Red (Sigma).
16. TES: 25 mM Tris-HCl, pH 8.0, 25 mM EDTA, pH 8.0, and 10.3% sucrose.
17. Kirby mixture (1X): 1% sodium tri-isopropylnaphthalene sulfonate, 6% sodium 4-amino-salycylate, 6% phenol mixture (v/v), and 50 mM Tris-HCl, pH 8.0.
18. Phenol mixture: 500 g phenol, and 0.5 g 8-hydroxyquinoline saturated with 50 mM Tris-HCl, pH 8.0.
19. Phenol/chloroform/isoamyl alcohol: 50 mL phenol mixture, 50 mL chloroform, and 1 mL isoamyl alcohol.
20. Chloroform/isoamyl alcohol: 24 mL chloroform, and 1 mL isoamyl alcohol.
21. TE: 10 mM Tris-HCl, pH 8.0, and 1 mM EDTA, pH 8.0.
22. Restriction enzymes, T4 DNA ligase.
23. Agarose (Sigma), ethidium bromide (Sigma) and DNA electrophoresis equipment.
24. TAE 50X running buffer: 57.1 mL glacial acetic acid, 100 mL 0.5 M EDTA, pH 8.0, 242 g Tris-HCl, and H_2O up to 1000 mL.
25. Loading buffer (6X): 0.25% (w/v) bromophenol blue, 40% (w/v), sucrose, and 0.25% (w/v) xylen-cianol.
26. Nylon membrane (Amersham International, UK).
27. Whatman filter paper no. 1 (Whatman, Kent, UK).
28. Denaturation solution: 0.5M NaOH, and 1.5 M NaCl.
29. Neutralization buffer: 0.5 M Tris-HCl, pH 7.5, 1.5 M NaCl, and 0.1 M EDTA.
30. 20X SSC: 3M NaCl and 0.15 M sodium citrate, pH adjusted to 7.0.
31. Prehybridization/hybridization solution: SSC 5X, sarkosyl 0.1%, SDS 0.02%, and blocking reagent 1%.
32. Random Primer DNA labeling kit (Roche Diagnostics, Barcelona, Spain).
33. *kan* probe.
34. *xyl* probe.
35. *cel* probe.
36. Probe 3: 3.3 kb *Bam*HI fragment of *B. lactofermentum* DNA containing 3 ORFs located downstream from *ftsZ* and labeled with digoxigenin.
37. Buffer 1: 150 mM NaCl, and 100 mM maleic acid, pH 7.5 (filter through 0.45-μm filter).
38. Buffer 2: Buffer 1 containing 2% blocking reagent, gently heat at 60°C, with constant mixing to dissolve blocking reagent.
39. Buffer 3: 100 mM Tris-HCl, 100 mM NaCl, and 250 mM $MgCl_2$; adjust to pH 9.5 with NaOH.
40. 2X wash solution: 2X SSC, containing 0.1% SDS.
41. 0.5X wash solution: 0.5X SSC, containing 0.1% SDS.
42. NBT: 4-nitro blue tetrazolium chloride.

43. BCIP: 5-Bromo-4-chloro-3 indolyl-phosphate.
44. SDS-PAGE (sodium dodecyl sulphate-polyacrylamide gel electrophoresis) equipment.
45. Ni-NTA-agarose (Qiagen, Hilden, Germany).
46. Equilibration buffer 1X: 50 mM sodium phosphate buffer, pH 8.0, and 300 mM NaCl.
47. Equilibration buffer 2X: 100 mM sodium phosphate buffer, pH 8.0, and 600 mM NaCl.
48. Washing buffer: 50 mM sodium phosphate buffer, pH 8.0, 300 mM NaCl, and 5 mM imidazol.
49. Elution buffer: 50 mM sodium phosphate buffer, pH 8.0, 300 mM NaCl, and 250 mM imidazol.

3. Methods

The methods described below outline (1) the culture media for coryneform bacteria, (2) the construction of the integrative plasmids, (3) plasmid transfer from *E. coli* to corynebacteria, (4) the screening of the transconjugants, and (5) purification of the recombinant protein from *B. lactofermentum.*

3.1. Culture Media for Corynebacteria

For many purposes (plasmid isolation, chromosomal DNA and RNA isolation, gene expression), corynebacteria are grown in different liquid or solid media from isolated colonies or from cell suspensions maintained in 20% glycerol. Liquid cultures are aerobically incubated at 30°C in an orbital shaker at 250 rpm in Erlenmeyer flasks. The most frequently used medium is TSB (tryptone soy broth) or TSA (tryptone soy agar).

TSB is a commercially available medium and is prepared by dissolving 30 g of tryptone soy broth in 1000 mL of distilled water. To prepare TSA, TSB is distributed in 100-mL quantities in 200-mL bottles containing 2 g of agar and is sterilized by autoclaving at 121°C for 20 min.

3.2. Construction of Conjugative Plasmids Designed to Insert Genes into the Chromosome of B. lactofermentum

The gene encoding the desired protein to be expressed and secreted in corynebacteria is cloned into the suicide plasmid pK18-3 *(23)*. The exogenous gene integrates by double recombination (*see* **Note 1**) in one of the three nonessential open reading frames (ORFs) located downstream from *ftsZ* in the genome of *C. glutamicum/B. lactofermentum* (*see* **Note 2**) *(58)*, and the resulting strain will lack any exogenous drug resistance marker.

Plasmid pK18-3 was constructed by cloning a 3.3-kb *Bam*HI fragment from the *B. lactofermentum* chromosomal DNA containing the three ORFs located downstream from *ftsZ* into the unique *Bam*HI restriction site of pK18mob (*see* **Note 3**) **(Fig. 1)**; because there are single restriction sites for *Nde*I and *Eco*RV

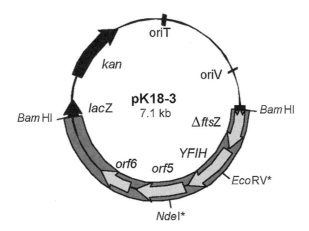

Fig. 1. Structure of plasmid pK18-3, a pK18mob derivative designed to deliver any gene into the *B. lactofermentum* chromosome. Single sites are marked with asterisks.

in pK18-3, they can be used for the cloning of any exogenous DNA fragment carrying the desired gene.

Requisites of the exogenous DNA (encoding the desired protein or desired fused protein) to be cloned in pK18-3:

1. Presence of a strong promoter, inducible or constitutive, able to be recognized by the *C. glutamicum* RNA polymerase (Ptac, Plac, Pkan, Pcsp).
2. If Plac or Ptac is the promoter used, *lacI* should be included in the construction, because *C. glutamicum* lacks *lacI*.
3. If the protein is to be secreted, a leader peptide (LP) from *Streptomyces* (LPamy, LPxyl) or from *C. glutamicum* (LPcsp) might be needed.
4. Exogenous DNA should be flanked by strong transcriptional terminators to control gene expression from the inducible promoter and to avoid interferences with chromosomal transcripts, which might make the recombinant strain unstable.
5. A tag should be added to the final protein to facilitate its purification by affinity chromatography.

As an example, the six-His-tagged Δ*xysA* gene from *S. halstedii* under the control of the promoter of the *kan* gene from Tn*5* and flanked by terminators was obtained from plasmid pXHis-Npro *(29)* and subcloned as a 2.6-kb *Bgl*II fragment (Klenow-filled) into the unique *Nde*I site (Klenow-filled) of pK18-3. The resulting plasmid was named pK18-3X **(Fig. 2)**.

Similarly, a truncated Δ*xysA* fused in-frame to *celA1* from *S. halstedii* **(34)** under the control of the promoter of the *kan* gene from Tn*5* and flanked by terminators was obtained from plasmid pXCNeo (S. A. I. Adham, unpublished)

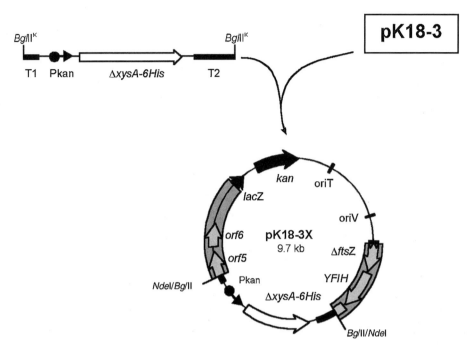

Fig. 2. Schematic representation of the construction of the conjugative suicide plasmid pK18-3X designed to integrate the His-tagged xylanase gene from *S. halstedii* into the *B. lactofermentum* chromosome.

and subcloned as a 3.2-kb *Bgl*II fragment (Klenow-filled) into the unique *Eco*RV site of pK18-3. The resulting plasmid was named pK18-3XC (**Fig. 3**).

3.3. Plasmid Transfer from E. coli to Corynebacteria by Mating

The procedure described here to transfer plasmid pK18-3X or pK18-3XC from *E. coli* S17-1 to *B. lactofermentum* R31 is similar to the protocol developed by Schafer et al. (**7**), except that in our procedure the donor-recipient ratio is 1:1 (*see* **Note 4**).

3.3.1. Preparation of Donor Cells

1. Introduce the desired mobilizable plasmid construction into *E. coli* S17-1 using any of the standard *E. coli* transformation protocols (**59**).
2. Inoculate 100 mL of LB medium (**60**) supplemented with the desired antibiotic to select for the mobilizable plasmid (in a 250-mL flask) with a single colony of transformed *E. coli* S17-1. Incubate at 37°C in a rotary shaker (200 rpm) until OD$_{600}$ = 1–1.5.
3. Harvest cells by centrifugation (in a bench centrifuge) at 1100g for 10 min. Discard supernatant.

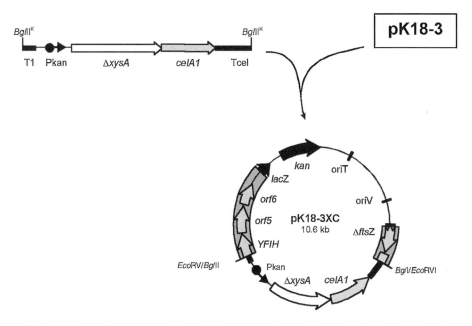

Fig. 3. Schematic representation of the construction of the conjugative suicide plasmid pK18-3XC designed to integrate the fused xylanase and cellulase genes from *S. halstedii* into the *B. lactofermentum* chromosome.

4. Resuspend pellet in 10 mL of LB medium and repeat **step 3**.
5. Resuspend cells in 5 mL of LB medium and repeat **step 3**.
6. Resuspend cells in LB medium to obtain about 3×10^8 cells/mL.

3.3.2. Preparation of Recipient Cells

1. Inoculate 100 mL of TSB medium with 1 mL of an overnight culture of *B. lactofermentum* R31. Incubate at 30°C in a rotary shaker (250 rpm) until $OD_{600} = 3–4$.
2. Harvest cells by centrifugation (in a bench centrifuge) at $1100g$ for 10 min. Discard supernatant.
3. Resuspend pellet in 10 mL of TSB medium and repeat **step 2**.
4. Resuspend cells in 5 mL of TSB medium and repeat **step 2**.
5. Resuspend cells in TSB medium to obtain about 3×10^8 cells/mL.
6. Treat with heat shock for 9 min at 49°C (not required for *B. lactofermentum* R31) and then place at 37°C before mating.

3.3.3. Mating

1. Mix 3×10^8 donor and recipient cells and centrifuge at room temperature $1100g$ for 5 min.
2. Carefully resuspend cells in 5 mL of TSB medium and centrifuge as in **step 1**.

3. Carefully resuspend cells in 0.5 mL of TSB medium and spread the mating mixture onto a 0.45-μm cellulose acetate filter (Millipore) placed on a prewarmed (to room temperature) TSA plate.
4. Incubate at 30°C for 20 h (mating).
5. Wash the cells from the filter with 1 mL of TSB and spread cell suspension (100–200 μL) on TSA plates with the appropriate antibiotic and nalidixic acid (30 μg/mL) (*see* **Note 5**). Use TSB medium to make dilutions if required.
6. Incubate plates at 30°C and score for resistant colonies after 24–48 h of incubation.

3.4. Screening for Single Recombination

The conjugative suicide plasmid pK18-3X and pK18-3XC were transferred separately by conjugation from *E. coli* S17-1 cells to *B. lactofermentum* R31 as described above, and *B. lactofermentum* kanamycin-resistant transconjugants were readily obtained (10^{-2} to 10^{-3} transconjugants per donor colony). In a first step, we scored colonies for kanamycin resistance (*see* **Note 6**).

3.4.1. Screening by Enzymatic Activity

The production of xylanase by *B. lactofermentum* transconjugants obtained with either plasmid pK18-3X or pK18-3XC was assayed on TSA medium containing 0.4% Remazol brilliant blue R-D-xylan (RBB-xylan, Sigma) (*see* **Note 7**). After 2 d of growth, a clear zone around the transconjugant colonies due to xylanase activity was observed. No clear zones were detected around colonies of *B. lactofermentum* R31 or *B. lactofermentum* carrying the parental vector pK18-3. These results were confirmed by growing transconjugants on TSA medium supplemented with 1% xylan. After 2 d of growth, the plates were flooded with Congo red (0.1%) (*see* **Note 8**) and washed with 1 *M* NaCl, and clear zones around the colonies due to xylan degradation were observed.

The cellulose activity of the transconjugants obtained with plasmid pK18-3XC was observed as halos around the colonies after 2 d of growth on MM or TSA, supplemented with 0.4% carboxy-methyl cellulose (CMC, Sigma), followed by Congo red staining. As above, no clear zones of cellulose degradation were detected around colonies of *B. lactofermentum* R31 or *B. lactofermentum* carrying vectors pK18-3 or pK18-3X.

3.4.2. Confirmation by DNA–DNA Hybridization

In order to confirm that the suicide conjugative plasmids pK18-3X or pK18-3XC were indeed integrated into the chromosome by single recombination or double recombination, chromosomal DNA isolation from transconjugants and Southern blot analysis are required.

3.4.2.1. METHOD FOR ISOLATING CORYNEBACTERIAL CHROMOSOMAL DNA

There are several methods for corynebacterial DNA isolation reported in the literature; here we shall describe the method used in our laboratory with *B. lactofermentum* and *C. glutamicum.*

This method is a modification of the Kirby mix procedure (developed by C. P. Smith) for the isolation of *Streptomyces* genomic DNA *(61)*. The average DNA fragment size obtained using this method is around 30 kb. The procedure is rapid (6 h) and allows 6 to 12 samples to be processed in parallel.

1. Inoculate 100 mL of TSB medium with 1 mL of an overnight culture of coryneform bacteria. Incubate at 30°C in a rotary shaker (220 rpm) until the end of the exponential phase (24–30 h).
2. Harvest cells by centrifugation at 12,100*g* for 10 min at room temperature. Discard supernatant.
3. Resuspend pellet in 10% sucrose and repeat **step 2.**
4. Resuspend cells in 3 mL of TES buffer and add lysozyme to a final concentration of 5 mg/mL. Incubate for 2–3 h at 30°C.
5. Add 3 mL of 2X Kirby mixture and shake for 1 min on a vortex mixer.
6. Add 6 mL phenol/chloroform/isoamyl alcohol and shake for 30 s by vortexing.
7. Centrifuge for 10 min at 12,100*g* to separate phases.
8. Transfer upper (aqueous) phase to a new tube containing 3 mL of phenol/chloroform/isoamyl alcohol and shake as in **step 6.** Centrifuge as in **step 7.**
9. Transfer the upper phase to another tube containing 3 mL of chloroform-isoamyl alcohol and shake as in **step 6.** Centrifuge as in **step 7.**
10. Transfer the upper phase to a fresh tube; add 1/10 volume of 3 *M* sodium acetate (pH 6.0) and 0.6 volume of isopropanol (for selective precipitation of DNA, leaving most of the RNA in solution); mix gently but thoroughly until DNA precipitates.
11. Recover DNA on a sealed Pasteur pipet and transfer to a battery of 10-mL tubes containing 5 mL of 70% ethanol, and finally to a 10-mL tube containing 5 mL of absolute ethanol.
12. Pour off the ethanol, dry the DNA in the laminar flow cabinet, and resuspend in 1 mL of TE buffer.

3.4.2.2. DNA DIGESTION, AGAROSE GEL ELECTROPHORESIS, TRANSFER TO SOLID SUPPORT, AND HYBRIDIZATION

Total DNA isolated from 20 *C. glutamicum* kanamycin-resistant transconjugants (10 from pK18-3X and 10 from pK18-XC) was digested with appropriate restriction enzymes (*Bam*HI in our example), and separated by agarose gel electrophoresis.

To prepare the gel, agarose was dissolved by heating in TAE buffer and DNA samples were mixed with loading buffer, heated at 65°C for 5–10 min, and then cooled by submerging them in a water bath at 4°C for 2–3 min. The DNA was loaded in the gel and electrophoresis was carried out in TAE buffer at 1.5 V/cm. DNA molecules were detected by staining in a solution of ethidium bromide (EtBr) at a final concentration of 0.5 µg/mL, which allows its visualization in a UV light transilluminator.

DNA fragments separated by agarose gel electrophoresis are transferred to a solid support according to Southern (*62*):

1. Denature DNA in gel by soaking in 1X denaturation solution for 30–60 min.
2. Neutralize by soaking gel in 1X neutralization buffer for 30–60 min.
3. Place gel on plastic wrap, overlay with pre-wetted (in water) positively charged nylon membrane.
4. Place a pre-wetted Whatman no. 1 filter paper over the nylon membrane; stack about 2 in. of paper towels on top with ~1–2 kg weight to press it down. Blot overnight.
5. Air-dry filter, UV cross-link (Stratalinker oven) and proceed to prehybridization step.
6. Prehybridize the filter in prehybridization solution and incubate, with gentle shaking, at 42°C for 1 h. It is important from this step forward to prevent the blots from air-drying and to minimize exposure to air at all times, since high backgrounds will result.
7. Label DNA probe (*kan, xysA, celA,* and probe 3) using the random priming method as follows: 2.5–25 ng of the DNA probe are diluted in 16 µL of H_2O, boiled for 10 min, and shocked by cooling for at least 4 min. Then, 4 µL of high prime digoxigenin is added to the mixture and incubated at 37°C overnight.
8. Add digoxigenin-labeled probe to fresh prehybridization solution and incubate at 42°C, 2–15 h, with gentle shaking. Pour off and save hybridization solution. This can be reused for additional blots.
9. Wash blot in 2X wash solution, twice, 5 min each, at room temperature.
10. Wash blot in 0.5X wash solution, twice, 15 min each, at 65°C.
11. Equilibrate blot 1 min in buffer 1, at room temperature.
12. Block membranes in buffer 2, 30–60 min, at room temperature. (Prepare anti-DIG antibody-alkaline phosphatase conjugate, 1/5000 to 1/10,000 dilution, in buffer 2.)
13. Pour off buffer 2, replace with buffer 2 + antibody; incubate 30 min at room temperature.
14. Pour off buffer 2 + antibody, and wash blot in buffer 1 twice, 15 min each, at room temperature.
15. Equilibrate blot for 2 min in buffer 3. Then add 90 µL NBT (100 mg/mL prepared in 70% N,N'dimethylformamide) and 70 µL BCIP (50 mg/mL in 100% N,N'dimethylformamide) to 20 mL of buffer 3 solution where the filter was incubated at 37°C until the appearance of DNA bands.

Following the method described above, total DNA isolated from 10 pK18-3X transconjugants (xylanase producers) was hybridized with the *kan* gene, with the *xysA* gene, and with probe 3. Two *Bam*HI chromosomal DNA fragments, one corresponding to the original fragment in the chromosome (3.3 kb) and a second one (5.9 kb) corresponding to the sum of the original fragment (3.3 kb) plus six-His-tagged *xysA* (2.6 kb), hybridized with probe 3; the 5.9-kb *Bam*HI DNA band hybridized with the *xysA* gene, and the 3.9-kb *Bam*HI DNA band hybridized with the *kan* gene, as expected.

Similarly, total DNA from 10 transconjugants (xylanase and cellulase producers) obtained with plasmid pK18-3XC was hybridized with the *kan* gene, with the *celA* gene, with the *xysA* gene, and with probe 3; DNA bands of the expected sizes were obtained in all cases.

The integrated plasmids were stable, and the xylanase and xylanase/cellulase were stably expressed without continued selection for kanamycin. It was also observed that *C. glutamicum/B. lactofermentum* can efficiently secrete two proteins from actinomycetes.

Because kanamycin-resistant strains have reduced applicability in vivo, we attempted to construct a double-recombinant strain expressing the xylanase gene and lacking any exogenous drug resistance marker.

3.5. Screening for Double Recombination

The starting strain for the screening for double recombination was a single colony of *B. lactofermentum* XylA$^+$Kan^R. This strain was incubated in TSB medium without kanamycin at 30°C for 200 generations in order to allow the second recombination event, which would excise the plasmid, rendering XylA$^+$ Kan^S colonies at a frequency of 1×10^{-4}. The second recombination event was confirmed by Southern blot hybridization as above. No signal was obtained when the *kan* gene was used as a probe. There was a positive signal of the expected size (5.9 kb, corresponding to the sum of the target fragment [3.3 kb] plus six-His-tagged *xysA* [2.6 kb]) when the *xysA* gene was used as a probe and with probe 3 (**Fig. 4**). The integrated *xysA* gene is stably maintained without the need for selective pressure.

3.6. Purification of His-Tagged Xylanase from B. lactofermentum

Because both the xylanase or the xylanase/cellulase genes were cloned downstream from Pkan and sandwiched between efficient transcriptional terminators, expression is constitutive from Pkan, reaching levels up to 1/10 of the levels obtained (*see* **Note 9**) when the gene was cloned in a multicopy plasmid *(29)*. More than 95% of the xylanase was extracellular and because of the 6 His-tag it could be readily purified from the culture supernatant using Ni-NTA-agarose, following the standard procedures of the manufacturer (Qiagen) as follows:

1. Centrifuge 200 µL of the resin at 1100*g* for 10 min; wash with 300 µL of 1X equilibration buffer.
2. Mix one volume of the supernatant containing the protein of interest with one volume of the 2X equilibration buffer (the amount depends on the quantity of

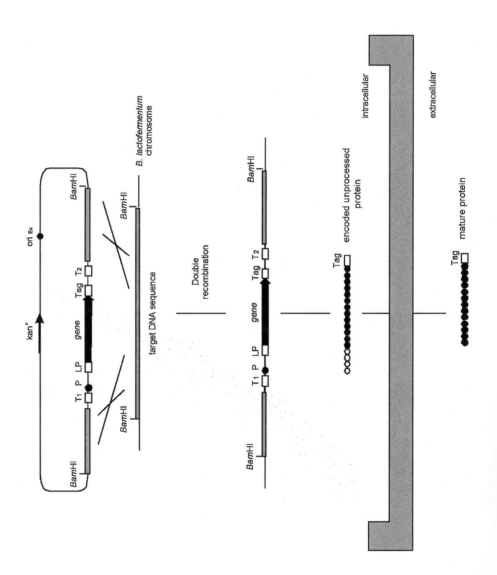

Fig. 4. Integration of a tagged-gene coding for a desired protein into the chromosome of *B. lactofermentum* by double recombination, and secretion of the tagged protein. T1 and T2: transcriptional terminators; P: promoter; LP: leader peptide; Tag: epitope tag.

protein and the retention capacity of the resin), and add 100 µL of the equilibrated resin.

3. Incubate the mixture at room temperature with a slow orbital shaking for 1 h.
4. Centrifuge at 23,600*g* for 1 min and discard the supernatant.
5. Add 1.6 mL of washing buffer and shake slowly for 15 min at room temperature.
6. Decant the supernatant and wash again for 15 min.
7. Add 100 µL elution buffer to the pellet of the second wash.
8. Shake for 1 min and centrifuge at 23,600*g* for 1 min.
9. Remove the supernatant (fraction E1) and add another 100 µL of the elution buffer.
10. Shake slowly in an orbital agitator for 1 h and centrifuge at 23,600*g* for 1 min.
11. Remove the supernatant (fraction E2).
12. Mix fractions E1 and E2 and filter them through a 0.2-µm filter to eliminate any remains of resin.

The His-tagged xylanase from *S. halstedii* JM8 was purified by affinity chromatography from the culture supernatant of the food-grade *B. lactofermentum* strain. The purified enzyme is active, almost 100% pure, and can be used as complement in animal food or for the production of enzymes with industrial applications *(63)*.

In this chapter we describe the method for integrating any gene coding for any hydrolytic enzyme or protein of interest into the genome of *B. lactofermentum* for the production of extracellular enzymes or protein of high added value. *B. lactofermentum* is itself considered a food-grade microorganism, but because no antibiotic resistance markers remain in these genetically manipulated corynebacteria, the resulting organisms are considered real food-grade microorganisms and safe for humans, animals, and the environment.

4. Notes

1. A method for inserting any DNA fragment into the chromosome of *B. lactofermentum* has been developed. The method relies on the suicide conjugative plasmid pK18-3, a pK18mob derivative containing (a) a gene conferring resistance to kanamycin as a selectable marker, (b) a chromosomal region necessary for its integration into the *B. lactofermentum* chromosome, and (c) unique *Nde*I and *Eco*RV restriction sites useful for cloning. Any gene that we wish to integrate (under the control of a promoter capable of functioning in *B. lactofermentum*) is cloned into

the unique restriction site of the plasmid, transformed into *E. coli,* and transferred by conjugation to *B. lactofermentum.*

2. The chromosomal region used for integration of the exogenous DNA (bearing genes encoding the enzyme of interest) is that corresponding to three ORFs of unknown function located downstream from *ftsZ* in the *B. lactofermentum* chromosome *(57)*. It has previously been demonstrated that interruption of any of the three ORFs is not deleterious for *B. lactofermentum* growth or cell division *(58)*.

3. pK18mob is an *E. coli* plasmid described by Schafer et al. *(11)* as a suicide vector for *B. lactofermentum* carrying a kanamycin resistance gene functional in both *E. coli* and *B. lactofermentum*. It contains the origin of replication of ColE1, the broad-host-range transfer machinery of plasmid RP4, the *lacZα* gene, and a multiple cloning site. This plasmid cannot replicate in *B. lactofermentum* but can be transferred by conjugation from *E. coli* S17-1. This plasmid has been used to interrupt several genes in *B. lactofermentum* or *C. glutamicum* *(11,64)*.

4. Plasmids to be transferred by conjugation from *E. coli* to coryneform bacteria are introduced by transformation into the donor strain *E. coli* S17-1 *(11)*. *E. coli* S17-1 is a mobilizing donor strain that has a derivative of plasmid RP4 integrated into the chromosome and provides the transfer functions *(tra)* for mobilization of plasmids carrying the mobilization fragment *(mob)* and the *oriT* of plasmid RP4. The method is useful for a variety of coryneform bacteria, but we used *B. lactofermentum* R31 as the recipient strain *(20)*, probably because of the lack of a DNA restriction system. The conjugation frequency of this strain does not increase after heat treatment prior to mating *(64)*. Transfer frequencies are expressed as the number of transconjugants per final donor colony, and values ranging from 10^{-2} to 10^{-4} are achieved with *B. lactofermentum* R31.

5. Most of the corynebacteria tested showed natural resistance to 30 µg/mL of nalidixic acid, whereas *E. coli* was sensitive.

6. Because chromosomal integration events occur rarely, plasmids used for integration do require the concomitant integration of a drug resistance marker in order to identify colonies of recombinant cells.

7. Remazol brilliant blue R-D-xylan (RBB-xylan, Sigma) is a colored substrate used for detecting xylanase activity. *B. lactofermentum* colonies expressing xylanase activity degrade RBB-xylan and the media around the colonies became colorless, whereas media around untransformed colonies or around colonies transformed with the empty vector remained blue.

8. The fluorophore Congo red typically binds to β-1,4 glucans *(65)* and β-1,4 xylans *(66)*. *B. lactofermentum* colonies expressing cellulase or xylanase activity will degrade cellulose or xylan. After staining with Congo red, the medium around those colonies became colorless, whereas the medium around untransformed colonies or around colonies transformed with the empty vector remained red.

9. The integration of foreign genes into the genome constitutes an interesting option for stably maintaining cloned genes without the need for selective markers. However, the level of expression of a given gene is expected to be lower than in a multicopy plasmid.

Acknowledgments

This work was supported by a grant from the European Community-Ministerio de Ciencia y Tecnología (1-FD1997-1134-C03-02). The authors thank Dr. Ramón I. Santamaría (Universidad de Salamanca, Spain) for providing plasmid pXC6. Thanks are also due to N. Skinner for supervising the English version of the manuscript.

References

1. Santamaría, R. I., Gil, J. A., Mesas, J. M., and Martín, J. F. (1984) Characterization of an endogenous plasmid and development of cloning vectors and a transformation system in *Brevibacterium lactofermentum. J. Gen. Microbiol.* **130,** 2237–2246.
2. Miwa, K., Matsui, H., Terabe, M., Nakamori, S., Sano, K., and Momose, H. (1984) Cryptic plasmids in glutamic acid producing corynebacteria. *Agric. Biol. Chem.* **48,** 2901–2903.
3. Deb, J. K. and Nath, N. (1999) Plasmids of corynebacteria. *FEMS Microbiol. Lett.* **175,** 11–20.
4. Santamaría, R. I., Martín, J. F., and Gil, J. A. (1987) Identification of a promoter sequence in the plasmid pUL340 of *Brevibacterium lactofermentum* and construction of new cloning vectors for corynebacteria containing two selectable markers. *Gene* **56,** 199–208.
5. Yeh, P., Oreglia, J., Prevots, F., and Sicard, A. M. (1986) A shuttle vector system for *Brevibacterium lactofermentum. Gene* **47,** 301–306.
6. Nesvera, J., Patek, M., Hochmannova, J., and Pinkas, P. (1990) Plasmid shuttle vector with two insertionally inactivable markers for coryneform bacteria. *Folia Microbiol. (Praha)* **35,** 273–277.
7. Schafer, A., Kalinowski, J., Simon, R., Seep-Feldhaus, A. H., and Puhler, A. (1990) High-frequency conjugal plasmid transfer from gram-negative *Escherichia coli* to various gram-positive coryneform bacteria. *J. Bacteriol.* **172,** 1663–1666.
8. Qian, H., Fan, W., Wu, J., and Zheng, Z. (1994) Plasmid transfer from *Escherichia coli* to coryneform bacteria by conjugation. *Chin. J. Biotechnol.* **10,** 55–60.
9. Simon, R. (1984) High frequency mobilization of Gram-negative bacterial replicons by the in vitro constructed Tn5-Mob transposon. *Mol. Gen. Genet.* **196,** 413–420.
10. Priefer, U. B., Simon, R., and Puhler, A. (1985) Extension of the host range of *Escherichia coli* vectors by incorporation of RSF1010 replication and mobilization functions. *J. Bacteriol.* **163,** 324–330.
11. Schafer, A., Tauch, A., Jager, W., Kalinowski, J., Thierbach, G., and Puhler, A. (1994) Small mobilizable multi-purpose cloning vectors derived from the *Escherichia coli* plasmids pK18 and pK19: selection of defined deletions in the chromosome of *Corynebacterium glutamicum. Gene* **145,** 69–73.
12. Cadenas, R. F., Martín, J. F., and Gil, J. A. (1991) Construction and characterization of promoter-probe vectors for corynebacteria using the kanamycin-resistance reporter gene. *Gene* **98,** 117–121.

13. Patek, M., Eikmanns, B. J., Patek, J., and Sahm, H. (1996) Promoters from *Corynebacterium glutamicum:* cloning, molecular analysis and search for a consensus motif. *Microbiology* **142**, 1297–1309.

14. Eikmanns, B. J., Kleinertz, E., Liebl, W., and Sahm, H. (1991) A family of *Corynebacterium glutamicum/Escherichia coli* shuttle vectors for cloning, controlled gene expression, and promoter probing. *Gene* **102**, 93–98.

15. Bardonnet, N. and Blanco, C. (1991) Improved vectors for transcriptional signal screening in corynebacteria. *FEMS Microbiol. Lett.* **68**, 97–102.

16. Ugorcakova, J., Bukovska, G., and Timko, J. (2000) Construction of promoter-probe shuttle vectors for *Escherichia coli* and corynebacteria on the basis of promoterless alpha-amylase gene. *Folia Microbiol. (Praha)* **45**, 114–120.

17. Cadenas, R. F., Fernández-González, C., Martín, J. F., and Gil, J. A. (1996) Construction of new cloning vectors for *Brevibacterium lactofermentum. FEMS Microbiol. Lett.* **137**, 63–68.

18. Soual-Hoebeke, E., Sousa-D'Auria, C., Chami, M., Baucher, M. F., Guyonvarch, A., Bayan, N., et al. (1999) S-layer protein production by *Corynebacterium* strains is dependent on the carbon source. *Microbiology* **145**, 3399–3408.

19. Adham, S. A., Rodríguez, S., Ramos, A., Santamaria, R. I., and Gil, J. A. (2003) Improved vectors for transcriptional/translational signal screening in corynebacteria using the *melC* operon from *Streptomyces glaucescens* as reporter. *Arch. Microbiol.* **180**, 53–59.

20. Santamaría, R. I., Gil, J. A., and Martín, J. F. (1985) High- frequency transformation of *Brevibacterium lactofermentum* protoplasts by plasmid DNA. *J. Bacteriol.* **162**, 463–467.

21. Liebl, W., Bayerl, A., Schein, B., Stillner, U., and Schleifer, K. H. (1989) High efficiency electroporation of intact *Corynebacterium glutamicum* cells. *FEMS Microbiol. Lett.* **53**, 299–303.

22. van der Rest, M. E., Lange, C., and Molenaar, D. (1999) A heat shock following electroporation induces highly efficient transformation of *Corynebacterium glutamicum* with xenogeneic plasmid DNA. *Appl. Microbiol. Biotechnol.* **52**, 541–545.

23. Adham, S. A., Campelo, A. B., Ramos, A., and Gil, J. A. (2001) Construction of a xylanase-producing strain of *Brevibacterium lactofermentum* by stable integration of an engineered *xysA* gene from *Streptomyces halstedii* JM8. *Appl. Environ. Microbiol.* **67**, 5425–5430.

24. Schafer, A., Kalinowski, J., and Puhler, A. (1994) Increased fertility of *Corynebacterium glutamicum* recipients in intergeneric matings with *Escherichia coli* after stress exposure. *Appl. Environ. Microbiol.* **60**, 756–759.

25. Yoshihama, M., Higashiro, K., Rao, E. A., Akedo, M., Shanabruch, W. G., Follettie, M. T., et al. (1985) Cloning vector system for *Corynebacterium glutamicum. J. Bacteriol.* **162**, 591–597.

26. Martín, J. F., Santamaría, R. I., Sandoval, H., del Real, G., Mateos, L. M., Gil, J. A., and Aguilar, A. (1987) Cloning systems in amino acid-producing corynebacteria. *Bio/Technology* **5**, 137–146.

27. Mazodier, P., Petter, R., and Thompson, C. (1989) Intergeneric conjugation between *Escherichia coli* and *Streptomyces* species. *J. Bacteriol.* **171**, 3583–3585.

28. Cadenas, R. F., Gil, J. A., and Martín, J. F. (1992) Expression of *Streptomyces* genes encoding extracellular enzymes in *Brevibacterium lactofermentum:* secretion proceeds by removal of the same leader peptide as in *Streptomyces lividans. Appl. Microbiol. Biotechnol.* **38**, 362–369.

29. Adham, S. A., Honrubia, P., Díaz, M., Fernández-Abalos, J. M., Santamaría, R. I., and Gil, J. A. (2001) Expression of the genes coding for the xylanase Xys1 and the cellulase Cel1 from the straw-decomposing *Streptomyces halstedii* JM8 cloned into the amino-acid producer *Brevibacterium lactofermentum* ATCC13869. *Arch. Microbiol.* **177**, 91–97.

30. Smith, M. D., Flickinger, J. L., Lineberger, D. W., and Schmidt, B. (1986) Protoplast transformation in coryneform bacteria and introduction of an alpha-amylase gene from *Bacillus amyloliquefaciens* into *Brevibacterium lactofermentum. Appl. Environ. Microbiol.* **51**, 634–639.

31. Vigal, T., Gil, J. A., Daza, A., García-González, M. D., and Martín, J. F. (1991) Cloning, characterization and expression of an alpha-amylase gene from *Streptomyces griseus* IMRU3570. *Mol. Gen. Genet.* **225**, 278–288.

32. Ruiz-Arribas, A., Fernandez-Abalos, J. M., Sanchez, P., Garda, A. L., and Santamaría, R. I. (1995) Overproduction, purification, and biochemical characterization of a xylanase (Xys1) from *Streptomyces halstedii* JM8. *Appl. Environ. Microbiol.* **61**, 2414–2419.

33. Paradis, F. W., Warren, R. A., Kilburn, D. G., and Miller, R. C., Jr. (1987) The expression of *Cellulomonas fimi* cellulase genes in *Brevibacterium lactofermentum. Gene* **61**, 199–206.

34. Fernández-Abalos, J. M., Sánchez, P., Coll, P. M., Villanueva, J. R., Pérez, P., and Santamaría, R. I. (1992) Cloning and nucleotide sequence of *celA1,* and endo-beta-1,4-glucanase-encoding gene from *Streptomyces halstedii* JM8. *J. Bacteriol.* **174**, 6368–6376.

35. Brabetz, W., Liebl, W., and Schleifer, K. H. (1991) Studies on the utilization of lactose by *Corynebacterium glutamicum,* bearing the lactose operon of *Escherichia coli. Arch. Microbiol.* **155**, 607–612.

36. Wells, J. A., Ferrari, E., Henner, D. J., Estell, D. A., and Chen, E. Y. (1983) Cloning, sequencing, and secretion of *Bacillus amyloliquefaciens* subtilisin in *Bacillus subtilis. Nucleic Acids Res.* **11**, 7911–7925.

37. Lilley, G. G., Riffkin, M. C., Stewart, D. J., and Kortt, A. A. (1995) Nucleotide and deduced protein sequence of the extracellular, serine basic protease gene *(bprB)* from *Dichelobacter nodosus* strain 305: comparison with the basic protease gene *(bprV)* from virulent strain 198. *Biochem. Mol. Biol. Int.* **36**, 101–111.

38. Billman-Jacobe, H., Wang, L., Kortt, A., Stewart, D., and Radford, A. (1995) Expression and secretion of heterologous proteases by *Corynebacterium glutamicum. Appl. Environ. Microbiol.* **61**, 1610–1613.

39. Shortle, D. (1983) A genetic system for analysis of staphylococcal nuclease. *Gene* **22**, 181–189.

40. Liebl, W., Sinskey, A. J., and Schleifer, K. H. (1992) Expression, secretion, and processing of staphylococcal nuclease by *Corynebacterium glutamicum. J. Bacteriol.* **174,** 1854–1861.

41. Ikura, K., Sasaki, R., and Motoki, M. (1992) Use of transglutaminase in quality-improvement and processing of food proteins. *Agric. Food Chem.* **2,** 389–407.

42. Ralf, P., Simone, D., Jens, T. O., Isabella, R. R., Sabine, W., and Hans-Lothar, F. (1998) Bacterial pro-transglutaminase from *Streptoverticillium mobaraense.* Purification, characterisation and sequence of the zymogen. *Eur. J. Biochem.* **257,** 570–576.

43. Peyret, J. L., Bayan, N., Joliff, G., Gulik-Krzywicki, T., Mathieu, L., Schechter, E., and Leblon, G. (1993) Characterization of the *cspB* gene encoding PS2, an ordered surface-layer protein in *Corynebacterium glutamicum. Mol. Microbiol.* **9,** 97–109.

44. Kikuchi, Y., Date, M., Yokoyama, K. I., Umezawa, Y., and Matsui, H. (2003) Secretion of active-form *Streptoverticillium mobaraense* transglutaminase by *corynebacterium glutamicum:* processing of the pro-transglutaminase by a cosecreted subtilisin-like protease from *Streptomyces albogriseolus. Appl. Environ. Microbiol.* **69,** 358.

45. Suzuki, M., Taguchi, S., Yamada, S., Kojima, S., Miura, K. I., and Momose, H. (1997) A novel member of the subtilisin-like protease family from *Streptomyces albogriseolus. J. Bacteriol.* **179,** 430.

46. Borremans, M., de Wit, L., Volckaert, G., Ooms, J., de Bruyn, J., Huygen, K., et al. (1989) Cloning, sequence determination, and expression of a 32-kilodalton-protein gene of *Mycobacterium tuberculosis. Infect. Immun.* **57,** 3123–3130.

47. Salim, K., Haedens, V., Content, J., Leblon, G., and Huygen, K. (1997) Heterologous expression of the *Mycobacterium tuberculosis* gene encoding antigen 85A in *Corynebacterium glutamicum. Appl. Environ. Microbiol.* **63,** 4392–4400.

48. Launois, P., DeLeys, R., Niang, M. N., Drowart, A., Andrien, M., Dierckx, P., et al. (1994) T-cell-epitope mapping of the major secreted mycobacterial antigen Ag85A in tuberculosis and leprosy. *Infect. Immun.* **62,** 3679–3687.

49. Radford, A. J., Hodgson, A. L., Rothel, J. S., and Wood, P. R. (1991) Cloning and sequencing of the ovine gamma-interferon gene. *Aust. Vet. J.* **68,** 82–84.

50. Billman-Jacobe, H., Hodgson, A. L., Lightowlers, M., Wood, P. R., and Radford, A. J. (1994) Expression of ovine gamma interferon in *Escherichia coli* and *Corynebacterium glutamicum. Appl. Environ. Microbiol.* **60,** 1641–1645.

51. Booth, R. J., Harris, D. P., Love, J. M., and Watson, J. D. (1988) Antigenic proteins of *Mycobacterium leprae.* Complete sequence of the gene for the 18-kDa protein. *J. Immunol.* **140,** 597–601.

52. Gosalbes, M. J., Esteban, C. D., Galán, J. L., and Pérez-Martínez, G. (2000) Integrative food-grade expression system based on the lactose regulon of *Lactobacillus casei. Appl. Environ. Microbiol.* **66,** 4822–4828.

53. Martín, M. C., Alonso, J. C., Suárez, J. E., and Álvarez, M. A. (2000) Generation of food-grade recombinant lactic acid bacterium strains by site-specific recombination. *Appl. Environ. Microbiol.* **66,** 2599–2604.

54. Pisabarro, A., Correia, A., and Martín, J. F. (1998) Characterization of the *rrnB* operon of the plant pathogen *Rhodococcus fascians* and targeted integrations of exogenous genes at *rrn* loci. *Appl. Environ. Microbiol.* **64,** 1276–1282.

55. Reyes, O., Guyonvarch, A., Bonamy, C., Salti, V., David, F., and Leblon, G. (1991) "Integron"-bearing vectors: a method suitable for stable chromosomal integration in highly restrictive corynebacteria. *Gene* **107,** 61–68.

56. Mateos, L. M., Schafer, A., Kalinowski, J., Martín, J. F., and Puhler, A. (1996) Integration of narrow-host-range vectors from *Escherichia coli* into the genomes of amino acid-producing corynebacteria after intergeneric conjugation. *J. Bacteriol.* **178,** 5768–5775.

57. Honrubia, M. P., Fernández, F. J., and Gil, J. A. (1998) Identification, characterization, and chromosomal organization of the *ftsZ* gene from *Brevibacterium lactofermentum. Mol. Gen. Genet.* **259,** 97–104.

58. Honrubia, M. P., Ramos, A., and Gil, J. A. (2001) The cell division genes *ftsQ* and *ftsZ,* but not the three downstream open reading frames YFIH, ORF5 and ORF6, are essential for growth and viability in *Brevibacterium lactofermentum* ATCC 13869. *Mol. Genet. Genomics* **265,** 1022–1030.

59. Hanahan, D. (1983) Studies on transformation of *Escherichia coli* with plasmids. *J. Mol. Biol.* **166,** 557–580.

60. Miller, J. H. (1972) *Experiments in Molecular Genetics.* Cold Spring Harbor Laboratory, Cold Spring Harbor, NY.

61. Kieser, T., Bibb, M. J., Buttner, M. J., Chen, B. F., and Hopwood, D. A. (2000) Practical *Streptomyces* genetics. The John Innes Foundation, Norwich.

62. Southern, E. M. (1975) Detection of specific sequences among DNA fragments separated by gel electrophoresis. *J. Mol. Biol.* **98,** 503–517.

63. Campbell, G. L. and Bedford, M. R. (1992) Enzyme applications for monogastric feeds: a review. *Can. J. Anim. Sci.* **72,** 449–466.

64. Fernández-González, C., Gil, J. A., Mateos, L. M., Schwarzer, A., Schafer, A., Kalinowski, J., Puhler, A., and Martín, J. F. (1996) Construction of L-lysine-overproducing strains of *Brevibacterium lactofermentum* by targeted disruption of the *hom* and *thrB* genes. *Appl. Microbiol. Biotechnol.* **46,** 554–558.

65. Teather, R. M. and Wood, P. J. (1982) Use of Congo red-polysaccharide interactions in enumeration and characterization of cellulolytic bacteria from the bovine rumen. *Appl. Environ. Microbiol.* **43,** 777–780.

66. Jeong, K. J., Park, I. Y., Kim, M. S., and Kim, S. C. (1998) High-level expression of an endoxylanase gene from *Bacillus* sp. in *Bacillus subtilis* DB104 for the production of xylobiose from xylan. *Appl. Microbiol. Biotechnol.* **50,** 113–118.

8

Enzymatic Synthesis of *(S)*-Phenylalanine and Related *(S)*-Amino Acids by Phenylalanine Dehydrogenase

Yasuhisa Asano

Summary

Phenylalanine dehydrogenase (L-phenylalanine: NAD$^+$ oxidoreductase, deaminating [EC 1.4.1.20], PheDH) from *Bacillus sphaericus* R79a, overproduced in *Escherichia coli* JM109/pBPDH1-DBL, is active toward 3-substituted pyruvic acids with bulky substituents in the reductive amination reaction. Optically pure *(S)*-phe and other *(S)*-amino acids are quantitatively synthesized from their oxo-analogs using PheDH, with a regeneration of NADH by formate dehydrogenase from *Candida boidinii* No. 2201. Acetone-dried cells of *B. sphaericus* R79a and *C. boidinii* No. 2201 are also effective for *(S)*-phe synthesis, providing a simple microbial method of synthesis.

Key Words: *(S)*-phenylalanine; phenylpyruvic acid; *Bacillus sphaericus* R79a; formate dehydrogenase; *Candida boidinii* No. 2201; regeneration of NADH.

1. Introduction

An enzymatic method to synthesize *(S)*-phe from phenylpyruvate was developed according to the success in phenylpyruvic acid synthesis by double carbonylation of benzylchloride in the presence of a cobalt catalyst *(1)*. Bacterial producers of a new enzyme, phenylalanine dehydrogenase (PheDH, EC 1.4.1.20), were isolated and characterized from *Sporosarcina ureae* R04 *(2,3)*, *Bacillus sphaericus* R79a *(3)*, and *Bacillus badius* IAM 1105 *(4)*. These *pdh* genes have been cloned and sequenced *(4,5,6,7)*, and overproduction of the enzymes has been achieved. *Escherichia coli* JM 109/pBPDH1-DBL expresses about 120-fold higher activity of PheDH (7,200 U/L) than the wild-type *B. sphaericus* R79a per liter of culture. The enzyme from *B. sphaericus* R79a was chosen to study the application of the enzyme to synthesize various *(S)*-amino acids, because it is very stable and shows broader substrate specificity than other PheDHs, acting on *(S)*-tyrosine as well as *(S)*-Phe (*see* **Note 1**). We used

From: *Methods in Biotechnology, Vol. 17: Microbial Enzymes and Biotransformations*
Edited by: J. L. Barredo © Humana Press Inc., Totowa, NJ

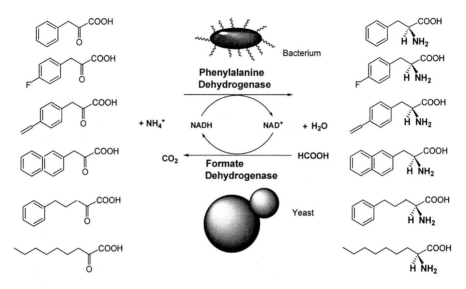

Fig. 1. Synthesis of *(S)*-amino acids from their 2-oxo analogs by PheDH and formate dehydrogenase.

the enzyme to catalyze the syntheses of *(S)*-amino acids using PheDH with the regeneration of NADH by formate dehydrogenase **(Fig. 1)**.

In this chapter, *(S)*-phe was synthesized by two methods: with partially purified PheDH and acetone-dried cells of *Candida boidinii* No. 2201, and acetone-dried cells of *B. sphaericus* R79a and *C. boidinii* No. 2201 *(8,9)*. Other *(S)*-amino acid derivatives were also synthesized from 2-oxo-4-phenylbutyrate, 2-oxo-5-phenylvalerate, 2-oxononanoate, 2-oxo-3-(2-naphthalene) propionic acid, 3-oxo-4-phenylbutyric acid, and the like, with partially purified PheDH and formate dehydrogenase.

2. Materials

1. Grignard reagents prepared from bromoethylbenzene, 1-bromo-3-phenylpropane, and 1-bromoheptane.
2. 10% HCl in acetic acid.
3. *B. sphaericus* R79a, accession number FERM P-8197 (National Institute of Technology and Evaluation, Kisarazu, Japan).
3. *C. boidinii* ATCC 32195 (ATCC, Manassas, VA); originally *C. boidinii* No. 2201 (AKU, Kyoto University, Japan).
4. *Pediococcus acidilactici* ATCC 8042 (ATCC).
5. *E. coli* JM109/pBPDH1-DBL, accession number FERM P-8873 (National Institute of Technology and Evaluation).

6. Medium A: 1% *(S)*-phe, 1% peptone (Kyokuto, Tokyo, Japan), 0.5% yeast extract, 0.2% K_2HPO_4, 0.1% NaCl, and 0.02% $MgSO_4 \cdot 7H_2O$, adjusted to pH 7.08.
7. Medium B: 1.3% methanol, 0.4% NH_4HCl, 0.1% K_2HPO_4, 0.1% KH_2PO_4, 0.05% $MgSO_4 \cdot 7H_2O$, and 0.2% yeast extract, pH 6.0.
8. Luria-Bertani (LB) medium: 10 g/L bacto-tryptone, 5 g/L bacto-yeast extract, 10 g/L NaCl, and 20 g/L agar. Adjust to pH 7.0 with 5 *N* NaOH. Autoclave at 121°C for 15 min.
9. Glycine buffer: 100 m*M* glycine-KCl-KOH, pH 10.4.
10. 0.1 *M* potassium phosphate buffer, pH 7.5.
11. 10 m*M* potassium phosphate buffer, pH 7.0.
12. Formate dehydrogenase purified from *C. boidinii* No. 2201, up to 2 U/mg according to the method of Kato et al. *(10)*.
13. Anion-exchange resin DEAE-Toyopearl 650M (Tosoh Corp., Tokyo, Japan).
14. Hydroxyapatite column, 4.5 × 8 cm.
15. Sephadex G-200 column, 2.2 × 123 cm.
16. Physiological saline: 0.85% NaCl.
17. Hydrophobic interaction resin Butyl-Toyopearl (Tosoh Corp.).
18. Sonicator Insonator 201M (Kubota, Tokyo, Japan).
19. Dialysis tubing, seamless cellulose tubing, small size 18 (Viskase, Willowbrook, IL).
20. Sodium phenylpyruvate (Sigma, St. Louis, MO).
21. 1 *M* Tris-HCl, pH 8.5.
22. 25 m*M* NAD^+ solution, store frozen.
23. Silicone tube, 30 cm.
24. Media for microbiological assay of amino acids (Takara Kosan, Tokyo, Japan).
25. Protein assay kit (Bio-Rad, Hercules, CA).
26. BSA solutions: 1 mg/mL and 0.1 mg/mL.
27. Thin-layer chromatography plates, thickness 0.25 mm, 5 cm ~ 8 cm.
28. Ninhydrin reagent, spray type: 0.7% solution in ethanol.
29. Cation exchange resin Dowex 50W-×8, 50–100 mesh, H^+ form, activated with 2 *N* HCl and 2 *N* NaOH.

3. Methods

The methods described below outline the chemical syntheses of the substrates.

3.1. Preparation of the Substrates

3.1.1. Preparation of 2-Oxo-4-Phenylbutyrate, 2-Oxo-5-Phenylvalerate, and 2-Oxononanoate

2-Oxo esters were prepared according to the method described by Weinstock et al. *(11)*.

1. Add Grignard reagents prepared from bromoethylbenzene, 1-bromo-3-phenylpropane, and 1-bromoheptane to diethyloxalate in dry THF to give ethyl 2-oxo-4-phenylbutyrate, ethyl 2-oxo-5-phenylvalerate, and ethyl 2-oxononoate, in 73, 73, and 68% yield, respectively.

2. Hydrolyze the ethyl esters in 10% HCl in acetic acid at 50°C for 3 h.
3. After extraction by ethylacetate, crystallize the oxo acids by adding n-hexane.
4. 2-Oxo-4-phenylbutyric acid, 2-oxo-5-phenylvaleric acid and 2-oxo-nonanoic acid were obtained from their esters in 58, 93, and 89% yields, respectively.

3.1.2. Preparation of Other Substrate Analogs

Ethyl 3-oxo-4-phenylbutyrate was synthesized in 75% yield from monoethylmalonate and phenylacetylchloride, according to the method described by Wierenga and Skulnick *(12)*.

1. Generate dianion of monoethylmalonate in THF with 2 equivalents of *N*-butyllithium from –30°C with slow warming to –5°C.
2. Cool the homogeneous reaction mixture to –65°C.
3. Add phenylacetylchloride and remove cooling to yield ethyl 3-oxo-4-phenylbutyrate.
4. Hydrolyze the ethyl ester in 10% HCl in acetic acid at 50°C for 3 h.
5. Work up the reaction mixture by a usual procedure and crystallize the oxo acids by adding n-hexane.

Hydrolysis of the ester by the same procedure as described above yielded 3-oxo-4-phenylbutyric acid. 1-Diazo-2-oxo-3-phenylpropane was synthesized in 64% yield from phenylacetylchloride and diazomethane. 2-Oxo-3-(2-naphthalene) propionic acid was prepared from 2-naphthalene acetic acid ethylester and diethyloxalate according to the procedure of Reimann and Voss *(13)*. 2-Oxo-3-(1-naphtyl)-3-ethoxycarbonylpropionic acid ethylester was synthesized from 2-naphthalene acetic acid ethylester and diethyloxalate with NaOH in ether. 2-Oxo-3-(2-naphthalene) propionic acid was obtained from the ester by treating with sulfuric acid. Other 2-oxo acids or their salts were purchased from commercial sources.

3.2. Preparation of the Biocatalysts

The methods described below outline the methods (1) to cultivate enzyme producers, (2) to measure the enzyme activity, (3) to prepare acetone-dried cells, and (4) to purify the enzymes.

3.2.1. Cultivation of Microorganisms

1. *B. sphaericus* R79a *(3)* was cultivated aerobically at 30°C for 20 h in a 2-L flask containing 500 mL of medium A.
2. *C. boidinii* No. 2201 was cultivated in medium B as described by Kato et al. *(10)*. The enzyme was purified from the cells cultivated for 24 h, by a procedure involving heat treatment at 55°C, column chromatographies on DEAE-Toyopearl, hydroxyapatite and Sephadex G-200.
3. The 1.3 Kb fragment of the *pdh* gene encoding PheDH with endogenous promoter and ribosomal recognition sequences was inserted into pUC9 to give

pBPDH1-DBL. This plasmid was transformed into *E. coli* JM109 by standard methods.

4. *E. coli* JM109/pBPDH1-DBL was cultivated with shaking at 37°C for 12 h in LB medium, supplemented with 50 μg/mL ampicillin *(5)*.

3.2.2. Enzyme Assay of PheDH and Definition of Unit *(3)*

1. NAD$^+$-dependent PheDH was assayed at 25°C by measuring reduction of NAD$^+$ at OD$_{340}$ in a cuvet placed in the beam of a 1-cm light path. The reaction mixture contained glycine buffer, 2.5 mM NAD$^+$, 10 mM *(S)*-phe, and PheDH in a total volume of 1.0 mL.
2. The enzyme activity for the reductive amination was measured at 25°C by the disappearance of NADH at OD$_{340}$ in a reaction mixture containing 100 μM Tris-HCl, pH 8.5, 0.1 μM NADH, 200 μM NH$_4$Cl, 10 μM sodium phenylpyruvate, and PheDH in a total volume of 1.0 mL.

One unit of enzyme activity was defined as the amount of enzyme which catalyzes the formation of 1 μmol of NADH per min in the oxidative deamination.

3.2.3. Enzyme Assay of Formate Dehydrogenase and the Definition of the Unit *(14)*

1. The enzyme activity was assayed by the reduction of NAD$^+$ monitored at OD$_{340}$ with a double-beam spectrophotometer, with sodium formate as a substrate. The enzyme activity was measured at 25°C in a reaction mixture containing 100 mM sodium formate, 0.1 M phosphate buffer, pH 7.5, 2.5 mM NAD$^+$, and formate dehydrogenase in a total volume of 1.0 mL.
2. A reaction mixture to give a linear change in OD$_{340}$ for at least 2 min was employed in the kinetic study, and OD$_{340}$ change for the initial 5 s was used for the calculation.

One unit of the enzyme is defined as the amount of enzyme that catalyzes the formation of 1 μmol of NADH per min.

3.2.4. Preparation of Acetone-Dried Cells of B. sphaericus R79a and C. boidinii No. 2201

Acetone-dried cells of *B. sphaericus* R79a and *C. boidinii* No. 2201 were prepared essentially as described by Izumi et al. *(15)*. A typical example of the preparation is described.

1. Wash *C. boidinii* No. 2201 cells, harvested from 6 L of culture, once with 10 mM phosphate buffer, pH 7.0, and suspend in 80 mL of the same buffer.
2. Add to the cell suspension 400 mL of acetone that has been cooled to –20°C.
3. Filter the cells, wash with the same volume of cold acetone, and leave for 30 min.
4. Store the cells at –20°C until use.

Acetone-dried cells from *B. sphaericus* R79a (2.0 g) and *C. boidinii* No. 2201 (1.3 g) per liter culture were typically obtained.

3.2.5. Purification of PheDH from B. sphaericus R79a (3)

1. Suspend cells (about 1 kg wet weight) from 100 L of culture in 0.1 M phosphate buffer.
2. Disrupt cells for 20 min (41 h in total) by the sonicator.
3. Remove the rests of the disrupted cells by centrifugation at 14,000g for 20 min.
4. Heat the cell-free extract for 10 min in a water bath at 50°C.
5. Cool the extract in ice and centrifuge at 14,000g for 20 min to remove denatured protein.
6. Add ammonium sulfate to 30% saturation to the supernatant and remove the precipitate formed by centrifugation at 14,000g.
7. Dissolve the precipitate formed on addition of ammonium sulfate to 60% saturation in a 10 mM phosphate buffer and dialyze in the same buffer with dialysis tubing.
8. Apply the enzyme solution on a DEAE-Toyopearl 650 M column (6.5 × 36 cm) equilibrated with 10 mM phosphate buffer.
9. After washing the column with 0.1 M phosphate buffer, elute the enzyme with the same buffer containing 0.1 M NaCl.
10. Combine the active fractions, dialyze, apply to the second DEAE-Toyopearl (4.5 × 9.5 cm), and elute in the same way.
11. Dialyze the enzyme and place it on a hydroxyapatite column (4.5 × 8 cm) equilibrated with 10 mM phosphate buffer.
12. Elute the enzyme with a linear concentration gradient of from 10 mM to 0.4 M phosphate buffer.
13. Pool the active fractions, concentrate by ultrafiltration, and place on a column of Sephadex G-200 (2.2 × 123 cm) equilibrated with 0.05 M phosphate buffer containing 0.1 M NaCl.
14. Combine the active fractions and concentrate by ultrafiltration.

3.2.6. Partial Purification of PheDH from E. coli JM109/pBPDH1-DBL

PheDH was partially purified from *E. coli* JM109/pBPDH1-DBL by a procedure involving ammonium sulfate fractionation, DEAE-Toyopearl, Butyl-Toyopearl, and Sephadex G-200 column chromatographies.

1. Inoculate a single colony of *E. coli* JM 109/pBPDH1-DBL to 20 mL of LB medium and incubate at 37°C for 12 h at 320 rpm.
2. Add the culture broth to 380 mL of LB medium containing 50 µg/mL ampicillin and IPTG (1 mM), and incubate aerobically at 37°C for 12 h at 200 rpm.
3. Harvest the cells by centrifugation at 12,000g and 4°C for 20 min.
4. Wash with physiological saline and centrifuge at 18,200g and 4°C for 10 min.
5. Add 5 mL of 0.1 M KPB buffer per 1 g washed cells and suspend.
6. Sonicate the cells at 4°C for 20 min.
7. Centrifuge (4°C, 10 min, 18,200g) to give the cell-free extract.

3.2.7. Assay of (S)-Phe

(S)-Phe and other natural amino acids were assayed microbiologically using *Pediococcus acidilactici* ATCC 8042, with assay media purchased from Takara

Kosan. *P. acidilactici* ATCC 8042 requires all the amino acids for the growth. Prepare a medium lacking only *(S)*-phe for the turbidmetric determination of *(S)*-phe. Monitor the formation of *(S)*-phe and other amino acids by TLC (n-butanol/acetic acid/water = 4/1/1).

3.3. Synthesis of (S)-Amino Acids by B. sphaericus PheDH (9)

In the study on the substrate specificity of the enzyme in the reductive amination of 2-oxo acids, various optically pure *(S)*-amino acids were quantitatively synthesized using PheDH and formate dehydrogenase.

The product from 2-oxo-3-*(RS)*-methylbutyrate and 2-oxo-3-*(RS)*-methyl-3-phenylpyruvate were identified as diastereomeric mixtures of isoleucine and *(S)-allo*-isoleucine, and *(S)*-2-amino-3-*(RS)*-methyl-3-phenylpropionic acid, respectively. *(S)*-2-amino-4-phenylbutyric acid (*(S)*-homophenylalanine) and other unnatural *(S)*-amino acids could be efficiently synthesized. *(S)*-Homophenylalanine is a building block of some of the angiotensin-converting-enzyme inhibitors. The solubility of these *(S)*-Phe homologs, such as *(S)*-tyrosine, *(S)*-2-amino-4-phenylbutyric acid, *(S)*-2-amino-5-phenylvaleric acid and so on, are so low that they are easily separated in crystalline forms from the reaction mixture by filtration. The filtered enzyme solution can be used for further repeated synthesis.

3.3.1. Enzymatic Synthesis of (S)-Phe by PheDH, and Acetone Dried Cells of C. boidinii

1. The reaction mixture contains 792 m*M* phenylpyruvate, 10 m*M* NAD+, 1.55 *M* sodium formate, 50 m*M* Tris-HCl, pH 8.5, 126 U PheDH, and acetone-dried cells of *C. boidinii* (39.4 U) in a total volume of 20 mL.
2. Mix partially purified PheDH by DEAE-Toyopearl column chromatography (126 U) and acetone-dried cells of *C. boidinii* (39.4 U), and incubate for 2 h at 30°C to synthesize *(S)*-phe.
3. Add sodium phenylpyruvate in portions for 8 times at intervals of every 15 min (0.792 *M* final concentration).

3.3.2. Synthesis of (S)-Phe by Acetone-Dried Cells of B. sphaericus R79a and C. boidinii No. 2201

1. The reaction mixture (30 mL) contains 1.85 m*M* sodium phenylpyruvate, 15 μ*M* NAD+, 12 m*M* ammonium formate, 1.5 m*M* Tris-HCl, pH 8.5, 300 mg acetone-dried cells of *C. boidinii* No. 2201, and 300 mg *B. sphaericus* R79a. Incubate at 30°C for 48 h.
2. Add the same amount of sodium phenylpyruvate (1.85 m*M*) at every 3, 7, 12, 25, and 30 h after the start of the reaction. Add 5.6 m*M* ammonium formate at 12 h.
3. Determine *(S)*-phe by the microbiological method (*see* **Subheading 3.2.7.**).

The concentration of *(S)*-phe reached 61.5 mg/mL with a yield of more than 99%. The reaction mixture becomes solidified. When used in combination with formate dehydrogenase, PheDH from *B. sphaericus* R-79a was effective in the enantioselective syntheses of various natural and unnatural *(S)*-amino acids from their oxo analogs. The enzyme showed relatively wide substrate specificity toward substituted pyruvic acids.

3.3.3. Purification of (S)-Phe by Ion Exchange Resin (Dowex)

1. Adjust the pH of the reaction mixture containing *(S)*-phe to pH 1.0 with 6 *N* HCl, to dissolve *(S)*-phe formed.
2. Take up the supernatant after centrifugation (4°C, 3000*g*, 15 min) to a separatory funnel, wash the water layer twice with ethyl acetate (50 mL), and adjust to pH 7.0 with 6 *N* NaOH.
3. Adsorb the solution to 10 mL of cation exchange resin Dowex 50W-×8.
4. Wash the column with water to pH 7.0, and elute *(S)*-phe with 50 mL of 1 *M* NH₄OH.
5. Combine fractions containing *(S)*-phe and concentrate by rotary evaporator.
6. Crystallize *(S)*-phe from hot water and 99% ethanol and recrystallize.

3.3.4. Synthesis of (S)-Amino Acids from Their 2-Oxo Analogs by Partially Purified PheDH and Formate Dehydrogenase

Various optically pure *(S)*-amino acids were quantitatively synthesized using partially purified PheDH from *E. coli* JM109/pBPDH1-DBL and formate dehydrogenase from *C. boidinii* No. 2201 (*see* **Note 2**). The following is an example of reductive amination of 2-oxo-5-phenylvalerate.

1. The reaction mixture (100 mL) contains 4 m*M* sodium 2-oxo-5-phenylvalerate, 20 m*M* ammonium formate, 8 m*M* NH₄OH-NH₄Cl buffer, pH 8.5, 0.1 m*M* NAD⁺, 5,760 U of partially purified PheDH from *E. coli* JM 109/pBPDH1-DBL, and 240 U of formate dehydrogenase. Incubate at 30°C for 60 h.
2. Add 3 mmol each of the sodium salt of the oxo acid to the reaction mixture every 10 h 5 times.
3. Add 20 mmol of ammonium formate at 40 h. Phenylpyruvate analogs substituted at the phenyl ring were relatively good substrates.

The enzyme utilized the compounds substituted at the 3-position of pyruvic acid with a longer or a bulkier group, although the relative velocity of the reductive amination reaction was low. The substitution at 3-position of phenylpyruvate with a bulkier group such as hexyl greatly lowered the reaction velocity. **Table 1** shows the yield of the *(S)*-amino acids thus synthesized. The products from 2-oxo-3-*(RS)*-methylbutyrate and 2-oxo-3-*(RS)*-methyl-3-phenylpyruvate were identified as a diastereomeric mixtures of *(S)*-isoleucine and *(S)*-*allo*-isoleucine, and *(S)*-2-amino-3-*(RS)*-methyl-3-phenylpropionic acid, respectively, after the purification procedure as described (*see* **Subheading 3.3.3.**). The product from 2-oxo-5-phenylvalerate solidified as the reaction proceeds (*see* **Note 3**).

Table 1
Synthesis of *(S)*-Amino Acids from 2-Oxo Acids by Using PheDH and Formate Dehydrogenase

Substrate	Product	Yield (%)
Phenylpyruvate	*(S)*-Phe	>99
4-Hydroxyphenylpyruvate	*(S)*-Tyrosine	>99
4-Fluorophenylpyruvate	*(S)*-4-Fluorophenylalanine	>99
2-Oxo-4-phenylbutyrate	*(S)*-2-Amino-4-phenylbutyric acid	99
2-Oxo-5-phenylvalerate	*(S)*-2-Amino-5-phenylvaleric acid	98
2-Oxo-3-methyl-3-phenylpropionate	*(S)*-2-Amino-3-DL-methyl-β-phenylpropionic acid	98
2-Oxononanoate	*(S)*-2-Aminononanoic acid	99

To prevent substrate inhibition, 2-oxo acids have been divided into portions and added to the reaction mixture not to exceed 50 mM.

4. Notes

1. Quantitative determination of *(S)*-phe in the plasma is important in diagnosing phenylketonuria. PheDH has been utilized for the microdetermination of *(S)*-phe in blood samples *(1,4,16)*. *B. badius* PheDH has narrower substrate specificity suitable for the microdetermination.

2. Benzoylformate, 2-oxo esters such as ethyl phenylpyruvate and ethyl 2-oxo-4-phenylbutyrate, 3-oxo esters such as ethyl 3-oxo-4-phenylbutyrate, 3-oxo acids such as 3-oxo-4-phenylbutyrate, and 2-oxoalcohols such as 2-oxo-3-phenyl-propanol, were inactive as substrates. This shows that free carboxylic acid moiety is required to be recognized as a substrate. The result that the enzyme utilized 2-oxo-4-phenylbutyrate and 2-oxo-5-phenylvalerate, but not 3-oxo-4-phenylbutyrate, shows it has a definite requirement for a distance between the carbonyl carbon and the carboxyl group of the substrates.

3. The enzyme has a wide pocket, which accommodates the hydrophobic substituent of the substrate: various 3-substituted pyruvic acid analogs with bulky groups could be relatively good substrates. It was also revealed that the enzyme does not differentiate the configuration of the substituent at the 3-position of pyruvic acid; for example, a diastereomeric mixture of *(S)*-2-amino-3-*(RS)*-methyl-3-phenylpropi-onate was synthesized from 2-oxo-3-*(RS)*-methyl-3-phenylpropionate.

References

1. Asano, Y. (1999) Phenylalanine dehydrogenase, in *Encyclopedia of Bioprocess Technology: Fermentation, Biocatalysis, and Bioseparation* (Flickinger, M. C. and Drew, S. W., eds.), John Wiley & Sons, Inc., New York, pp. 1955–1963.
2. Asano, Y. and Nakazawa, A. (1985) Crystallization of phenylalanine dehydrogenase from *Sporosarcina ureae. Agric. Biol. Chem.* **49,** 3631–3632.

3. Asano, Y., Nakazawa, A., and Endo, K. (1987) Novel phenylalanine dehydrogenases from *Sporosarcina ureae* and *Bacillus sphaericus*—purification and characterization. *J. Biol. Chem.* **262**, 10,346–10,354.

4. Asano, Y., Yamada, A., Endo, K., Hibino, Y., Ohmori, M., Numao, N., and Kondo, K. (1987) Phenylalanine dehydrogenase from *Bacillus badius*—purification, characterization and gene cloning. *Eur. J. Biochem.* **168**, 153–159.

5. Okazaki, N., Hibino, Y., Asano, Y., Ohmori, M., Numao, N., and Kondo, K. (1988) Cloning and nucleotide sequencing of phenylalanine dehydrogenase gene of *Bacillus sphaericus. Gene* **63**, 337–341.

6. Yamada, A., Dairi, T., Ohno, Y., Huang, X.-L., and Asano, Y. (1995) Nucleotide sequencing of phenylalanine dehydrogenase gene from *Bacillus badius* I AM 11059. *Biosci. Biotech. Bioch.* **59**, 1994–1995.

7. Asano, Y., Endo, K., Nakazawa, A., Hibino, Y., Ohmori, M., Numao, N., and Kondo, K. (1987) *Bacillus* phenylalanine dehydrogenase produced in *Escherichia coli*—its purification and application to L-phenylalanine synthesis. *Agric. Biol. Chem.* **51**, 2621–2623.

8. Asano, Y. and Nakazawa, A. (1987) High yield synthesis of L-amino acids by phenylalanine dehydrogenase from *Sporosarcina ureae. Agric. Biol. Chem.* **51**, 2035–2036.

9. Asano, Y., Yamada, A., Kato, Y., Yamaguchi, K., Hibino, Y., Hirai, K., and Kondo, K. (1990) Enantioselective synthesis of *(S)*-amino acids by phenylalanine dehydrogenase from *Bacillus sphaericus:* use of natural and recombinant enzymes. *J. Org. Chem.* **55**, 5567–5571.

10. Kato, N., Kano, M., Tani, Y., and Ogata, K. (1974) Purification and characterization of formate dehydrogenase in a methanol-utilizing yeast, *Kloeckera* sp. No. 2201. *Agric. Biol. Chem.* **38**, 111–116.

11. Weinstock, L. M., Currie, R. B., and Lovell, A. V. (1981) A general, one-sep synthesis of α-keto esters. *Synth. Commun.* **11**, 943–946.

12. Wierenga, W. and Skulnick, H. I. (1979) General, efficient, one-step synthesis of β-keto esters. *J. Org. Chem.* **44**, 310–311.

13. Reimann, E. and Voss, D. (1976) Synthese von 3-(1-naphthyl)-alanin durch katakytische reductive aminierung. *Arch. Pharm.* **309**, 978–983.

14. Asano, Y., Sekigawa, T., Inukai, M., and Nakazawa, A. (1988) Purification and properties of formate dehydrogenase from *Moraxella* sp. strain C-1. *J. Bacteriol.* **170**, 3189–3193.

15. Izumi, Y., Mishra, S. K., Ghosh, B. S., Tani, Y., and Yamada, H. (1983) NADH production from NAD$^+$ using formate dehydrogenase system with cells of a methanol-utilizing bacterium. *J. Ferment. Technol.* **61**, 135–142.

16. Naruse, H., Ohashi, Y. Y., Tsuji, A., Maeda, M., Nakamura, K., Fujii, T., et al. (1992) A method of PKU screening using phenylalanine dehydrogenase and microplate system. *Screening* **1**, 63–66.

9

Enzymes in Modern Detergents

Susumu Ito, Tohru Kobayashi, Yuji Hatada, and Koki Horikoshi

Summary

Huge amounts of alkaline enzymes are used in the detergent industry, and they have been widely incorporated into heavy-duty laundry and automatic dishwashing detergents. The alkaline enzymes used in modern detergents are protease, cellulase, α-amylase, lipase, and mannanase. In this chapter, methods for screening alkaline enzyme-producing alkaliphilic *Bacillus* strains, enzyme assays, purification, properties, and genetics of enzymes are described.

Key Words: Serine protease; subtilisin; cellulase; endoglucanase; α-amylase; pectinase; lipase; mannanase; alkaliphile; *Bacillus,* laundry detergent; dishwashing detergent.

1. Introduction

The 1999 world market for industrial enzymes was estimated at more than $1.6 billion. The market for detergent enzymes, including protease, cellulase, α-amylase, and lipase, occupies approx 40% of the world market. After the switch from tripolyphosphate to zeolite in laundry detergents in the early 1980s, consumers became dissatisfied with the cleaning performance of detergents. Further, especially in Europe and the United States, consumers were using lower washing temperatures to save energy. The enzymes were extensively introduced into detergents to compensate for low detergency.

1.1. Alkaline Protease

Protease (extract of the pancreatic gland) was the first enzyme incorporated into a detergent (presoak-type), in 1931. Due to the high alkalinity of the detergent, pancreatic enzymes did not work well. An alkaline protease (subtilisin, EC 3.4.21.14) from *Bacillus* strains was exploited for use in detergents from the early 1960s. Since then, many high-alkaline proteases suitable for detergents have been found, such as Alcalase® and Savinase® (Novozymes), Maxacal® and

From: *Methods in Biotechnology, Vol. 17: Microbial Enzymes and Biotransformations*
Edited by: J. L. Barredo © Humana Press Inc., Totowa, NJ

Purafect® (Genencor), KAP (Kao), and Blap® (Henkel) *(1,2)*. A serious problem with such subtilisins is their inactivation by chemical oxidants. Novozymes and Genencor replaced Met residues with nonoxidizable amino acids by protein engineering (Durazyme®, Maxapem®, and Purafect Oxp®) for use in detergents with bleach *(2)*. Recently, oxidatively stable serine proteases such as E-1 and KP-43 have been found in alkaliphilic *Bacillus* spp. *(3,4)*.

1.2. Alkaline Cellulase

The alkaline cellulase (carboxymethylcellulase; CMCase) (endo-1,4-β-glu-canase, 1,4-β-D-glucan 4-glucanohydrolase; EC 3.2.1.4) from alkaliphilic *Bacillus* strain sp. KSM-635 (Egl-635) was first incorporated into detergents by Kao in 1987 *(5,6)*. After that, Novozymes launched a detergent using a cellulase complex from *Humicolla insolence* (Celluzyme®) and, later, one of the cellulase components in Celluzyme® (Carezyme®) *(1)*. Recently, a thermostable alkaline cellulase (Egl-237) from alkaliphilic *Bacillus* sp. strain KSM-S237 was found *(7)*, and its recombinant enzyme is now incorporated into heavy-duty laundry detergents. The first use of an alkaline lipase (Lipolase®; Novozymes) in laundry detergents was by Lion (Japan) in 1989.

1.3. Alkaline Amylase

Liquefying α-amylases (1,4-α-D-glucan glucanohydrolases, EC 3.2.1.1), particularly *Bacillus licheniformis* enzyme (BLA) (Termamyl®, Novozymes; Maxamyl®, Genencor), have wide industrial applications. In the early 1990s, α-amylases were included in both laundry and automatic dishwashing detergents in Europe and also into heavy-duty laundry detergent in Europe and the US *(8)*. There were several problems in the use of α-amylases in detergents—for instance, low reaction rate at alkaline pH, inactivation by oxidants, and Ca^{2+} ion-dependent thermal stability. To overcome such problems, new α-amylases have been created by protein engineering and/or screened from nature (Novozymes, WO 94/02597, 1993; WO 95/10603, 1995; WO 96/23873, 1996; Genencor, WO 94/18314, 1994; Kao, WO 98/44126, 1998). Recently, a novel calcium-free α-amylase (AmyK38) has been found from alkaliphilic *Bacillus* sp. KSM-K38, which is highly resistant to excess H_2O_2 and EDTA *(9)*.

Recently, an alkaline mannanase (Novozymes), which can remove gum stains of cosmetics and/or foods, was introduced into a laundry detergent (Procter & Gamble, WO 99/09128 ~ WO 99/09133, 1998). Alkaline mannanase was first discovered by Horikoshi and his colleagues (Research and Development Corporation of Japan, JP H03–065754, 1991). Although alkaline pectinolytic enzymes have not yet been used in detergents, many patents concerning detergent formulation with the enzymes have been documented.

Pectinolytic enzymes are also promising for use in detergents. Nowadays, these alkaline enzymes are essential for heavy-duty laundry and automatic dishwashing detergents all over the world.

2. Materials

2.1. Alkaline Protease

1. Saline solution: 0.85% (w/v) NaCl aqueous solution.
2. Keratin agar (w/v): 1% glucose, 0.2% yeast extract (Difco, Detroit, MI), 1% wool keratin (Kanto Chemical, Tokyo, Japan), 1% CMC, 0.1% KH_2PO_4, 0.02% $MgSO_4·7H_2O$, 1% Na_2CO_3 (sterilized separately, *see* **Note 1**), and 1.5% agar.
3. Liquid medium (w/v): 2% glucose, 1% meat extract (Wako Pure Chemical, Osaka, Japan), 1% soybean meal (Ajinomoto, Tokyo, Japan), 0.1% KH_2PO_4, and 1% Na_2CO_3 (sterilized separately, *see* **Note 1**).
4. Casein solution: 1 % (w/v) Hammerstein casein (Merck, Darmstadt, Germany) is dissolved in 50 mM borate buffer, pH 10.5, and then the pH is adjusted to 10.5 with 1–10 N NaOH.
5. TCA solution: 0.11 M trichloroacetic acid, 0.22 M sodium acetate, and 0.33 M acetic acid.
6. Filter paper grade 2 (Whatman, Kent, UK).
7. Reagent A: 1% (w/v) sodium potassium tartrate, 1% cupric sulfate, and 2% sodium carbonate in 0.1 N NaOH (1:1:100, v/v). Each solution should be mixed just before use.
8. Phenol reagent (Kanto Chemical).
9. Tetracycline (Sigma, St. Louis, MO) is dissolved in ethanol, and equal volume of sterilized distilled water is added to the solution. The solution is filter-sterilized using a 0.45-μm membrane.
10. Membrane YM-5 (Amicon, Bedford, MA).
11. Dialysis solution: 5 mM Tris-HCl, pH 8.0, and 2 mM $CaCl_2$.
12. DEAE-Bio-Gel A (Bio-Rad, Hercules, CA).
13. Equilibration buffer for a column of DEAE-Bio-Gel A: 10 mM Tris-HCl, pH 8.0, and 2 mM $CaCl_2$.
14. CM-Bio-Gel A (Bio-Rad).
15. Equilibration buffer for a column of CM-Bio-Gel A: 10 mM borate buffer, pH 9.6, and 2 mM $CaCl_2$.
16. *Bacillus* sp. strain KSM-K16 (FERM BP-3376) (Fermentation Research Institute, Tsukuba, Japan).
17. DNACELL (D'aiichi Pure Chemicals, Tokyo, Japan).
18. Restriction enzymes, alkaline phosphatase, and T4 DNA ligase.
19. DM3 agar (w/v): 0.5 M disodium succinate, 0.5% casamino acid, Technical (Difco), 0.5% yeast extract, 0.35% KH_2PO_4, 0.15% K_2HPO_4, 0.5% glucose, 0.4% $MgCl_2·6H_2O$, 0.01% bovine serum albumin (Sigma), and 0.8% agar.
20. Selection agar (w/v): 1% glucose, 0.5% yeast extract, 0.5% NaCl, 1% skim milk (sterilized separately; Difco), and 1.5% agar.
21. Plasmid pHY300PLK (Yakult, Tokyo, Japan).

22. *Bacillus subtilis* ISW1214 (Yakult).
23. Taq dideoxy termination cycle sequencing kit (Applied Biosystems, Foster City, CA).

2.2. Alkaline Cellulase

1. Carboxymethyl cellulose (CMC) : Sunrose A01MC (degree of substitutions 0.58 (Nippon Paper Industries, Tokyo, Japan).
2. CMC agar (w/v): 2% CMC, 1% meat extract (LAB-LEMCO powder; Oxoid, Hampshire, UK), 1% Bacto peptone (Difco), 1% NaCl, 0.1% KH_2PO_4, 0.5% Na_2CO_3 (sterilize separately), 0.005% trypan blue (sterilize separately; Merck) (*see* **Note 2**), and 1.5% agar.
3. Liquid medium: 0.1% CMC (*see* **Note 3**), 0.1% yeast extract, 2% polypepton S (Nippon Pharmaceuticals, Osaka, Japan), 1% meat extract, 0.5% sodium glutamate, 0.15% K_2HPO_4, 0.01% $CaCl_2 \cdot 2H_2O$, 0.02% $MgSO_4 \cdot 7H_2O$, and 0.5% Na_2CO_3 (sterilized separately).
4. Substrate solution: 2.5% CMC, 0.5 M glycine–NaOH buffer, pH 9.0, and distilled water (4:3:2, v/v).
5. DNS reagent: 0.5% (w/v) dinitrosalicylic acid dissolved in 30% sodium potassium tartrate, and 0.4 N NaOH. The reagent can be used for several months when stored at 4°C.
6. Membrane PM-10 (Amicon).
7. Buffer A: 10 mM Tris-HCl, pH 7.5, and 5 mM $CaCl_2$.
8. DEAE-Toyopearl 650S (Tosoh, Tokyo, Japan).
9. Bio-Gel A 0.5 m (Bio-Rad).
10. *Bacillus* sp. strain KSM-635 (FERM BP-1485) (Fermentation Research Institute).
11. *Bacillus* sp. strain KSM-S237 (FERM BP-7875) (Fermentation Research Institute).
12. GFX PCR DNA and gel band purification kit (Pharmacia Biotech, Uppsala, Sweden).
13. Restriction enzymes, alkaline phosphatase, and T4 DNA ligase.
14. *Escherichia coli* HB101 (Takara Bio, Kyoto, Japan).
15. Plasmid pUC18 (Takara Bio).
16. Ampicillin (Sigma).
17. Luria-Bertani (LB) medium: 1% (w/v) Bacto-tryptone (Difco), 0.5% yeast extract, 1% NaCl, and 1.5% agar.
18. Soft agar (w/v): 0.5% CMC, 1% NaCl, 2 mg/mL lysozyme, 50 mM glycine-NaOH buffer, pH 9.0, and 0.8% agar.
19. Congo red (Wako Pure Chemical).
20. Microprep plasmid purification kit (Amersham, Piscataway, NJ).
21. BigDye terminator cycle sequencing ready reaction kit (Applied Biosystems).
22. *Pwo* DNA polymerase (Boehringer Mannheim, Mannheim, Germany).

2.3. Alkaline Amylase

1. Starch agar (w/v): 1% soluble starch (Wako Pure Chemical), 0.4% starch azure (Sigma), 0.2% Bacto-tryptone, 0.1% yeast extract, 0.2% KH_2PO_4, 0.1% $MgSO_4 \cdot 7H_2O$, 0.1% $CaCl_2 \cdot 7H_2O$, 0.001% $FeSO_4 \cdot 7H_2O$, 0.0001% $MnCl_2 \cdot 4H_2O$, 1% Na_2CO_3 (sterilized separately), and 1.5 % agar (pH 10.0).

2. Liquid medium (w/v): 1% soluble starch, 0.2% Bacto-tryptone, 0.1% yeast extract, 0.2% KH_2PO_4, 0.1% $MgSO_4 \cdot 7H_2O$, 0.1% $CaCl_2 \cdot 2H_2O$, 0.001% $FeSO_4 \cdot 7H_2O$, 0.0001% $MnCl_2 \cdot 4H_2O$, and 1% Na_2CO_3 (sterilized separately).

3. *Bacillus* sp. strain KSM-AP1378 (FERM BP-3048) (Fermentation Research Institute).

4. Substrate solution: 1.0% (w/v) solution of soluble starch (from potato; Sigma) in 50 mM Tris-HCl buffer, pH 8.5, or 50 mM glycine-NaOH buffer, pH 10.0.

5. DNS reagent: the same as described in **Subheading 2.2.5.**

6. Buffer A: 10 mM Tris-HCl buffer, pH 7.5, and 2 mM $CaCl_2$.

7. DEAE-Toyopearl 650M (Tosoh).

8. CM-Toyopearl 650S (Tosoh).

9. Restriction enzymes, alkaline phosphatase, and T4 DNA ligase.

10. BigDye terminator cycle sequencing ready reaction kit (Applied Biosystems).

11. *Pwo* DNA polymerase (Boehringer Mannheim).

12. Primer A: 5′-TNGAYGCNGTNAARCAYATHAA-3′.

13. Primer B: 5′-CGNCANTGNAARCANCTRTTRGTRCT-3′.

14. Primer C: 5′-AGCCAATCTCTCGTATAGCTGTA-3′.

15. Primer D: 5′-GTACAAAAACACCCTATACAATG-3′.

16. Primer E: 5′-AATGGWACWATGATGCAKTA-3′.

17. Primer F: 5′-CATTTGGCAAATGCCATTCAAA-3′.

18. Primer G: 5′-AAAATTGATCCACTTCTGCAG-3′.

19. Primer H: 5′-CAGCGCGTAGATAATATAAATTTGAAT-3′.

20. Primer I: 5′-AAGCTTCCAATTTATATTGGGTGTAT-3′.

21. PCR production purification kit (Boehringer Mannheim).

3. Methods

3.1. Alkaline Protease

3.1.1. Screening for Alkaline Protease-Producing Bacteria

The high-alkaline KAP-producing *Bacillus* sp. strain KSM-K16 (patent strain FERM BP-3376) was screened as follows *(10)*:

1. A soil sample (0.5 g) was suspended in 10 mL of saline solution and incubated at 80°C for 30 min (*see* **Note 4**).

2. A portion of the solution was spread on keratin agar plates (*see* **Note 5**).

3. Plates were incubated at 30°C for 3–5 d. Alkaline protease producing-bacteria formed a keratin-soluble zone around the colonies.

4. The selected bacteria were propagated, with shaking, at 30°C for 2 d in liquid medium.

3.1.2. Enzyme assays

1. For measurement of caseinolytic activity, the reaction mixture (1 mL) composed of 1% casein solution was incubated at 40°C for 5 min.

2. Then, 0.1 mL of a suitably diluted enzyme solution was added and incubation continued at 40°C for 10 min.

3. The reaction was stopped by adding 2 mL of TCA solution, and the mixture was allowed to stand for 15 min at room temperature (*see* **Note 6**).
4. The solution was passed through No. 2 filter paper to remove denatured protein.
5. Acid-soluble materials in the filtrate were quantified by the method of Lowry et al. *(11)*. To 0.5 mL of the filtrate was added 2.5 mL of reagent A. After incubation at 30°C for 10 min, 0.25 mL of the phenol reagent (twofold dilution with distilled water) was added to the mixture and further incubated at 30°C for 30 min. Absorbance in the mixture was measured at 660 nm. One unit (U) of the enzyme activity was defined as the amount of protein that produced acid-soluble protein equivalent to 1 μmol of L-tyrosine per minute.

3.1.3. Purification of Alkaline Proteases

KAP was purified to homogeneity as follows *(10)*:

1. A centrifuged supernatant (1 L) was concentrated on the membrane YM-5, and the concentrate was dialyzed against dialysis solution.
2. The retentate was loaded onto a column of DEAE-Bio-Gel A (2.5 × 16 cm) equilibrated with equilibration buffer. The column was washed with the same buffer, and nonadsorbed fractions showing protease activity were combined and concentrated to 20 mL by ultrafiltration.
3. The concentrate was applied to a column of CM-Bio-Gel A (2.5 × 16 cm) equilibrated with the equilibration buffer. The column was washed, and proteins were eluted with a 900-mL linear gradient of 0–100 m*M* KCl in the same buffer. The main protease activity (M-protease) was eluted between 35 m*M* and 75 m*M* KCl. The active fractions were pooled and dialyzed against the equilibration buffer, and concentrated to 7.7 mL by ultrafiltration.
4. The concentrate was applied to a column of CM-Bio-Gel A (1.5 × 16 cm) equilibrated with borate buffer. The proteins were eluted with a 300-mL linear gradient of 0–100 m*M* triethanolamine-HCl (TEA-HCl) in the same buffer (*see* **Note 7**). The main active fractions (M-protease) were eluted between 50 m*M* and 80 m*M* of TEA-HCl and concentrated by ultrafiltration. The concentrate was dialyzed against equilibration buffer. The retentate was concentrated by ultrafiltration.

KAP (M-protease) was purified to homogeneity as judged by nondenaturing polyacrylamide gel electrophoresis (PAGE) *(12)* (*see* **Note 8**). The overall yield of purification was 18.6%. Oxidatively stable alkaline proteases such as E-1 from *Bacillus* sp. strain D-6 (JCM9154) *(3)* and KP-43 from *Bacillus* sp. strain KSM-KP43 (patent strain FERM BP-6532) *(4)* were purified in a similar way with M-protease.

3.1.4. Cloning and Sequencing of Gene for Alkaline Proteases

The gene of *Bacillus* sp. strain KSM-K16 coding for M-protease was cloned by the shotgun method *(13)*.

1. The genomic DNA of *Bacillus* sp. KSM-K16 was prepared by a standard method *(14)*, and then digested with *Hin*dIII at 37°C for 2 h. After agarose gel

electrophoresis, the digests (3.5–4.5 kb) were electroeluted from the gel in a DNACELL.

2. The electroeluted DNA (2 μg) was ligated with T4 ligase into pHY300PLK (1 μg) that had been treated with *Hin*dIII and alkaline phosphatase. *B. subtilis* ISW1214 cells were transformed with the ligation mixture and then grown on DM3 agar *(15)* plus 5 μg/mL tetracycline. After incubation at 37°C for 2 d, the transformants were spread on selection agar with 15 μg/mL tetracycline and cultured at 37°C for 16 h.

3. The colonies that formed around a clear zone were picked up. Plasmid DNA was isolated by the alkaline extraction procedure *(16)*. Nucleotide sequencing of double-stranded DNA was done directly by the dideoxy chain-termination method on a model 370A DNA sequencer, using a Taq dideoxy termination cycle sequencing kit.

The gene for M-protease consisted of a single open reading frame (ORF) of 1143 bp (380 amino acids [aa]) encoding a prepro-peptide (111 aa) and a mature protein (269 aa, 26,723 Da). The deduced aa sequence of the enzyme showed high homology to high-alkaline proteases from alkaliphilic *Bacillus* strains with more than 80% identity, but showed moderate homology to those of true subtilisins, BPN′ and Carlsberg, with less than 60% identity *(13)* (*see* **Note 9**).

3.2. Alkaline Cellulase

3.2.1. Screening for Alkaline Cellulase-Producing Bacteria

The Egl-635-producing *Bacillus* sp. strain KSM-635 (patent strain FERM BP-1485) was screened as follows *(5)*:

1. A soil sample (0.5 g) was suspended in 10 mL of saline and incubated at 80°C for 30 min.
2. A portion of the solution was spread on CMC agar.
3. Agar plates were incubated at 30°C for 3–5 d. Alkaline cellulase-producing bacteria formed translucent zones around colonies.
4. The selected bacteria were propagated, with shaking, at 30°C for 2–3 d in liquid medium.

3.2.2. Enzyme Assays

1. The substrate solution (0.9 mL) was incubated at 40°C for 5 min.
2. Then, 0.1 mL of a suitably diluted enzyme solution was added and incubation continued at 40°C for 20 min.
3. The reaction was stopped by adding 1 mL of the DNS reagent, and the solution was placed in a boiled water bath for 5 min.
4. The solution was rapidly cooled in an ice bath, and then 4 mL of distilled water was added. The color developed was measured at 535 nm. One unit of the enzyme (CMCase) activity was defined as the amount of protein that produced 1 μmol of reducing sugars as glucose.

3.2.3. Purification of Alkaline Cellulases

Egl-635 from *Bacillus* sp. strain KSM-635 was purified as follows *(17)*:

1. Three liters of the cell-free culture supernatant was concentrated on a PM-10 membrane and dialyzed against buffer A.
2. The retentate was applied to a column of DEAE-Toyopearl 650S (4.4 × 40 cm) equilibrated with buffer A plus 0.2 *M* NaCl. The adsorbed proteins were eluted by a 6-L linear gradient of 0.2–0.4 *M* NaCl in the buffer. The fractions containing CMCase activity were eluted around 0.25 *M* NaCl. The active fractions were concentrated by ultrafiltration.
3. The concentrate was loaded onto a column of Bio-Gel A0.5m (2.5 × 90 cm) equilibrated with buffer A plus 0.1 *M* NaCl. The CMCase activity was separated into two peaks; the first peak of the activity (E-H) contained 70% of the total amount of CMCase activity loaded on the column. The second peak of the activity was designated E-L.

E-H and E-L were purified to homogeneity, as judged by both nondenaturing PAGE and SDS-PAGE. The degrees of purification and total recovery were 2.1-fold and 56% for E-H, and 3.7-fold and 12% for E-L. The specific activities of E-H and E-L toward CMC were 39.8 and 59.2 U/mg protein, respectively. The molecular masses of the enzymes were 130 kDa for E-H and 103 kDa for E-L by SDS-PAGE *(17)*.

3.2.4. Cloning and Sequencing Gene for Alkaline Cellulases

The genes for Egl-635 and Egl-237 were cloned by the shotgun method *(18,19)*. Here the procedures for cloning and sequencing of the gene of *Bacillus* sp. strain KSM-S237 (patent strain FERM BP-7875) coding for Egl-237 *(19)* are described:

1. The genomic DNA of *Bacillus* sp. strain KSM-S237 was prepared by a standard method *(14)*. The genomic DNA was partially digested with *Eco*RI at 37°C for 2.5 h.
2. The digests were fractionated by agarose gel electrophoresis and purified by a GFX PCR DNA and gel band purification kit. The purified digests (1 µg, 2–9 kb) were ligated with T4 ligase at 16°C for 20 h into pUC18 that had been treated with *Eco*RI and alkaline phosphatase.
3. *E. coli* HB101 cells were transformed with the constructed plasmid and grown on LB agar supplemented with 50 µg/mL ampicillin at 37°C for 24 h. CMCase-positive clones were selected by the soft-agar-overlay method. After the overlaid soft agar had been incubated at 37°C for 3 h, a 1% (w/v) solution of Congo red was poured onto the soft agar. The positive clones form a clear zone around the colonies, and washing the agar with 1 *M* NaCl makes the halos clearer.
4. The plasmid was extracted from a positive clone and purified by a Microprep plasmid purification kit. The nucleotide sequence of the insert was determined by a

BigDye terminator cycle sequencing ready reaction kit and a model 377 DNA sequencer.

5. The nucleotide sequence coded Met[1]-Ser[420] had an *Eco*RI site on the cloned gene. To obtain the entire gene, primers, 5′-GATGCAACAGGCTTATATTTA-GAG-3′ designed from a sequence upstream of the possible promoter region and 5′-AAATTACTTCATCATTCTATCAC-3′ designed from a sequence downstream of the gene for Egl-1139 from *Bacillus* sp. strain 1139 *(20)* *(see* **Note 10**), were synthesized. PCR was done in a DNA thermal cycler (model 480; Perkin-Elmer) using the two primers and the *Bacillus* sp. strain KSM-S237 genomic DNA. The reaction program was 2 min at 94°C, followed by 30 cycles of 1 min at 94°C, 1 min at 55°C, and 3 min at 72°C, and a final 5 min at 72°C using *Pwo* DNA polymerase. The amplified fragment (3.1 kb) was purified and sequenced.

The entire gene for Egl-237 harbored a 2472-bp ORF encoding 824 aa, including a 30-aa signal sequence *(19)*. Egl-237 showed moderate homology to Egl-635 *(18)* with 53.0% identity and also high homology to Egl-64 from *Bacillus* sp. strain KSM-64 (patent strain FERM BP-2886) *(21)* and Egl-1139 from *Bacillus* sp. strain no.1139 *(20)* with 91.8 and 89.1%, respectively.

3.3. Alkaline Amylase

3.3.1. Screening for Alkaline α-Amylase-Producing Bacteria

AmyK producer *Bacillus* sp. strain KSM-AP1378 (patent strain FERM BP-3048) was screened as follows *(22)*:

1. A soil sample (0.5 g) was suspended in 10 mL of saline and incubated at 80°C for 15 min.
2. A portion of the solution was spread on starch agar plates.
3. Agar plates were incubated at 30°C for 2–5 d. Alkaline α-amylase-producing bacteria formed a clear zone around colonies.
4. The selected bacteria were propagated, with shaking, at 30°C for 2 d in liquid medium.

3.3.2. Enzyme Assays

1. The substrate solution (0.5 mL) plus 0.4 mL of distilled water was incubated at 50°C for 5 min.
2. Then, 0.1 mL of suitably diluted enzyme solution was added and incubation continued at 50°C for 10 min.
3. The reaction was stopped by adding 1 mL of the DNS reagent, and the solution was placed in a boiled water bath for 5 min.
4. The solution was rapidly cooled in an ice bath, and then 4 mL of distilled water was added. The color developed was measured at 535 nm. One unit (U) of the enzyme activity was defined as the amount of protein that produced 1 μmol of reducing sugars as glucose.

3.3.3. Purification of Alkaline α-Amylases

AmyK was purified as follows *(22)*:

1. Ammonium sulfate was added to the centrifugal supernatant, and the precipitates formed at 60% saturation were collected. The precipitates were dissolved in a small volume of buffer A, and then dialyzed against the same buffer.
2. The retentate was applied to a column of DEAE Toyopearl 650M (10 × 15 cm) equilibrated with buffer A. The column was washed with the same buffer and non-adsorbed fractions that contained amylase activity were combined and concentrated by ultrafiltration (PM-10 membrane).
3. The concentrate was put on a column of CM-Toyopearl 650S (2.5 × 50 cm) equilibrated with buffer A. The column was washed and proteins were eluted with a 2-L linear gradient of 0–0.5 *M* NaCl in the same buffer. The active fractions were combined and concentrated by ultrafiltration. The concentrate was dialyzed against buffer A.

AmyK was purified to homogeneity on SDS-PAGE (*see* **Note 11**), 6.1-fold with an overall yield of 35%. AmyK has a molecular mass of 53 kDa. The specific activities toward soluble starch were approximately 5000 U/mg for AmyK, values being several times greater than that of the industrial α-amylase, BLA.

3.3.4. Cloning Genes Encoding Alkaline α-Amylases

The gene of *Bacillus* sp. strain KSM-AP1378 coding for AmyK was cloned and sequenced as follows *(22)*:

1. Primers A and B, designed from the two common regions, DAVKHIK and DVT-FVDNHD, in typical liquefying α-amylases, were synthesized.
2. PCR was performed using the genomic DNA (1 μg) of *Bacillus* sp. strain KSM-AP1378 as template and primers A and B (0.2 μg each). The reaction program was 30 cycles of 1 min at 94°C, 1 min at 55°C, and 2 min at 72°C. Consequently, a 0.3-kb fragment was amplified (fragment A).
3. To determine the complete sequence of the AmyK gene, inverse PCR was done using a self-circularized *Xba*I-digested genomic DNA of *Bacillus* sp. strain KSM-AP1378 as template, and primers C and D designed from fragment A.
4. A 0.7-kb fragment was amplified, which encoded a deduced aa sequence identical to the C-terminal sequence of typical liquefying α-amylases and also a stop codon TAA. Primer E designed from the N-terminal aa sequence of purified AmyK from a culture broth of *Bacillus* sp. strain KSM-AP1378 was synthesized. PCR was performed to amplify a 0.7-kb fragment using the genomic DNA as template, and primers C and E.
5. The second inverse PCR was performed using a self-circularized *Hind*III-digested genomic DNA as template and primers F and G. An amplified 0.8-kb fragment encoded a putative regulatory region and aa sequence identical to the N-terminal sequence of AmyK.

6. A 1.8-kb fragment containing the entire AmyK gene was amplified using genomic DNA as template and primers H and I.

The gene for AmyK consisted of a single ORF of 1548 bp (516 aa) that encoded a signal peptide of 31 aa and a mature protein (485 aa, 55,391 Da). The deduced aa sequence of the mature enzyme showed moderate similarity to other those of α-amylases from *B. licheniformis* (BLA: 68.9%), *B. amyloliquefaciens* (BAA: 66.7%), and *B. stearothermophillus* (BSA: 68.6%) *(22)*.

4. Notes

1. When Na_2CO_3 is autoclaved together with other components, a browning reaction occurs, and some nutrients are also destroyed at alkaline pH. The pH of the medium must not be adjusted with NaOH, KOH, and K_2CO_3 because the productivity of alkaline enzymes decreases.
2. Alternatively, Congo red assay on a soft agar (described in **Subheading 3.2.4.**) can be used for the detection of cellulase-producing bacteria.
3. The production of cellulase strongly depends on the source of manufacturers, lot numbers, and degrees of substitution.
4. Targets of detergent-enzyme producers in this section were alkaliphilic *Bacillus* strains. The organisms usually survive as heat-resistant spores in soil samples.
5. Alkaline proteases that can degrade insoluble proteins, such as keratin, show good detergency. CMC is added to agar to prevent precipitation of keratin powders to the bottom of the plate.
6. Acid-denatured proteins grew gradually to large precipitates during incubation and were easily removed by filtration.
7. TEA-HCl in elution buffer increases the ionic strength and decreases the pH.
8. Taber and Sherman *(12)* used 18 mM glycine/34 mM 2,6-dimethylpyridine as the electrode buffer. The buffer is quite irritating to the eyes and nose during electrophoresis. Alternatively, 25 mM Tris/192 mM glycine buffer (pH 8.3) can be used. Toluidine blue N is used as a marker dye in protein samples.
9. Subtilisins are weak to chemical oxidants when examined with oligopeptidyl *p*-nitroanilide as substrate in borate buffer.
10. Homology search of the sequence Met[1]-Ser[420] in Egl-237 revealed very high similarity to that of Egl-1139. Also, the upstream and downstream regions of both Egls show high similarity.
11. *Bacillus* α-amylases often do not move into gels during nondenaturing PAGE. Then we certified the degrees of purification of the enzymes by SDS-PAGE.

References

1. Houston, J. H. (1997) Detergent enzymes' market, in *Enzymes in Detergency* (van Ee, J. H., Misset, O., and Baas, E. J., eds.), Marcel Dekker, New York, pp. 11–21.
2. Bott, R. (1997) Development of new proteases for detergents, in *Enzymes in Detergency* (van Ee, J. H., Misset, O., and Baas, E. J., eds.), Marcel Dekker, New York, pp. 75–91.

3. Saeki, K., Okuda, M., Hatada, Y., Kobayashi, T., Ito, S., Takami, H., and Horikoshi, K. (2000) Novel oxidatively stable subtilisin-like serine proteases from alkaliphilic *Bacillus* spp.: Enzymatic properties, sequences, and evolutionary relationships. *Biochem. Biophys. Res. Commun.* **279,** 313–319.

4. Saeki, K., Hitomi, J., Okuda, M., Hatada, Y., Kageyama, Y., Takaiwa, M., et al. (2002) A novel species of alkaliphilic *Bacillus* that produces an oxidatively stable alkaline serine protease. *Extremophiles* **6,** 65–72.

5. Ito, S. (1997) Alkaline cellulases from alkaliphilic *Bacillus:* Enzymatic properties, genetics, and application to detergents. *Extremophiles* **1,** 61–66.

6. Hoshino, E. and Ito, S. (1997) Application of alkaline cellulases that contribute to soil removal in detergents, in *Enzymes in Detergency* (van Ee, J. H., Misset, O., and Baas, E. J., eds.), Marcel Dekker, New York, pp. 149–174.

7. Hakamada, Y., Koike, K., Yoshimatsu, T., Mori, H., Kobayashi, T., and Ito, S. (1997) Thermostable alkaline cellulase from an alkaliphilic isolate, *Bacillus* sp. KSM-S237. *Extremophiles* **1,** 151–156.

8. UpaDek, H. and Kottwitz, B. (1997) Application of amylases in detergents, in *Enzymes in Detergency* (van Ee, J. H., Misset, O., and Baas, E. J., eds.), Marcel Dekker, New York, pp. 203–212.

9. Hagihara, H., Igarashi, K., Hayashi, Y., Endo, K., Ikawa-Kitayama, K., Ozaki, K., et al. (2001) Novel α-amylase that is highly resistant to chelating reagents and chemical oxidants from the alkaliphilic *Bacillus* isolate KSM-K38. *Appl. Environ. Microbiol.* **67,** 1744–1750.

10. Kobayashi, T., Hakamada, Y., Adachi, S., Hitomi, J., Yoshimatsu, T., Koike, K., et al. (1995) Purification and some properties of an alkaline protease from alkalophilic *Bacillus* sp. KSM-K16. *Appl. Microbiol. Biotechnol.* **43,** 473–481.

11. Lowry, O. H., Rosebrough, N. J., Farr, A. L., and Randall, J. R. (1951) Protein measurement with the Folin phenol reagent. *J. Biol. Chem.* **242,** 265–275.

12. Taber, H. W. and Sherman, F. (1964) Spectrophotometric analyzers for disk electrophoresis: Studies of yeast cytochrome *c. Ann. NY Acad. Sci.* **121,** 600–615.

13. Hakamada, Y., Kobayashi, T., Hitomi, J., Kawai, S., and Ito, S. (1994) Molecular cloning and nucleotide sequence of the gene for an alkaline protease from the alkalophilic *Bacillus* sp. KSM-K16. *J. Ferment. Bioeng.* **78,** 105–108.

14. Saito, H. and Miura, K. (1963) Preparation of transforming deoxyribonucleic acid by phenol treatment. *Biochim. Biophys. Acta* **72,** 619–629.

15. Chang, S. and Cohen, S. N. (1979) High frequency transformation of *Bacillus subtilis. Mol. Gen. Genet.* **168,** 111–115.

16. Birnboim, H. C. and Doly, J. (1979) A rapid alkaline extraction procedure for screening recombinant plasmid DNA. *Nucleic Acids Res.* **7,** 1513–1523.

17. Yoshimatsu, T., Ozaki, K., Shikata, S., Ohta, Y., Koike, K., Kawai, S., and Ito, S. (1990) Purification and characterization of alkaline endo-1,4-β-glucanases from alkalophilic *Bacillus* sp. KSM-635. *J. Gen. Microbiol.* **136,** 1973–1979.

18. Ozaki, K., Shikata, S., Kawai, S., Ito, S., and Okamoto, K. (1990) Molecular cloning and nucleotide sequence of a gene for alkaline cellulase from *Bacillus* sp. KSM-635. *J. Gen. Microbiol.* **136,** 1327–1334.

19. Hakamada, Y., Hatada, Y., Koike, K., Yoshimatsu, T., Kawai, S., Kobayashi, T., and Ito, S. (2000) Deduced amino acid sequence and possible catalytic residues of a thermostable, alkaline cellulase from an alkaliphilic *Bacillus* strain. *Biosci. Biotechnol. Biochem.* **64,** 2281–2289.

20. Fukumori, F., Kudo, T., Narahashi, Y., and Horikoshi, K. (1986) Molecular cloning and nucleotide sequence of the alkaline cellulase gene from the alkalophilic *Bacillus* sp. strain 1139. *J. Gen. Microbiol.* **132,** 2329–2335.

21. Sumitomo, N., Ozaki, K., Kawai, S., and Ito, S. (1992) Nucleotide sequence of the gene for an alkaline endoglucanase from an alkalophilic *Bacillus* and its expression in *Escherichia coli* and *Bacillus subtilis. Biosci. Biotechnol. Biochem.* **56,** 872–877.

22. Igarashi, K., Hatada, Y., Hagihara, H., Saeki, K., Takaiwa, M., Uemura, T., et al. (1998) Enzymatic properties of a novel liquefying α-amylase from an alkalophilic *Bacillus* isolate and entire nucleotide and amino acid sequences. *Appl. Environ. Microbiol.* **64,** 3282–3289.

10

Microbial Proteases

Chandran Sandhya, K. Madhavan Nampoothiri, and Ashok Pandey

Summary

Proteases represent one of the three largest groups of industrial enzymes and account for about 60% of the total worldwide sale of enzymes. They are degradative enzymes of central importance because they can be employed in a number of industries to create change in product taste, texture, and appearance, as well as in waste recovery. They are also important in medical and pharmaceutical applications. Microorganisms represent an excellent source of proteases owing to their broad biochemical diversity and their susceptibility to genetic manipulation. In the microbial fermentation process, optimization of culture media is important to yield an economically viable amount of proteases. Taking into consideration the need for large-scale production of proteases, a brief outline of production techniques, recovery, purification, and characterization is discussed here.

Key Words: Protease; production; fermentation; recovery; purification.

1. Introduction

Enzymes are highly efficient environment-friendly protein catalysts, synthesized by living systems. They have significant advantages over chemical catalysts, of which the most important are specificity, high catalytic activity, ability to work at moderate temperatures, and the ability to be produced in large amounts. The current demand for better utilization of renewable resources and pressure on industry to operate within environmentally compatible limits stimulated development of new enzyme-catalyzed industrial processes. Proteases (proteolytic enzymes) represent one of the three largest groups of industrial enzymes and account for about 60% of the total worldwide sale of enzymes. These enzymes play a critical role in many physiological and pathological processes such as protein catabolism, blood coagulation, cell growth and migration, tissue arrangement, morphogenesis in development, inflammation, tumor growth and metastasis, activation of zymogens, release of hormones and pharmacologically active peptides from precursor proteins, and transport of secretory proteins across membranes (1). Besides this, they have extensive applications

From: *Methods in Biotechnology, Vol. 17: Microbial Enzymes and Biotransformations*
Edited by: J. L. Barredo © Humana Press Inc., Totowa, NJ

in food industry, laundry detergents, leather treatment, bioremediation processes, and the pharmaceutical industry, among others. They also play a major role in nutrition due to their depolymerizing activity. Among the major protease producers of the world, Novo Industries (Denmark) occupies 40% of the market share of proteases. It manufactures three different proteases—Aquaderm, NUE, and Pyrase—for use in soaking, dehairing, and bating, respectively.

Proteases, being ubiquitous, are found in a wide diversity of sources such as plants, animals, and microorganisms *(1)*. Increased interest in microbial proteases resulted from the inability of plant and animal proteases to meet current world demand; they are preferred to other sources because they possess almost all the characteristics desired for their biotechnological applications. Proteins are degraded by microorganisms, and they utilize the degradation products as nutrients for their growth. Numerous proteases have been reported to be produced by microorganisms depending on the species of the producers or the strains even belonging to the same species. Some are also produced by the same strain under various cultural conditions. Human immunodeficiency virus (HIV), a causative of AIDS, presents an aspartic protease essential for the retroviral life cycle and it has been a good target for the chemotherapy with specific inhibitors. **Table 1** gives a list of some industrially important protease-producing microorganisms.

Proteases are enzymes of class 3, the hydrolases, and subclass 3.4, the peptide hydrolases or peptiodases. They are subdivided into two major groups, i.e., exopeptidases and endopeptidases, depending on their site of action and based on the functional group present at the active site, they are further classified into four prominent groups, i.e., serine proteases, aspartic proteases, cysteine proteases, and metalloproteases. Depending on the pH at which they are active, proteases are also classified into acid, alkaline, and neutral proteases. Most commercial serine proteases, mainly neutral and alkaline, are produced by organisms belonging to the genus *Bacillus.* Similar enzymes are also produced by other bacteria such as *Thermus caldophilus* and *Desulfurococcus mucosus, Streptomyces, Aeromonas,* and *Escherichia* genera. Fungi produce several serine proteases. Among them, these enzymes are produced by various strains of *Aspergillus oryzae.* Cysteine proteases are not so widely distributed as was seen with serine and aspartic proteinases.

In the near future, the significant potential of these enzymes for diverse applications will necessitate large-scale enzyme production facilities, so an outline on various aspects of microbial production is necessary for advanced research in these enzymes.

1.1. Isolation of Proteolytic Organisms

In the development of an industrial fermentation process, the first step is to isolate strains capable of producing the desired product in commercial yields.

Table 1.
Industrially Important Proteases from Microbes

Source	Species	Industrial application
Fungi	*Aspergillus oryzae*	Food/Pharmaceutical
	Mucor sp.	Dairy
	Conidiobolus coronatus	Detergent
	Conidiobolus coronatus NCIM 1238	Resolution of racemic mixtures of D,L-phenylalanine and glycine
	Tritirachium album	Detergent
	Penicillium sp.	Detergent
	Aspergillus flavus	Dehairing
Bacteria	*Bacillus licheniformis*	Detergent
	Bacillus amyloliquefaciens	Dehairing/leather
	Bacillus subtilis	Contact lens cleansing agent/leather
	Bacillus licheniformis (Alcalase)	Synthesis of biologically active peptides
	Bacillus sp. (P-001A)	Production of biomass from natural waste
	Streptococcus sp.	Dairy/cheese production
	Bacillus stearothermophillus	Detergents and heavy-duty laundry powders

This will follow screening programs to test a large number of strains to identify high producers having novel properties. While designing a medium for screening proteases, the medium should contain its inducers and be devoid of constituents that may repress protease synthesis.

1.2. Production of Proteases

The amount of protease produced varies greatly with the strain and media used. So in order to develop economically viable yields of protease, it is important to optimize fermentation media for their growth and production. Media for protease production mostly contain carbohydrate sources like starch, ground barley, or lactose, and nitrogen sources like soybean meal, casein, or corn steep liquor. Carbohydrate consumption is high during fermentation and it is often the practice to add carbohydrate continuously in order to keep the concentration low at all times. This is to avoid a high concentration of carbohydrate, which is known to repress enzyme production. Free amino acid often represses protease production, while peptides and proteins induce protease synthesis.

Protease production has been extensively studied by submerged fermentation (SmF) as well as solid-state fermentation (SSF) processes.

1.2.1. Submerged Fermentation

Submerged fermentation comprises a large variety of stirred or nonstirred microbial processes, where biomass is completely surrounded in the liquid culture medium. The proteases of *Mucor michei, Endothia parasitica,* and all species of *Bacillus* are produced in submerged fermentation. Many of the medium components are common, inexpensive agricultural products that can be supplied in reliable, uniform quality. The media are usually supplemented with 10–15% dry substance and high protein content. Carbohydrate may be fed as glucose, lactose, sucrose, or starch hydrolysate.

1.2.2. Solid-State Fermentation

Recently SSF has generated much interest, because it is a simpler process and can use wastes or agroindustrial substrates such as defatted soybean cake, wheat bran, rice bran, and the like. The environmental conditions in solid-state fermentation can stimulate the microbe to produce enzymes with different properties from those of same organism under submerged conditions. There have been increasing attempts to produce different types of proteases (acid, neutral, and alkaline) through the SSF route. It is interesting to note that although a number of substrates have been employed for cultivating different microorganisms, wheat bran has been the preferred choice in most of the studies *(2)*. A comparative study on production of alkaline protease in SmF and SSF showed that total protease activity present in 1 g bran (SSF) was equivalent to that in 100 mL broth (SmF) *(3)*. Generally protease synthesis is inhibited by the carbon source indicating the presence of catabolite repression of protease biosynthesis.

The important parameters that influence large-scale production of protease are accumulation density, bed height, agitation, and so forth. In trays, the mold grows poorly and produces less protease. This may be due to evaporation, which tends to dry out the media and cause growth inhibition. The shear effects on the surface of the lumps in the rolling drum fermentor may prevent growth and protease production. Packed-bed fermenters with temperature, humidity, and aeration control are features for large-scale operations. However, flasks were chosen for continued laboratory studies because of the yield and convenience of their preparation.

1.3. Recovery and Purification of Proteases

Product recovery during bioprocessing is difficult due to dilute and labile products mixed with macromolecules of similar properties. Therefore the first step is to separate cell biomass and insoluble nutrient ingredients from the supernatant. Purification is usually achieved by precipitation and chromatographic procedures.

1.3.1. Ammonium Sulfate Precipitation

In order to differentiate between diverse proteins present in the crude extract, it may be necessary to adopt procedures to separate out the desired protein. Salt fractionation techniques are used to prepare protein fractions precipitated by successively increased concentrations of the salt (sulfates, sulfites, phosphates). When high concentrations of salt are present, proteins tend to aggregate and precipitate out of solution. Ammonium sulfate is the salt of choice because it combines many features such as salting out effectiveness, pH versatility, high solubility, low heat of solution, and low price. After completion of protein precipitation, ammonium sulfate can be removed by dialysis, where protein solution is contained within a membrane whose pore size prevents protein from escaping and that permits solute exchange with the surrounding solution.

1.3.2. Gel Filtration Chromatography

Gel filtration chromatography separates proteins according to their size. With the proper filtration matrix, the protein would emerge from the column in the initial protein fractions and all the smaller proteins would still remain in the column, allowing a considerable purification. Depending on molecular characteristics as well as physical properties, proteases can be further separated by other chromatographic techniques such as ion exchange chromatography and affinity chromatography.

1.4. Measurement of Protease Activity

A colorimetric assay using azocasein (chemically modified protein) as a substrate can be used to determine the protease activity in crude and purified samples.

1.5. Effect of Different Kinetic Parameters on Protease Activity and Its Stability

The catalytic activity of enzymes is dependent on the native structure; any slight variations may result in significant changes in activity. Studies on effect of substrate are important since enzymes possess a cleft or depression in structure that is lined by certain amino acid residues concerned with the specificity of enzymes. Enzymes are also sensitive toward environmental effects such as temperature, pH, presence of ions in the medium, and so on.

1.6. Gelatin Zymography

This technique involves the electrophoresis of secreted proteases through discontinous polyacrylamide gels containing enzyme substrate. After electrophoresis, removal of SDS from the gel by washing in 2.5% Triton X-100 solution allows enzymes to renature and to degrade the protein substrate. Staining of the gel with Coomassie blue allows the bands of proteolytic activity to be detected on clear bands of lysis against a blue background.

1.7. Immunoblot Analysis

Immunoblotting is an extremely powerful technique for identifying a single protein in a complex mixture following separation based on its molecular weight, size, and charge. Immunoblotting can be divided into two steps: transfer of the protein from the gel to the matrix (nitrocellulose membrane is preferred because it is relatively inexpensive and blocking nitrocellulose from nonspecific antibody binding is fast and simple) and decoration of the epitope with the specific antibody.

2. Materials

2.1. Isolation of Proteolytic Organisms

1. Soil sample.
2. Nutrient agar medium: 5.0 g/L peptic digest of animal tissue, 1.5 g/L beef extract, 1.5 g/L yeast extract, 5.0 g/L NaCl, and 15 g/L agar. Final pH at 25°C is 7.4.
3. 1% casein.

2.2. Production of Proteases

1. Culture medium for SmF: casamino acids (variable), glucose (variable), 1.0 g/L KH_2PO_4, 3.0 g/L K_2HPO_4, 2.0 g/L Na_2SO_4, and 0.1 g/L $MgSO_4 \cdot 7H_2O$.
2. Microorganism.
3. Orbital shaker.
4. Cooling centrifuge.
5. Agroindustrial residues (substrate).
6. Salt solution: 2.0 g/L KNO_3, 0.5 g/L $MgSO_4 \cdot 7H_2O$, 1.0 g/L K_2HPO_4, 0.44 g/L $ZnSO_4 \cdot 7H_2O$, 1.12 g/L $FeSO_4 \cdot 7H_2O$, and 0.2 g/L $MnSO_4 \cdot 7H_2O$. Adjust to pH 5.0.
7. Incubation chamber.

2.3. Recovery and Purification of Proteases

1. 1% NaCl.
2. 80% $(NH_4)_2SO_4$.
3. 0.2 M citrate phosphate buffer, pH 6.5.
4. Magnetic stir plate and stir bar.
5. Centrifuge.
6. Dialysis tubing.
7. Chromatography equipment.
8. Appropriate gel matrix.

2.4. Measurement of Protease Activity

1. Azocasein 1% (w/v) (Sigma, St. Louis, MO).
2. 0.2 M citrate phosphate buffer, pH 6.5.

3. Crude enzyme (dialyzed in water for 12 h at 4°C).
4. Trichloroacetic acid (TCA) (10% v/v).
5. 1 *M* NaOH.
6. Spectrophotometer.

2.5. Gelatin Zymography

1. Minigel apparatus.
2. Power supply (capacity 200 V, 500 mA).
3. Zymogram resolving gel: 7.5% polyacrylamide, 0.1% sodium dodecyl sulfate (SDS), and 0.15% w/v copolymerized gelatin.
4. Laemmli sample buffer (5X): 60 m*M* Tris-HCl, pH 6.8, 25% glycerol, 2% SDS, 14.4 m*M* 2-mercaptoethanol, and 0.1% bromophenol blue.
5. 2.5% Triton X-100.
6. Zymogram development solution: 0.2 *M* citrate phosphate buffer, pH 6.5.
7. Staining solution: 45% methanol, 10% acetic acid, and 0.1% Coomassie brilliant blue R-250.
8. Destaining solution: 40% v/v methanol, 10% v/v acetic acid, and 50% distilled water.

2.6. Immunoblot Analysis

1. Electroblotting apparatus.
2. Power supply.
3. Nitrocellulose membranes (Bio-Rad, Hercules, CA).
4. Whatman 3 MM paper (Whatman, Kent, UK).
5. Transfer buffer: 20 m*M* Tris, 150 m*M* glycine, and 20% methanol.
6. Blocking solution: 3% bovine serum albumin in Tris-buffered saline.
7. Tris-buffered saline (TBS): 10 m*M* Tris-HCl, pH 7.5, and 150 m*M* NaCl.
8. Dilution buffer: 1% gelatin, 200 m*M* Tris-HCl, pH 7.5, 500 m*M* NaCl, and 0.05% Tween-20.
9. Antiguinea-pig-horseradish-peroxidase complex (Bio-Rad).
10. Color developing solution: Add 10 mL methanol to 1 mL of chloronaphthol solution (30 mg/mL in methanol), and made up to 50 mL with TBS. Then add 30 µL 30% H_2O_2.

3. Methods

3.1. Isolation of Proteolytic Organisms

The organisms inhabiting protein-rich soil tend to utilize more amounts of proteinaceous material by producing higher amounts of proteolytic enzymes. To isolate proteolytic microorganisms, the following protocol can be used:

1. Suspend about 1 g of the soil sample in 5 mL sterile distilled water and mix vigorously.
2. Streak aliquots of the clear suspension onto nutrient agar medium containing 1% casein.
3. Incubate for 48 h at 37°C.

4. Observe protease production as a clearing zone around the colony in protease positive isolates (*see* **Note 1**).
5. Transfer individual colonies to fresh nutrient agar plates.
6. Purify protease-producing colonies through repeated streaking.
7. Store on agar slants.

3.2. Production of Proteases

3.2.1. Submerged Fermentation

Culture media can be prepared either by adding precise amounts of pure inorganic or organic chemicals to distilled water (chemically defined media) or by employing crude digests of substances such as casein, beef, soybeans, yeast cells, and so on (*see* **Note 2**).

1. Autoclave the culture medium in an appropriate container (shake flask, benchtop fermenter, etc.).
2. Inoculate the sterile culture medium with a desired protease-producing microorganism. Adjust the concentration of inoculum by serial dilution.
3. Incubate at a desired temperature and rpm on a rotary shaker for the required period.
4. Harvest the supernatant by centrifugation at 4°C (12,000g) and use it as crude enzyme extract (*see* **Note 3**).

3.2.2. Solid-State Fermentation

1. Prepare SSF medium in desired container (250 mL Erlenmeyer flask, tray fermenter, column fermenter, etc.).
2. Moisten a definite amount of substrate with salt solution and water to a desired initial moisture content.
3. Autoclave at 121°C for 20 min.
4. Inoculate with a spore suspension (desired number of spores per gram dry substrate) and incubate at desired temperature for desired period (*see* **Note 4**).

The following protocol can be used for alkaline protease production by *Rhizopus oryzae:*

1. Prepare 250 mL conical flask containing 10 g wheat bran.
2. Moisten (140%) with salt solution of pH 5.5.
3. Autoclave at 121°C for 20 min.
4. Inoculate with ~2 × 10^5 spores/g wheat bran.
5. Incubate at optimum conditions of temperature (32°C) and relative humidity (90–95%).

Alkaline protease production was 341 U/g wheat bran *(4)*. Using 1% (w/v) defatted soybean meal as substrate in a 20-L fermenter, submerged culturing of *Bacillus amyloliquefaciens* ATCC 23844 produced 800,000 U alkaline protease, whereas in solid culturing 250,000 U/g was produced *(3)*.

Effect of the height of the culture medium on large-scale protease production showed that a height up to 4.0 cm was suitable, and an increase to 5.5 cm led to a 20% reduction in the enzyme yield *(5)*. The optimum conditions for protease production by *R. oryzae* consisted of an accumulation density of 0.3125 g/cc and a bed height of 1 cm without aeration. Aikat and Bhattacharya compared protease production by *R. oryzae* on wheat bran in a stacked-plate fermentor (SPF) and packed-bed fermenter (PBF) incorporating the liquid culture medium recycle strategy. It has been found that in PBF the activity was four times higher than that from SPF. Rapid changes including a product washout effect occurred in PBF but not in the SPF *(6)*.

3.3. Recovery and Purification of Proteases

In submerged cultures, culture supernatants are used as enzyme source, while in SSF, the crude extract is obtained by adding 50 mL of 1% NaCl solution to the fermented matter and mixing for 20 min at room temperature (28°C). Solids are removed first by filtering through muslin cloth and then by centrifuging at 10,000g for 5 min. (*see* **Note 5**).

3.3.1. Ammonium Sulfate Fractionation

1. Place beaker of 100 mL crude enzyme extract in a cooling bath on top of a magnetic stir plate.
2. While agitating on a magnetic stirrer, slowly add 56.8 g ammonium sulfate to attain 80% saturation.
3. Continue stirring for 30 min.
4. Centrifuge at 10,000g for 10 min.
5. Decant supernatant and resuspend precipitate in minimum amount of citrate-phosphate buffer.
6. Remove ammonium sulfate by dialysis (*see* **Note 6**).

3.3.2. Gel Filtration Chromatography

1. Choose the appropriate gel matrix. They are often made of either cross-linked dextran, polyacrylamide, or agarose beads. The dextran and polyacrylamide matrices separate proteins of small to moderate molecular weights, while the agarose matrices have large pores and are able to separate much larger protein complexes.
2. Determine column dimensions.
3. Determine the experimental buffer conditions.
4. Set up the chromatography equipment.
5. Prepare the gel matrix and pack the column.
6. Run the chromatography.
7. Test the fractions for protease activity.
8. Analyze the results (*see* **Note 7**).

The crude alkaline protease extract from a mutant of *Bacillus polymyxa* was purified by 40–80% ammonium sulfate fractionation followed by application to a DEAE cellulose column. The fractions obtained were applied to a gel filtration column, Sephadex G-100, to give 9.9% recovery of 99.4-fold purified enzyme *(7)*. Purification of a novel salt-tolerant protease from *Aspergillus* sp., produced in solid-state fermentation by ultrafiltration, lyophilization, anion-exchange chromatography on DEAE–Sepharose-CL-6B, preparative isoelectic-focusing electrophoresis, and gel filtration of sephacryl-S-200HR, gave a 13.8-fold in 12.7% yield *(8)*.

3.4. Measurement of Protease Activity

1. The colorimetric assay is carried out at 45°C with 250 µL 1% w/v azocasein in 0.2 *M* citrate phosphate buffer, pH 6.5.
2. Start the reaction by the addition of 150 µL of the crude extract.
3. After incubation for 1 h, inactivate the enzyme by addition of 1.2 mL of TCA solution (10% v/v).
4. Neutralize the solution using 1 *M* NaOH solution.
5. Read OD_{440} against a blank of the inactivated crude extract (100°C for 10 min) *(9)*.

One unit of proteolytic enzyme activity is defined as the amount of enzyme that produced absorbance difference during 1 h incubation at 45°C per milliliter of solution of crude extract *(see* **Note 8**).

3.5. Effect of Different Kinetic Parameters on Protease Activity and Its Stability

3.5.1. Effect of Substrate

Different substrates can be used to determine the specificity of enzymes (1% concentration) under optimized assay conditions. Three different substrates (azocoll, azoalbumin, and azocasein) were chosen for determination of the specificity of protease from *Penicillium* sp. produced by SSF *(10)*. The results showed azocasein as the best substrate for maximum protease production, and apparent Km value (2.6 mg/mL) was determined using azocasein as a substrate at different concentrations and at different times.

3.5.2. Effect of pH

Effect of pH can be studied by diluting the crude enzyme in different buffers of varying pH. For stability experiments using buffers having a wide range of pH, the crude enzyme is incubated in initial conditions for 1 h at 37°C and the residual protease activity is determined according to standard conditions. Protease from *Penicillium* sp. was found to be quite stable at pH 6.0–8.0.

Table 2
Effect of Metallic Ions (0.1 mM) on Enzymatic Activity

Cation	Activity (U/mL)
Control	51.6
Mg^{2+}	22.7
Na^+	53.0
Ca^{2+}	55.0
Co^{2+}	37.4
Zn^{2+}	2.2

3.5.3. Effect of Temperature on Activity of Protease and Its Stability

The enzyme extract is incubated for 1 h at different temperatures, then cooled and analyzed for protease activity to determine the effect of temperature on protease activity. The thermal stability of the enzyme is assayed by incubating at different temperatures and different times. Effect of temperature on protease activity showed that at temperatures higher than 50°C, the residual activity decreased significantly. Experiments on stability of enzymes showed that at 28°C, 100% activity remained after 3 d, whereas at 40°C, 98% of the residual activity remained for 30 d.

3.5.4. Effect of Ions and Oxidizing Agents

Crude enzyme produced by *Penicillium* sp. was incubated with different ionic solutions at 0.1 mM concentration ($CaCl_2$, $MnSO_4$, $ZnSO_4$, etc.). Results presented in **Table 2** show the effect of metallic ions on enzyme activity *(10)*. To study the effect of oxidizing agents, H_2O_2 was employed in different concentrations and residual proteolytic activity was determined. The results obtained showed good stability in peroxidase solution of 5% (v/v) for 1 h of incubation at 28°C, retaining 80% of its activity. This result is very significant since commercial detergents generally contain 10% (v/v) hydrogen peroxide.

3.6. Gelatin Zymography

This method will help to determine the number of proteases present in the crude extract.

1. Apply samples (total protein 1–3 μg) to electrophoresis in zymogram resolving gel *(11)*.
2. Mix samples with Laemmli sample buffer (5X) containing 2% SDS but not reducing agent.

3. Carry out electrophoresis by standard methods at 100 V and 4°C using a vertical minigel system.
4. After electrophoresis, wash the gels twice in 2.5% Triton X-100 solution for 30 min to remove SDS (*see* **Note 9**).
5. Carry out the proteolytic reaction in zymogram development solution at 45°C for 18 h.
6. Stain the gel in staining solution. Clear zones can be observed on a blue background after several washings in the destaining solution.

Protease profiles of cell-free extracts of three archaeal strains analyzed by gelatin zymography showed that the isolate, 898-BS24, produced just a single protein band (molecular mass >150 kDa), while isolate 898-BS17 displayed five discrete activity bands (molecular masses of 38–88 kDa). Besides a single band of 51 kDa, a broad zone of clearing from the position of 134 kDa to a distance corresponding to 43 kDa was detected on the gel with extract of isolate 898-BS1, so that individual proteolytic bands would not be distinguished. This heterogeneity of the proteolytic bands observed on substrate gels may be due to autolytic degradation of one or more proteases, or to the presence of several proteases, or to the occurrence of variants generated by post-translational modifications of one or more protease species *(12)*.

3.7. Immunoblot Analysis

Immunoblot analysis is carried out to identify protease in a complex mixture following separation based on its molecular weight, size, and charge.

1. Carry out electrophoresis of protein samples by standard methods in 10% polyacrylamide gels.
2. Blot the gel onto nitrocellulose membranes with an electroblotting apparatus (16 h, 40 V, 4°C) by using transfer buffer.
3. After electrophoretic blotting, immerse the nitrocellulose membranes in blocking solution for 30 min.
4. Expose nitrocellulose membranes to diluted guinea pig serum (1:200 in dilution buffer) containing antibodies against protease for 2 h at room temperature (*see* **Note 10**).
5. Wash the membranes with an antiguinea-pig-horseradish-peroxidase complex (1:2000 dilution) for 1 h at room temperature.
6. Finally, immerse the membranes in the color developing solution *(13)*.

Cell-free extracts from *Pyrococcus furiosus* contain five proteases, two of which (S66 and S102) are resistant to SDS denaturation. Immunoblot analysis with guinea pig sera containing antibodies against protease S66 showed that S66 is related neither to S102 nor to other proteases. This result also suggests that S66 might be the hydrolysis product of a 200 kDa precursor that does not have proteolytic activity *(14)*.

4. Notes

1. It has been reported that *Bacillus licheniformis, Aspergillus oryzae,* and *Aspergillus sojae* produce very narrow zones of hydrolysis on casein agar despite being good producers.

2. An important disadvantage of using complex media is the loss of control over the precise nutrient specifications of the medium. Different carbon sources (glucose, starch, maltose, lactose, glycerol, etc.), and nitrogen sources (casein, yeast extract, peptone, tryptone, etc.) have been reported for protease production by SmF.

3. There are several biological and physiochemical factors influencing submerged fermentation: (a) Carbon and nitrogen sources: These two factors represent the energetic source that will be available for the growth of the microorganism. The carbon source could be a monosaccharide such as glucose or a complex molecule such as starch. For selecting a proper energy or carbon source there could be two points of view: (1) select the carbon source according to the microorganism to be employed and the product to obtain or (2) select the adequate microorganism in order to employ a particular substrate component. The ratio between the mass of carbon and nitrogen (C/N) is most crucial for a particular process to obtain a specified product. When dealing with fungi, it is necessary to take into account how the C/N ratio could induce or delay sporulation. (b) Temperature: Microorganisms are classified into extremothermophiles, thermophiles, mesophiles, and psychrophiles, depending on the temperature range in which the microrganism grow. During submerged fermentation, a large amount of heat is liberated, which is proportional to the metabolic activity of the microorganism. (c) pH: Each microorganism possesses a range of pH for its growth and activity with an optimum value within this range. Due to homogeneous nature of the fermentation medium in submerged fermentation, pH control is easy. (d) Aeration and agitation: Aeration and agitation have a great influence in submerged fermentation. For aerobic-type fermentation, aeration is essential. Agitation will help the homogenization and aeration of the mixture, and the biological oxygen demand of the process. It enhances the phenomenon of mass transfer and will also help in the uniform distribution of organisms. All these factors have to be considered while designing a medium for SmF.

4. Factors to be controlled in SSF are: (a) Carbon and nitrogen source: These two factors determine the growth of microorganisms. Carbon and nitrogen ratio is also crucial for obtaining a desired product. (b) Temperature: Biological processes are developed in a relatively very narrow range of temperature. It determines effects such as protein denaturation, enzymatic inhibition, promotion, or inhibition on the production of a particulate metabolite, etc. (c) Moisture and water activity: Moisture is a factor intimately related with the definition of SSF and with the characteristics of the biological material. A great majority of cells are characterized by a moisture content of 70–80%. (d) pH: Control of pH in SSF has a vital role. pH variability during SSF processes is obtained by substrate formulation considering the buffering capacity of the different components employed or by the use of a buffer formulation with components that have no deleterious influence on biological activity. (e) Particle size: Particle size is an important characteristic related to

system capacity to interchange with microbial growth and heat and mass transfer during SSF process. Reduction in particle size provides a larger surface area for microbial growth. But mycelium formation is the main cause for void fraction variation, which provokes canalization during the process. Therefore, the particle size should have an optimum range. Very tiny particles could produce matrix contraction or compaction, enhanced channeling problems, increasing mass and heat transfer problems, and the like. However, while studying SSF systems, one also faces some limitations. This includes poor knowledge of automation in SSF processes. Yet another major problem is on scale-up strategies as the system offers complications due to intense heat generation and nonhomogeneity. The thermal characteristics of the organic materials and low moisture content in SSF create especially difficult conditions for heat transfer.

5. Adsorption of the enzyme to fermented matter has been attributed to ionic bond, hydrogen bond, and van der Waal's forces. Sodium chloride (1%) has been found to be the most suitable protease extractant for *Mucor bacilliformis, Rhizopus oligosporus, Penicillium* sp., etc.

6. During ammonium sulfate precipitation, stirring must be regular and gentle. To ensure maximal precipitation, it is best to start with a protein concentration of at least 1 mg/mL. The salt can subsequently be removed from the protein preparation by dialysis.

7. Sephacryl S-200 has a fractionation range from 5 kDa to 250 kDa, while Sephacryl S-100 has a fractionation range from 1 kDa to 100 kDa, and so on. The choice of gel filtration matrix should be based on the molecular weight of the protease. Matrix flow rate can also be considered when selecting a gel filtration matrix. Faster flow rates will allow a separation to proceed more rapidly, while slower flow rates sometimes offer better resolution of peaks.

8. In azocasein, the orange sulfanilamide groups are covalently linked to peptide bonds of milk protein casein. During incubation for 1 h, proteases hydrolyze peptide bonds, liberating shorter peptides and amino acids from the chain. Trichloroacetic acid (TCA) is then added to precipitate macromolecules, including the enzyme and undigested azocasein, which will be removed by centrifugation. Short peptide chains and free amino acids liberated are not precipitated by TCA and thus remain in solution, which is orange in color. The intensity of color measured spectrophotometrically can be used to determine protease activity.

9. In addition of detecting enzyme activity, it can be used to provide information about the molecular weight of an enzyme and so help identifying the enzyme. Different strategies have been applied for SDS removal from the gel after electrophoresis. These include washing the gel in aqueous buffers in the absence or in the presence of anion exchange resins, adding organic solvents, or using competing nonionic detergents or chaotropic agents (e.g., urea or guanidine hydrochloride).

10. To prepare the antibodies, a protein band corresponding to protease was excised from polyacrylamide gels after electrophoretic separation and the gel piece was crushed and homogenized in 50 mM sodium phosphate buffer. This preparation was used to immunize the guinea pig.

References

1. Rao, M. B., Tanksale, A .P., Ghatge, M. S., and Deshpande, V. V. (1998) Molecular and biotechnological aspects of microbial proteases. *Microbiol. Mol. Biol. R.* **62**, 597–635.
2. Pandey, A., Soccol, C. R., Rodriguez-Leon, J. A., and Nigam, P. (2001) Production of enzymes by solid-state fermentation, in *Solid-State Fermentation in Biotechnology Fundamentals and Applications,* Asiatech Publishers, New Delhi, pp. 98–110.
3. George, S., Raju, V., Subramanian, T. V., and Jayaraman, K. (1997) Comparative study of protease production in solid state fermentation versus submerged fermentation. *Bioprocess Eng.* **16**, 381–382.
4. Tunga, R., Banerjee, R., and Bhattacharya, B. C. (1998) Optimizing factors affecting protease production under solid state fermentation. *Bioprocess Eng.* **19**, 187–190.
5. Chakraborty, R. and Srinivasan, M (1993) Production of a thermostable alkaline protease by a new *Pseudomonas* sp. by solid state fermentation. *J. Microb. Biotechnol.* **8**, 7–16.
6. Aikat, K. and Bhattacharya, B. C. (2001) Protease production in solid state fermentation with liquid medium recycling in a stacked plate reactor and in packed bed reactor by a local strain of *Rhizopus oryzae. Process Biochem.* **36**, 1059–1068.
7. Meenu, M., Santhosh, D., and Randhir, S. (2002) Purification and characterisation of alkaline protease from a mutant of *Bacillus polymyxa. Ind. J. Microbiol.* **42**, 155–159.
8. Su, N. W. and Lee, M. H. (2001) Purification and characterisation of a novel salt tolerant protease from *Aspergillus* sp. FC-10, a soy sauce koji mold. *J. Ind. Microbiol. Biotechnol.* **26**, 253–258.
9. Leighton, T. J., Doi, R. H., Warren, R. A. J., and Kelln, R. A. (1973) The relationship of serine protease activity to RNA polymerase modification and sporulation in Bacillus subtilis. *J. Mol. Biol.* **76**, 103–122.
10. Germano, S., Pandey, A., Osaku, C. A., Rocha, S. N., and Soccol, C. R. (2003) Characterization and stability of proteases from *Penicillium* sp. produced by solid-state fermentation. *Enzyme Microb. Tech.* **32**, 246–251.
11. Makowski, G. S. and Rampsy, M. L. (1997) Gelatin zymography, in *Protein Structure: A Practical Approach* (Creighton, T. E., ed.), Oxford University Press, New York, pp. 21–23.
12. Kocabiyik, S. and Erdem, B. (2002) Intracellular alkaline proteases produced by thermoacidophiles: detection of protease heterogeneity by gelatin zymography and polymerase chain reaction (PCR), *Bioresource Technol.* **84**, 29–33.
13. Towbin, H., Staehelin, T., and Gordon, J. (1979) Electrophoretic transfer of proteins from polyacrylamide gels to nitrocellulose sheets: procedure and some applications. *Proc. Natl. Acad. Sci. USA* **76**, 4350–4354.
14. Blumentals, I. I., Robinson, A. S., and Kelly, R. M. (1990) Characterisation of sodium dodecyl sulphate-resistant proteolytic activity in the hyperthermophilic archaebacterium *Pyrococcus furiosus. Appl. Environ. Microbiol.* **56**, 1992–1998.

11

Proteases Produced by Halophilic Bacteria and Archaea

Encarnación Mellado, Cristina Sánchez-Porro, and Antonio Ventosa

Summary

Proteases from halophilic microorganisms present the advantage of being stable at high salinities, constituting interesting enzymes from a biotechnological point of view. To maintain osmolarity in saline environments the microorganisms adopt mainly two strategies, one followed by most moderately halophilic bacteria, accumulating organic compatible solutes in the cytoplasm, and the second followed by the halobacteria (extremely halophilic aerobic archaea), accumulating inorganic salts in the cytoplasm. In this chapter, we describe the methods for the production of proteases by a representative organism from each of the two main groups inhabiting the saline habitats: the moderately halophilic bacterium *Pseudoalteromonas* sp. CP76 and the archaeon *Natrialba magadii*. The production process involves (1) the culture of the microorganisms under optimal conditions for the production of the extracellular proteases, and (2) the recovery and purification of the enzymes from the culture supernatant.

Key Words: Halophiles; extremophiles; bacteria; archaea; extracellular enzymes; proteases.

1. Introduction

Hypersaline environments are typical extreme habitats that are characterized by their high salt concentration *(1)*. Halophilic microorganisms, which require salt for growth, possess haloadaptation mechanisms to grow and survive in such saline habitats. The most widely distributed halophilic organisms in these environments are the extremely halophilic microorganisms, which are able to grow optimally in media containing from 15% to 25% NaCl, and moderately halophilic microorganisms, growing optimally in media containing between 3% and 15% NaCl *(2)*. Extreme halophiles are well-represented among the *Halobacteriaceae,* although some extremely halophilic bacteria have been described *(3)*. Moderately halophilic microorganisms are phylogenetically very diverse, including a great variety of microorganisms: some methanogenic archaea and a large number of Gram-positive and Gram-negative bacteria *(4)*.

From: *Methods in Biotechnology, Vol. 17: Microbial Enzymes and Biotransformations*
Edited by: J. L. Barredo © Humana Press Inc., Totowa, NJ

Microorganisms inhabiting hypersaline environments are expected to have specific proteins presenting features distinguished from those proteins from organisms inhabiting nonsaline environments. These halophilic enzymes have to maintain their catalytic properties at high salt concentrations.

Proteases constitute a group of enzymes of important applications in many industrial processes *(5)*. Several extracellular proteases have been reported from halobacteria *(6–8)*. These enzymes are characterized by their dependency on high salt concentration for activity and stability. Among them, a serine protease from the haloalkaliphile *Natrialba magadii* has been deeply characterized *(9)*. On the other hand, the production of extracellular enzymes by moderately halophilic bacteria has not been so extensively investigated. However, in recent years, efforts have been made for exploring the potential of this versatile group of halophilic bacteria. In this sense, a number of extracellular enzymes produced by moderately halophilic bacteria, proteases, amylases, DNases, pullulanases, and lipases have been reported *(10)*. The only protease deeply characterized from this group of halophilic bacteria is the protease CP1 produced by *Pseudoalteromonas* sp. CP76 *(11)*.

Moderately halophilic bacteria show important advantages for being used as a source of halophilic enzymes, such as the few nutritional requirements, the ability to grow in a wide range of salinities and being very easy to culture (most of them being able to use a great variety of compounds as the sole source of carbon and energy). These characteristics exhibited by moderately halophilic bacteria make them a group with great potential biotechnological applications *(4)*.

The optimization of the production process of these halophilic enzymes has allowed the recovery of enough amount of the proteins for a further characterization and potential biotechnological use. The interest in these halophilic enzymes lies in their possibility to be used in industrial processes requiring optimal activities over a very wide range of salinities, from almost no saline conditions to high salt concentrations *(8)*. The description of the production process of two proteases produced by selected members of moderately and extremely halophilic microorganisms will help in the description of specific methods for the production of halophilic enzymes in order to be used in biotechnological systems.

2. Materials

1. *Pseudoalteromonas* sp. strain CP76 (CECT 5782) (CECT, Valencia, Spain).
2. *Natrialba magadii* (formely *Natronobacterium magadii*) (ATCC 43099) (ATCC, Manassas, VA).
3. SW-7.5 medium: 60.7 g/L NaCl, 5.2 g/L $MgCl_2$, 7.2 g/L $MgSO_4$, 0.27 g/L $CaCl_2$, 1.5 g/L KCl, 0.045 g/L $NaHCO_3$, 0.0019 g/L NaBr, and 5 g/L yeast extract.

4. Yeast extract (Difco Laboratories, Detroit, MI).
5. Carbohydrates: lactose, sucrose, glycerol, maltose, mannitol, fructose, and glucose.
6. Casamino acids (Difco).
7. Saline medium SM1: 200 g/L NaCl, 18.5 g/L Na_2CO_3, 20 g/L yeast extract, 3 g/L sodium citrate, 2 g/L KCl, 1 g/L $MgSO_4 \cdot 7H_2O$, 3.6×10^{-4} g/L $MnCl_2 \cdot 4H_2O$, and 2×10^{-5} g/L $FeSO_4 \cdot 7H_2O$.
8. Buffer A: 20 mM Tris-HCl, pH 8.5.
9. Buffer B: 20 mM Tris-HCl, pH 8.5, and 1 M NaCl.
10. Buffer C: 20 mM Tris-HCl, pH 8.5, and 0.1 M NaCl.
11. Buffer D: 50 mM Tris-HCl, pH 8.5.
12. Buffer 1: 50 mM Tris-HCl, pH 8.0, and 4.5 M NaCl.
13. Buffer 2: 50 mM Tris-HCl, pH 8.0, 4.5 M NaCl, and 15% (v/v) ethanol.
14. Buffer 3: 50 mM Tris-HCl, pH 8.0, and 3 M NaCl.
15. SDS-PAGE (sodium dodecyl sulfate-polyacrylamide gel electrophoresis) equipment.
16. Coomassie blue reagent: 0.01% Coomassie Brilliant Blue G-250 in 4.7% ethanol and 85% (v/v) phosphoric acid.
17. Chromatography equipment.
18. Q-Sepharose column (2.5 × 20 cm) (Amersham, Freiburg, Germany).
19. Bacitracin-Sepharose 4B column (Sigma, St. Louis, MO).
20. Superdex S-200 prep column (1.5 × 96 cm) (Amersham).
21. FPLC equipment (Amersham).
22. Sephacryl S-200 gel filtration column (Amersham).
23. Gelatin-PAGE.
24. Staining solution: 0.6% amido black in water-ethanol-acetic acid (60:30:10).
25. Ultrafiltration unit (cut-off 10,000 Da) (Amicon, Bedford, MA).
26. Dialysis buffer: 0.1 M sodium borate buffer, pH 8.0, and 4 M betaine.

3. Methods

The methods described below outline the production of two extracellular proteases by halophilic microorganisms: (1) the protease CP1 by the moderately halophilic bacterium *Pseudoalteromonas* sp. CP76, and (2) a protease secreted by the extremely halophilic archaeon *Natrialba magadii*.

Among other processes, microbial proteases are used in the leather industry for selective hydrolysis in the soaking, dehairing, and bating stages of leather processing. The use of halophilic proteases functionally active in different salt concentrations and in alkaline conditions would improve this process.

3.1. Protease CP1

Protease CP1, an extracellular enzyme showing proteolytic activity, produced by the moderately halophilic bacterium *Pseudoalteromonas* sp. CP76, constitutes an interesting halophilic enzyme for application in different biotechnological processes used in the detergent, pharmaceutical, food, and feed industries. This protease not only presents optimal activity at a wide range of salinities (from 0 to 4 M NaCl), but it also shows an excellent thermostability (optimal activity at 55°C) and activity at alkaline pH values (optimal activity at pH 8.5) *(11)*.

3.1.1. Culture of Pseudoalteromonas sp. CP76 for Optimal Production of Protease CP1

To obtain the highest extracellular protease activity the following method has been developed for the culture of *Pseudoalteromonas* sp. CP76:

1. Seed a Petri dish with complex medium SW-7.5 (7.5% total salts final concentration) *(12)*.
2. Inoculate the complex saline liquid medium SW-7.5 with a single colony.
3. Incubate the culture at 37°C (the highest proteolytic activity was measured at this temperature).
4. Collect the cells during the stationary growth phase (24 h) in order to reach the highest production of CP1.

The effect of culture conditions on protease production was assayed by growing *Pseudoalteromonas* sp. CP76 in SW-7.5 medium to which different carbohydrates as lactose, sucrose, glycerol, maltose, mannitol, fructose, glucose (50 mM), NH$_4$Cl (50 mM), or Casamino acids (1%) were added *(11)*. The results of this assay showed that the protease CP1 produced by *Pseudoalteromonas* sp. CP76 is strongly influenced by the composition of the culture medium. To obtain an optimal production of CP1 the saline medium is supplemented with sucrose, fructose, and glycerol.

3.1.2. Protein Purification

Protease CP1 is purified to homogeneity from the concentrated culture supernatant of *Pseudoalteromonas* sp. CP76 by anion-exchange chromatography followed by gel filtration procedure:

1. After cultivation of *Pseudoalteromonas* sp. CP76 as described above (**Subheading 3.1.1.**), centrifuge the culture at 12,000g for 30 min at 4°C.
2. Dialyze the cell-free supernatant (1 L) against buffer A (*see* **Note 1**).

3.1.2.1. ANION-EXCHANGE CHROMATOGRAPHY

1. Apply the supernatant onto a Q-Sepharose column (2.5 × 20 cm) equilibrated in buffer A until no absorbance at 280 nm is detectable.
2. Wash the column with 100 mL buffer A.
3. Elute proteins with buffer B.
4. Pool the active fractions and concentrate by ultrafiltration.
5. Dialyze overnight against 100 volumes of buffer C.

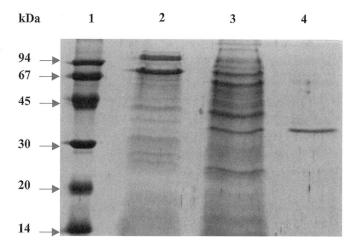

Fig. 1. SDS-PAGE analysis of protease CP1-enriched fractions obtained during the different purification steps. Lane 1: molecular mass markers; lane 2: dialyzed culture supernatant preparation; lane 3: Q-Sepharose column chromatography; lane 4: purified protease after Superdex S-200 column. Proteins were detected with Coomassie blue (0.1%) (reprinted from ref. *11*, with permission).

3.1.2.2. Gel Filtration

1. Apply the concentrated sample (about 2 mL) to a fast protein liquid chromatography Superdex S-200 prep column (1.5 × 96 cm) connected to a FPLC system equilibrated in buffer C.
2. Elute the protease with buffer C. The collection of 1-mL fractions will be performed at a flow rate of 1 mL/min.
3. Analyze the different fractions containing enzyme by 12% SDS-PAGE. This purification procedure is an effective method for the recovery of protease activity, yielding a pure homogeneous protein as evidenced by detection of a single band (**Fig. 1**).

3.1.3. Characterization

The purified protease CP1 has to be characterized in order to determine that the enzyme conserves its optimal activity at temperature and pH values, as well as its tolerance to different concentrations of NaCl. The purified protease can be assayed for proteolytic activity using casein or azocasein as substrates (*see* **Note 2**).

3.1.3.1. Proteolytic Activity with Casein

A modification of the method by Kunitz *(13)* is described in order to determine the proteolytic activity of protease CP1:

1. Prepare samples (duplicates) containing 400 µL of 0.5% casein (w/v) in buffer D and incubate in a water bath with 100-µL enzyme sample at 55°C and pH 8.5 for 30–60 min.
2. Stop the enzyme reaction by the addition of 500 µL of 10% trichloroacetic acid (w/v) and keep at room temperature for another 10 min.
3. Centrifuge the reaction mixture at 3000g.
4. Measure OD_{280} of the supernatant against a blank (nonincubated) sample.

One unit of protease is defined as the amount of enzyme yielding the equivalent of 1 µmol of tyrosine per minute under the defined assay conditions.

Protein concentration was determined according to Bradford *(14)*:

1. Prepare a solution of bovine serum albumin (BSA) of 100 µg in 0.1 mL 1 M NaOH.
2. Create a series of dilutions (10–100 µg in 0.1 mL) in 1 M NaOH.
3. Add 5 mL Coomassie blue protein reagent to 0.1 mL of the standard prepared dilutions and mix.
4. Measure OD_{595} after 2 min and before 1 h. A reagent blank prepared from 0.1 mL 1 M NaOH and 5 mL protein reagent was used.
5. Construct a standard curve of µg protein vs OD_{595}.
6. Dissolve the samples (1–100 µL) in 1 M NaOH and treat as above indicated to calculate the protein concentration.

3.1.3.2. PROTEOLYTIC ACTIVITY WITH AZOCASEIN

1. The reaction mixture contains 0.5% (w/v) azocasein, 0.1 M sodium borate buffer, pH 8.0, 1.5 M NaCl, and the solution of the enzyme.
2. Incubate the mixture at 45°C and stop by adding 1 volume of cold 10% (v/v) trichloroacetic acid.
3. Cool for 15 min on ice and centrifuge at 3000g and 4°C for 15 min.
4. Measure OD_{335} in order to detect the acid-soluble products in the supernatant.

One unit of enzyme activity is defined as the amount that yields an increase in OD_{335} of 1 in 1 h at 45°C.

3.1.3.3. ZYMOGRAM

1. Load the samples in a 12.5% SDS-PAGE gel containing 0.1% (w/v) copolymerized gelatin (gelatin-PAGE).
2. After electrophoresis, rinse the gel in 0.25% (v/v) Triton X-100 at 4°C for 1 h to remove SDS.
3. Incubate the gel 10–30 min under optimal assay conditions to detect the proteolytic activity (*see* **Subheading 3.1.3.4.**).
4. Stain the gel with staining solution for 1 h.
5. Distain the gel in water-ethanol-acetic acid (60:30:10).

Proteolytic activity is visualized in the gel by clearing zones resulting from gelatin hydrolysis (**Fig. 2**).

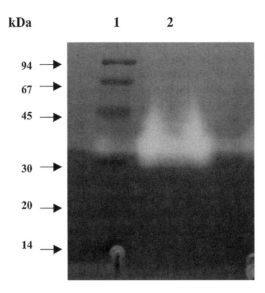

Fig. 2. Gelatin zymogram of the purified protease. Lane 1: molecular mass markers; lane 2: purified protease after Superdex S-200 column (from *ref.* **11**, with permission).

3.1.3.4. Optimal Conditions Recommended for Use of Protease CP1

Protease CP1 is active between 20°C and 75°C, but optimum activity is found at 55°C. The pH range exhibited for the enzyme is 6.0 to 10.0, with an optimum at pH 8.5. The maximal activity of protease CP1 is detected in the range 0 to 1 M NaCl, but at higher salinities (up to 4 M NaCl) activity is still detected. Addition of 0.8 M urea to the protease incubation mixture slightly increases the activity of the enzyme (*see* **Note 3**).

3.2. Protease from Natrialba magadii

This extracellular enzyme has a native molecular mass of 130 KDa, and is very dependent on high salt concentrations for activity and stability (maximal activity at 1–1.5 M NaCl or KCl), presenting optimum activity at 60°C in 1.5 M NaCl and pH 8.0–10.0 *(9)*.

3.2.1. Culture of Natrialba magadii for Optimal Production of Extracellular Protease

To obtain the highest extracellular protease activity the following method has been developed for the culture of *Natrialba magadii:*

1. Seed a Petri dish with complex medium SM1.
2. Inoculate the complex liquid medium SM1.

3. Incubate in darkness on rotary shaker at 150 rpm at 37°C.
4. Collect the cells during the stationary growth phase (80 h).

3.2.2. Protein Purification

After cultivation of *Natrialba magadii* as described earlier (**Subheading 3.2.1.**), the protease is purified to homogeneity by ethanol precipitation, affinity chromatography, and gel filtration. The purification process is performed at room temperature.

3.2.2.1. ETHANOL PRECIPITATION

1. Add 1 volume of cold absolute ethanol to the culture medium of *N. magadii* and maintain during 1 h on ice to allow the protein precipitation.
2. Pellet the precipitated material at 5000*g* for 10 min.
3. Resuspend the precipitated material in 1/10 volume of buffer 1.
4. Centrifuge again at 5000*g* for 10 min to eliminate insoluble material.

3.2.2.2. AFFINITY CHROMATOGRAPHY

1. Equilibrate a bacitracin-Sepharose 4B column (25 cm^3) in buffer 1.
2. Load the ethanol-precipitated protein onto the equilibrated column (*see* **Note 4**).
3. Elute the protease with buffer 2.

3.2.2.3. GEL FILTRATION

1. Concentrate the fractions displaying proteolytic activity by ultrafiltration (cut-off 10,000 Da).
2. Equilibrate a Sephacryl S-200 gel filtration column (74 × 1.6 cm) with buffer 3.
3. Apply the fractions onto the equilibrated column. Column flux 0.5 mL/min.
4. Elute the enzyme with the same buffer used to equilibrate the column.
5. Evaluate the purified fractions by 10% SDS-PAGE in presence of 4 *M* betaine (*see* **Note 5**).

3.2.3. Protease Characterization

The optimal conditions established by Giménez et al. (*9*) for protease activity are 60°C (measured in 3 *M* KCl and pH 8.0) (*see* **Note 6**), 1–1.5 *M* NaCl or KCl, and a pH range from 8.0 to 10.0. All these assays are performed determining the azocaseinolytic activity (*see* **Note 2**).

4. Notes

1. To optimize the purification process of CP1 the supernatant must be dialyzed prior to the two purification steps in order to remove all extra salt contained in the culture growth medium. This chapter describes a dialysis method effective for the purification of protease CP1.
2. Two methods could be used for the detection of the proteolytic activity, although azocasein is the method standardized for the protease of the archaeon *N. magadii*.

3. Studies on the influence of different inhibitors on the activity of protease CP1 revealed that the enzyme activity is not enhanced with the addition of different ions (Zn^{2+}, Ca^{2+}, Mg^{2+}, or Mn^{2+}) to the reaction mixture as was expected due to the results obtained using EDTA as chelating agent.
4. The contaminating proteins are removed with the same buffer used to equilibrate the bacitracin-Sepharose column (Tris-HCl, pH 8.0) but containing 5 M NaCl.
5. Concentrations of 4 M betaine prevent autolysis of halophilic proteases under conditions of low ionic strength. Before loading the gel, the samples are dialyzed against dialysis buffer.
6. It is important to note that the highest proteolytic activity is established at 60°C, but the enzyme is not stable for more than 1 h at this temperature. In this sense, the optimal activity concerning NaCl concentration and pH are determined at 45°C.

Acknowledgments

This work was supported by grants from the Spanish Ministerio de Ciencia y Tecnología (BMC2003-1344 and REN2003-01650), the Quality of Life and Management of Living Resources Programme of the European Commission (QLK3-CT-2002-01972) and Junta de Andalucía. The authors thank María José Bonete for the critical reading of the manuscript.

References

1. Oren, A. (2002) Diversity of halophilic microorganisms: environments, phylogeny, physiology, and applications. *J. Ind. Microbiol. Biotechnol.* **28,** 56–63.
2. Grant, W. D., Gemmell, R. T., and McGenity, T. J. (1998) Halophiles, in *Extremophiles: Microbial Life in Extreme Environments* (Horikoshi, K., Grant, W. D., eds.), Wiley-Liss, New York, pp. 93–132.
3. Anton, J., Oren, A., Benlloch, S., Rodriguez-Valera, F., Amann, R., and Rossello-Mora, R. (2002) *Salinibacter ruber* gen. nov., sp. nov., a novel, extremely halophilic member of the Bacteria from saltern crystallizer ponds. *Int. J. Syst. Evol. Microbiol.* **52,** 485–491.
4. Ventosa, A., Nieto, J. J., and Oren, A. (1998) Biology of moderately halophilic aerobic bacteria. *Microbiol. Mol. Biol. Rev.* **62,** 504–544.
5. Rao, M. B., Tanksale, A. M., Ghatge, M. S., and Deshpande, V. V. (1998) Molecular and biotechnological aspects of microbial proteases. *Microbiol. Mol. Biol. Rev.* **62,** 597–635.
6. Oren, A. (1994) Enzyme diversity in halophilic archaea. *Microbiologia* **10,** 217–228.
7. Margesin, R. and Schinner, F. (2001) Potential of halotolerant and halophilic microorganisms for biotechnology. *Extremophiles* **5,** 73–83.
8. Mellado, E. and Ventosa, A. (2003) Biotechnological potential of moderately and extremely halophilic microorganisms, in *Microorganisms for Health Care, Food and Enzyme Production* (Barredo, J. L., ed.), Research Signpost, Kerala, India, pp. 233–256.
9. Giménez, M. I., Studdert, C. A., Sánchez, J. J., and De Castro, R. E. (2000) Extracellular protease of *Natrialba magadii:* purification and biochemical characterization. *Extremophiles* **4,** 181–188.

10. Mellado, E., Sánchez-Porro, C., Martín, S., and Ventosa, A. (2004) Extracellular hydrolytic enzymes produced by moderately halophilic bacteria, in *Halophilic Microorganisms* (Ventosa, A., ed.), Springer-Verlag, Berlin, pp. 285–295.

11. Sánchez-Porro, C., Mellado, E., Bertoldo, C., Antranikian, G., and Ventosa, A. (2003) Screening and characterization of the protease CP1 produced by the moderately halophilic bacterium *Pseudoalteromonas* sp. strain CP76. *Extremophiles* **7,** 221–228.

12. Ventosa, A., Quesada, E., Rodríguez-Valera, F., Ruiz-Berraquero, F., and Ramos-Cormenzana, A. (1982) Numerical taxonomy of moderately halophilic Gram-negative rods. *J. Gen. Microbiol.* **128,** 1959–1968.

13. Kunitz, M. (1947) Crystalline soybean trypsin inhibitor. II. General properties. *J. Gen. Physiol.* **30,** 291–310.

14. Bradford, M. M. (1976) A rapid and sensitive method for quantitation of microgram quantities of protein utilizing the principle of protein-dye binding. *Anal. Biochem.* **72,** 248–254.

12

Microbial Pectinases

Rupinder Tewari, Ram P. Tewari, and Gurinder S. Hoondal

Summary

The biotechnological potential of pectinolytic enzymes from microorganisms has drawn a great deal of attention for use as biocatalysts in a variety of industrial processes. The role of acidic pectinases in extraction and clarification of fruit juices is well established. Recently, these have emerged as important industrial enzymes with wide-ranging applications in textile processing, degumming of plant bast fibers, pretreatment of pectic wastewater, papermaking, and coffee and tea fermentation. This chapter describes key features for screening, production (liquid and solid-state fermentation), purification, characterization, applications, and gene cloning of pectinases.

Key Words: Pectinases; *Bacillus;* enrichment; production; purification; characterization; gene cloning.

1. Introduction

Pectic substances are a heterogenous group of high-molecular-weight complex, acidic structural polysaccharides *(1)* that consists largely of D-galactopyranosyluronic acids that are α(1→4) glycosidically linked to polygalacturonic acid with a small amount of L-rhamnose (β 1→2 linked) and various side chains consisting of L-arabinose, D-galactose, and β-D-xylose *(2)*. The enzymes that hydrolyze pectic substances are broadly classified into two categories: depolymerizing enzymes and saponifying enzymes or pectin esterases. Depolymerizing enzymes are further classified based on the following criteria: hydrolytic or transeliminative cleavage of glycosidic bonds; endo- or exo-mechanism of the splitting reaction, i.e., whether breakdown starts from within the polymer or from the end of the polymer; and preference for pectic acid or pectin as substrate. Pectin esterases catalyze de-esterification of the methoxyl group of pectin, forming pectic acid, whereas pectin lyase catalyzes breakdown of pectin by β-elimination cleavage, resulting in galacturonic acid with an unsaturated bond between C4 and C5 at the nonreducing end. Polygalacturonases catalyze the hydrolysis

From: *Methods in Biotechnology, Vol. 17: Microbial Enzymes and Biotransformations*
Edited by: J. L. Barredo © Humana Press Inc., Totowa, NJ

of α-1,4-glycosidic linkages in pectic acid, whereas pectate lyase catalyzes cleavage of α-1,4-glycosidic linkages in pectic acid by transelimination reaction. These have been classified as acidic, neutral, and alkaline pectinases based on their optimal activity. The substrates for acidic and alkaline pectinases are esterified pectin and pectic acid, respectively. For neutral pectinases either pectin or pectic acid can serve as substrate.

Alkaline and thermotolerant pectinases, mainly produced by *Actinomycetes* and *Bacillus (4–6)*, have great importance in industrial processes such as degumming of fiber crops *(7–9)* and pretreatment of pectic wastewater from fruit and vegetable processing units *(10,11)*. Acidic pectinases, used mostly in fruit juices and winemaking industries, are derived from fungal sources, especially *Aspergillus* sp. *(3)*. There are a number of reports available in the literature regarding screening and optimization of cultural conditions favoring maximal production of alkaline and thermotolerant pectinases using solid-state fermentation (SSF) and submerged fermentation (SmF) techniques *(4–6,12–14)* from *Bacillus* sp. These pectinases have been successfully purified to homogeneity *(4,6,14)*. In this chapter, methods for screening, production, purification, characterization, and applications of alkalophilic and thermotolerant pectinase will be illustrated using *Bacillus* sp. DT7 *(4)* as reference. Characterization and application of acidic pectinases will also be discussed.

2. Materials

1. Samples for the isolation of microorganisms producing pectinases (*see* **Note 1**).
2. Solid substrates: wheat bran, rice bran, and apple pomace can be obtained from flour mills, cereal-shellers, and apple-processing industrial units.
3. Plant fibers: The bast fibers of ramie *(Boehmeria nivea)*, buel *(Grewia optiva)*, and sun hemp *(Cannabis sativa)* can be procured from places where these are cultivated.
4. Pectic wastewater from fruit- and vegetable-processing units.
5. Yeast extract pectin (YEP) medium: 1% yeast extract and 0.25% pectin, pH 7.2, adjusted with 0.1 *N* NaOH.
6. Cetrimide solution: 0.1% (w/v) aqueous solution.
7. Pectin (Sigma, St Louis, MO).
8. Polygalacturonic acid (PGA) (Sigma).
9. Galacturonic acid (GA) (Sigma).
10. Sephadex G-25 (Pharmacia, Uppsala, Sweden).
11. Sephadex G-150 (Pharmacia).
12. DEAE Sephacel (Pharmacia).
13. Dinitrosalicylic acid (Sigma)—DNSA reagent: 2 g/L NaOH, 40 g/L sodium potassium tartarate, 0.1 g/L sodium sulfite, 0.4 g/L phenol, and 2.0 g/L DNSA.
14. Neurobion multivitamin solution (Merck, Darmstadt, Germany).
15. Luria-Bertani (LB) broth: 1% tryptone, 0.5% NaCl, and 0.5% yeast extract, pH 7.2, adjusted with 0.1 *N* NaOH.

16. Luria Agar (LA): 1% tryptone, 0.5% NaCl, 0.5% yeast extract, and 1.5% agar, pH 7.2, adjusted with 0.1 *N* NaOH.
17. TE buffer: 10 m*M* Tris-HCl, pH 8.0, and 1 m*M* EDTA, pH 8.0.
18. CTAB/NaCl solution: 10% CTAB (cetyltrimethylammonium bromide), and 0.7 *M* NaCl.
19. Phenol:chloroform:isoamyl alcohol: 24 mL phenol, 24 mL chloroform, and 1 mL isoamyl alcohol.
20. Restriction enzymes and T4 ligase.
21. Plasmid pUC18 *Bam*HI dephosphorylated (Stratagene, La Jolla, CA).
22. 5X loading dye: Bromophenol blue 0.01% (w/v).
23. Ethidium bromide solution: 1 μg/mL ethidium bromide in distilled water.
24. *E. coli* competent cells.
25. 0.01 *M* Tris-HCl, pH 8.0.
26. 0.01 *M* Tris-HCl, pH 7.5.
27. Reagent A: 1% (w/v) $CuSO_4 \cdot 5H_2O$.
28. Reagent B: 2% (w/v) sodium potassium tartrate.
29. Reagent C: 2% (w/v) anhydrous sodium carbonate in 0.1 *N* NaOH.
30. Working solution: 1 mL reagent A, 1 mL reagent B, and 98 mL reagent C.
31. Folin Ciocalteu's reagent (Sigma).
32. Polyacrylamide gel electrophoresis (PAGE) equipment.
33. Amicon fliter assembly 10-kDa cut-off membrane.
34. Chromatography columns.
35. Refrigerated centrifuge.
36. Spectrophotometer.
37. Scanning electron microscope, model JSM 6100 SM JEOL (JEOL, Tokyo, Japan).
38. Ion sputter, model JFC-1100 (JEOL).
39. Water bath.
40. Incubation chamber.

3. Methods

The methods described below outline (1) the screening of alkalophilic, thermotolerant pectinase-producing *Bacillus* sp., (2) scale-up of enzyme production (liquid-batch fermentation and solid-state fermentation), (3) purification of the enzyme, (4) pectinase characterization, (5) applications of the pectinase, and (6) gene cloning of alkaline pectinase.

3.1. Enrichment of Pectinase-Producing Microorganisms (see Note 2)

1. Collect soil samples (approx 25–50 g) in sterile plastic bags (or Petri dishes) using autoclaved spoons wrapped in aluminum foils.
2. Store the samples in a cold room (4°C) until they are processed (*see* **Note 3**).
3. Transfer nearly 5 g of sample, under aseptic conditions, into 50 mL of sterilized YEP medium contained in 500-mL conical flask.
4. Incubate the flasks on a rotary shaker (100 rpm) at 37°C for about 24 h.
5. Aseptically take out 2 mL of the liquid culture and transfer to a 500-mL YEP flask.

6. Incubate the flask as described in **step 4.**
7. Repeat the enrichment process two more times.

3.2. Screening for Pectinase-Producing Bacteria

1. Streak the enriched samples onto duplicate YEP agar plates.
2. Incubate overnight at 37°C.
3. Pectinase-producing bacterial colonies are detected by flooding the plates with cetrimide solution. A clear halo zone around the colonies against a white background (of the medium) indicates the ability of an isolate to produce pectinase.

3.3. Colorimetric Assay of Pectinase Activity

1. Grow the isolated strain in YEP medium (18 h, 37°C, 200 rpm).
2. Centrifuge the culture at 10,000g for 15 min at 4°C.
3. Analyze the supernatant for enzyme activity and protein content.
4. Incubate suitably diluted cell-free supernatant (100 µL) with 100 µL of substrate (PGA, 1.0% w/v) at 60°C for 5 min.
5. Add 500 µL of DNSA reagent as modified by Aguillar and Huitron *(16)* and cover the tops of the tubes with glass marbles (to prevent evaporation).
6. Place the tubes in a boiling water bath for 15 min.
7. Adjust the volume of the solutions to 5.0 mL with deionized distilled water.
8. Measure spectrophotometrically the intensity of the brownish color of the solutions developed (OD$_{530}$) (*see* **Note 4**).
9. Estimate the concentration of reducing groups from the standard curve obtained from the solution of known concentrations (10–1000 µg/mL) of D-galacturonic acid (GA).

One unit (U) of enzyme is defined as the amount of enzyme that catalyzes the formation of 1.0 µmol of galacturonic acid per minute under the assay conditions.

3.4. Optimum pH for Activity of Pectinase

1. Assay pectinase activity in the cell-free supernatant (as described in **Subheading 3.3**) in a range of pH values (pH 5.0–10.0) by preparing substrate (PGA, 1.0%, w/v) in 0.01 M of different buffers, such as citrate phosphate buffer, pH 5.0–6.5; phosphate buffer, pH 6.0–7.9; Tris-HCl buffer, pH 7.5–8.5; and glycine NaOH buffer, pH 9.0–10.0.
2. Mix suitably diluted enzyme (100 µL) with equal volume of PGA (1%, w/v) and keep at 37°C to determine optimum pH for its activity.

3.5. Optimum Temperature for Activity of Pectinase

Assay pectinase activity (*see* **Subheading 3.3**) in the cell-free supernatant at different temperatures (30–90°C) at optimized pH by incubating suitably diluted enzyme sample (100 µL) for 5 min with PGA (1%, w/v).

3.6. Culture Identification

The morphology and general physiological features of the isolated strains are examined per *Bergey's Manual of Systematic Bacteriology (15)*. The standard identification scheme is as follows.

3.6.1. Growth Characteristics

Inoculate YEP broth with actively growing culture of isolates followed by incubation at 37°C under stationary conditions. The following observations should be made every day for 7 days: (1) Formation of pellicle at the surface or mucoid formation or sedimentary growth. (2) Development of any characteristic smell. (3) Cells are capsulated or noncapsulated.

3.6.2. Physiological and Biochemical Characteristics

Physiological characteristics of isolates are studied with respect to Gram staining, shape, size, cell diameter of the bacteria, presence of spores, and motility of the isolate. All the characteristics are checked by standard methods *(15)*. Biochemical characteristics can be studied with respect to production of acid/gas from carbohydrates, utilization of organic compounds, anaerobic growth, growth at different hydrogen ion concentration, growth at different temperature, catalase production, oxidase production, nitrate reduction, indole production, Methyl red (MR) and Voges-Proskauer (VP) test, and hydrolysis of urea per *Bergey's Manual of Systematic Bacteriology (15)*.

3.7. Storage of Culture

1. Maintain the running cultures on YEP slants and store at 4°C.
2. For long-term storage a thick suspension of actively growing cells in 10% glycerol can be maintained at –20°C or at –80°C. The cultures can also be lyophilized for very long periods of preservation as well as transportation.

3.8. Optimization of Growth Medium for Maximal Pectinase Production

The methods describe below outline optimization of (1) nutrient media, (2) yeast extract and pectin, (3) initial pH of the media, (4) incubation temperature, (5) agitation, (6) inoculum size, (7) salts, (8) carbon and nitrogen sources, and (9) vitamins in submerged fermentation (SmF) and solid-state fermentation (SSF).

3.8.1. Inoculum Preparation

Inoculate YEP medium (25 mL/250 mL flask) with pectinase positive culture and incubate on rotary shaker (100 rpm) at 37°C overnight. This seed culture is inoculated into YEP medium at 1% inoculum concentration.

3.8.2. Enzyme Production Using SSF

1. Prepare solid substrate (wheat bran, rice bran, etc.) by moistening 5.0 g of it in 250 mL conical flasks with variable moisture contents (40–80%) *(12)* (*see* **Note 5**).
2. Inoculate these flasks with 2.0 mL of overnight-grown culture.
3. Mix the flasks well so that the inoculum is evenly distributed.
4. Incubate the flasks at 37°C for 48 h.
5. Under aseptic conditions, periodically withdraw about 0.5 g solid substrate culture from the flasks and monitor its pectinase production.

3.8.3. Extraction of Pectinase

1. Extract the enzyme pectinase from solid substrate by adding 25 mL of 0.01 *M* Tris-HCl, pH 8.0, to each flask and shake the flasks thoroughly so as a slurry is formed.
2. Keep the flasks at 4°C for 30 min under static conditions to facilitate the enzyme extraction.
3. Centrifuge the slurry at 8,000*g* for 15 min at 4°C, and collect the clear supernatant.
4. Again add 25 mL of 0.01 *M* Tris-HCl, pH 8.0, to the centrifuged solid residues to extract the remaining enzyme.
5. Pool both enzyme-containing supernatants and assay pectinase activity **(Subheading 3.3)**.
6. Determine pectinase activity in the supernatant as U/g of solid substrate used.

3.8.4. Nutrient Media

Various nutrient media can be tested for pectinase production. Some of them are the following: Yeast extract (YE), Luria-Bertani broth (LB), Nutrient broth (NB), brain heart infusion (BHI), corn steep liquor (CSL, 1% w/v), soybean meal (1.0% w/v), and minimal media (M-9) lacking glucose (*see* **Note 6**).

1. Incubate the inoculated flasks (50 mL medium in 500-mL conical flasks) at 37°C (100 rpm, 48 h).
2. Withdraw 2 mL of culture at different time intervals under aseptic conditions.
3. Centrifuge the culture at 10,000*g* for 10 min.
4. Assay for pectinase activity by the method described in **Subheading 3.3.** *(16)*.

3.8.5. YE and Pectin

Use different concentrations of YE (0.25–4.0%, w/v) and pectin (0–0.5%, w/v) in the YEP medium (50 mL medium in 500-mL conical flasks) used for pectinase production.

3.8.6. Initial pH of Medium

Adjust initial pH of 50 mL YEP medium in 500-mL conical flasks used for pectinase production to various pHs (5.0–9.0) using 0.1 *N* NaOH or 1 *N* Na_2CO_3.

3.8.7. Incubation Temperature

The effect of incubation temperature on enzyme production can be studied by incubating the YEP medium inoculated with test bacterial culture at different temperatures (25–50°C, 100 rpm, 48 h)

3.8.8. Agitation

Incubate 50 mL of inoculated YEP medium at different speeds (static–250 rpm, 37°C) of the incubating shaker.

3.8.9. Inoculum Size

Inoculate 50 mL of YEP medium in 500-mL conical flasks with various inoculum sizes (0.5–5.0%, v/v) (*see* **Note 7**) followed by incubation at 37°C for 48h.

3.8.10. Salts

Supplement different metal ions (0.05–3.0 m*M*) such as: $CaCl_2 \cdot 2H_2O$, $MgSO_4 \cdot 7H_2O$, $CuCl_2 \cdot 2H_2O$, $CoCl_2 \cdot 2H_2O$, $MnSO_4 \cdot 4H_2O$, H_3BO_3, $ZnCl_2$, Fe (III) citrate, $Na_2MoO_4 \cdot 2H_2O$, $FeSO_4$, KCl, and NaCl in YEP medium used for pectinase production by SmF. To study the effect of these salts on pectinase production by SSF, incorporate these salts in the water used for adjusting moisture level of the solid substrate.

3.8.11. Carbon Sources

Supplement various sugars (0.25–3.0%, w/v) such as glucose, mannitol, glycerol, maltose, sucrose, lactose, cellobiose, fructose, xylose, starch, galactose, arabinose, rhamnose, galacturonic acid, and sodium acetate, in YEP medium along with 0.25% pectin to optimize the best carbon source for pectinase production. For SSF dissolve these carbon sources to desired concentrations in distilled water. Use this water containing the carbon sources as moistening agent.

3.8.12. Nitrogen Sources

Supplement various organic and inorganic nitrogen sources, namely casein, soybean meal, skim milk powder, peptone, tryptone, glycine, urea, ammonium chloride, ammonium nitrate, ammonium sulfate, and ammonium citrate (0.25–3.0%, w/v) in YEP medium in order to optimize a suitable nitrogen source for pectinase production under shaking conditions (150 rpm), at 37°C for 48 h. For SSF dissolve these nitrogen sources to desired concentrations in distilled water. Use this water containing the nitrogen sources as moistening agent.

3.8.13. Vitamins

Vitamin solutions are known to enhance pectinase production. The effect of multivitamin solution (neurobion) on pectinase production can be studied by incorporating this solution (0.05–0.2%, v/v) into the YEP medium. For SSF add different concentrations (3–18 μL/mL moistening agent) of neurobion dissolved in sterile distilled water. Moisten the autoclaved solid medium with this water to the desired moisture level. Use this medium to study the effect of multivitamin solution on pectinase production.

3.9. Purification of Pectinase

The purification of pectinase is described in **Subheading 3.9.1.–3.9.4.** This includes (1) the preparation of cell-free supernatant, (2) ammonium sulfate precipitation, (3) concentration of protein, and (4) chromatography.

3.9.1. Preparation of Cell-Free Supernatant

1. Grow the isolated strain in 1 L of YEP (18 h, 37°C, 200 rpm).
2. Centrifuge the culture at 10,000g for 15 min at 4°C.
3. Analyze the supernatant for enzyme activity and protein content.

3.9.2. Ammonium Sulfate Precipitation

1. Precipitate the protein in cell-free supernatant containing pectinase with $(NH_4)_2SO_4$ to two cut-offs (0–40 and 40–100% saturation) in a refrigerator by addition of small amounts of ammonium sulfate with constant stirring *(4)*.
2. Centrifuge the mixture at 11,000g for 20 min at 4°C to collect the protein precipitates.
3. Dissolve the precipitates in a minimum amount of 0.01 M Tris-HCl, pH 7.5.

3.9.3. Concentration

1. Pass the soluble fraction through a membrane filter with a cut-off of 10 kDa using a filter assembly, e.g., Amicon filter units. This procedure results in the concentration of the samples to nearly 3.0 mL as well as removal of ammonium sulfate.
2. Wash the membrane with 1 mL of 0.01 M Tris-HCl, pH 7.5, to recover adsorbed enzyme and pool the two fractions.
3. Add 0.01% of sodium azide to the sample and store it at 4°C.

3.9.4. Chromatography

3.9.4.1. Ion-Exchange Chromatography

Ion-exchange chromatography is carried out by using DEAE Sephacel, which is a weak anion exchanger with total ionic capacity of 0.11–0.14 mmol/gel, and pH working range of 2.0–9.0. The ion-exchange group is diethyl aminoethyl.

1. Apply the concentrated enzyme sample onto an ion-exchange column (35 × 0.55 cm, 10 mL bed volume) containing DEAE Sephacel equilibrated with 0.01 M Tris-HCl, pH 7.5, and elute the protein using a 0–1.0 M NaCl linear gradient made in equilibrating buffer.
2. Collect fractions of 1.5 mL.
3. Measure OD_{280} using equilibrating buffer as blank.
4. Assay all the fractions for pectinase activity (**Subheading 3.3**).
5. Determine protein content by Lowry's method *(17)*. Take 1 mL suitably diluted sample, add 5 mL working solution, and mix. Keep for 10 min at room temperature and add 0.5 mL Folin Ciocalteu's reagent 1:1 diluted with distilled water. Mix by vortexing. Incubate the tubes at 37°C and record the OD_{660} on a spectrophotometer against a blank processed in similar manner with distilled water instead of protein sample.
6. Pool the fractions showing pectinase activity, concentrate, and save for further analysis.

3.9.4.2. GEL FILTRATION CHROMATOGRAPHY

1. Prepare a gel filtration column of following dimensions: 35 × 1.5 cm, bed volume 60 mL using Sephadex G-150. Equilibrate the column with 0.01 M Tris-HCl, pH 7.5. Apply 3.0 mL of the concentrated and partially purified enzyme onto the column. Run the column using same buffer at a flow rate of 20 mL/h.
2. After the void volume (20 mL), collect 1.5-mL volume fractions.
3. Assay all the fractions for pectinase activity (**Subheading 3.3**) and protein content by Lowry's method (**Subheading 3.9.4.1.**) *(17)*.
4. Pool pectinase-positive fractions and concentrate them.
5. Load approx 30 μL of the sample onto SDS-PAGE (10%), to determine the protein profile of the purified sample.
6. Separate the proteins by the gel at 30 mA for stacking gel and 50 mA for separating gel. Stain the gel either by Coomassie blue or silver staining standard methods.

3.10. Characterization of Purified Pectinase

3.10.1. Optimum pH

1. Assay purified pectinase activity at different pH values (pH 5.0–10.0) using different buffers (0.01 M) such as citrate phosphate buffer, pH 5.0, phosphate buffer, pH 6.0–7.9, Tris-HCl buffer, pH 7.5–8.5, and glycine NaOH buffer, pH 9.0–10.0.
2. Carry out the assay (**Subheading 3.3**) at 60°C to determine optimum pH for its activity.

3.10.2. pH Stability

1. Keep the pectinase preparation buffered at different pHs from 4.0–12.0 at room temperature for different time intervals up to 4 h (*see* **Note 8**).
2. Assay pectinase activity at 60°C.

3.10.3. Optimal Temperature

To study the effect of temperature on purified pectinase preparation, assay the enzyme **(Subheading 3.3)** at different temperatures from 20–90°C with 5°C intervals at pH 8.0.

3.10.4. Thermostability

1. Keep the enzyme and substrate mixture at different temperatures ranging from 45–90°C for different time intervals up to 5 h.
2. Assay pectinase activity **(Subheading 3.3)** at 60°C and pH 8.0.

3.10.5. Chelating Agents, Surfactants, Metal Ions, and Salts

The effect of chelating agents and surfactants including mercaptoethanol, urea, ascorbic acid, glycine, cysteine, EDTA (1 mM), SDS (1%, w/v), Tween (20, 40, 60 and 80; 0.1%, v/v), and Triton X-100 (0.1%, v/v) can be studied by directly incorporating them in to the enzyme substrate system.

The effect of metal ions and salts on purified pectinase activity can be studied by directly incorporating them into the enzyme substrate system at a final concentration of 1 mM. Metal ions and salts to be examined for their effect include Ca^{2+}, Mg^{2+}, Pb^{2+}, Hg^{2+}, Co^{2+}, Cu^{2+}, Zn^{2+}, Fe^{3+}, Cd^{2+}, Ni^{2+}, Ba^{2+}, and Mn^{2+}.

3.10.6. Type of Pectinolytic Activity

Different methods may be used to determine the type of pectinase produced by the isolates.

3.10.6.1. PECTINESTERASES

1. Adjust 20 mL of 1% pectin solution containing 0.1 M NaCl to pH 7.5 (placed in constant temperature water bath maintained at 30°C).
2. Add 0.1–0.5 mL of the enzyme solution.
3. Adjust the pH immediately to 7.5 with 0.1 N or 0.02 N NaOH and note the time (*see* **Note 9**).
4. Enzyme activity can be also expressed as milligrams of CH_3OH liberated in 30 min/mL.

3.10.6.2. PECTIN LYASE

1. Add 1 mL of suitable diluted sample to 5 mL of pectin solution (1%, w/v).
2. Adjust the volume of test sample to 10 mL with distilled water.
3. Incubate the samples at 40°C for 2 h.
4. Add 0.6 mL of zinc sulfate (9%, w/v) and 0.6 mL of sodium hydroxide (0.5 M).
5. Centrifuge the samples at 3000g for 10 min.
6. Add 5 mL of clear supernatant to a mixture of 3 mL of 0.04 M thiobarbituric acid, 2.5 mL of 0.1 M HCl, and 0.5 mL of distilled water.

7. Heat the mixture in a boiling water bath for 30 min.
8. After cooling to room temperature, measure OD_{550}.
9. Estimate the concentration of reducing groups from the standard curve obtained from the solution of known concentrations (10–1000 μg) of pectin.
10. One unit (U) of activity may be defined as the amount of enzyme that causes a change in absorbance of 0.01 under the conditions of assay.

3.10.6.3. POLYGALACTURONASES

1. Add 1.0 mL of 0.5% (w/v) PGA to 1.0 mL of enzyme solution.
2. Incubate at desired temperature (20–90°C) for 3–10 min.
3. Add 5 mL of DNSA reagent and keep the solution in a water bath for 10 min.
4. Cool the solution.
5. Measure OD_{530} in a 1-cm light path cuvet (*see* **Note 10**), against the product of a blank in which 1 mL of buffer substitutes for enzyme solution.
6. Estimate the concentration of reducing groups from the standard curve obtained from the solution of known concentrations (10–1000 μg) of galacturonic acid.
7. One unit (U) of activity may be defined as the amount of enzyme that releases 1 μmol of reducing groups per minute under the assay conditions.

3.10.6.4. PECTATE LYASE

1. Incubate 250 μL of suitably diluted purified enzyme preparation and 250 μL of PGA (1%, w/v) mixture at 60°C for 10 min.
2. Take 100 μL of sample from this mixture and add to 500 μL of 0.5 *N* HCl in a test tube.
3. Add 1.0 mL of 0.01 *M* thiobarbituric acid (dissolved in distilled water) and heat the mixture in a boiling water bath for 30 min.
4. Cool it and restore to volume with distilled water. Centrifuge at 10,000*g* for 15 min.
5. Determine the absorption spectrum of the supernatant using a spectrophotometer in the range 480–580 nm.
6. The occurrence of a peak at 550 nm is due to the hydrolytic products formed by pectate lyase enzyme.

3.10.7. Michaelis-Menten Constant (K_m) and V_{max} Values

The K_m and V_{max} values of purified pectinase can be determined from Lineweaver-Burk's plot between substrate (PGA) concentrations (1–10 mg/mL) and enzyme activity.

3.11. Cloning of Gene(s) for Pectinase Production from Bacillus *sp.* DT7

3.11.1. Preparation of Genomic DNA

1. Inoculate *Bacillus* DT7 in 5 mL of LB and incubate at 37°C overnight under shaking conditions (100 rpm).
2. Centrifuge the culture (8,000*g*, 10 min, 4°C). Discard the supernatant and save the cells.

3. Suspend the cells in 950 μL of TE buffer. Add 50 μL of 10% SDS. Mix thoroughly and incubate at 37°C for 1 h. This will result in weakening of the cell wall of the bacteria.
4. Add 200 μL of 5 *M* NaCl and 150 μL of CTAB/NaCl solution. Mix the contents and incubate at 65°C for 20 min. This will result in complete lysis of the cells.
5. Add 1.4 mL of phenol:chloroform:isoamyl alcohol (24:24:1). Mix the contents. This will result in denaturation of proteins.
6. Centrifuge the tubes at 10,000*g* for 5 min at 4°C. Denatured proteins will appear as insoluble debris at the interphase of solvents.
7. Transfer the upper aqueous phase (which contains DNA) to another tube and repeat **step 5.**
8. Add 0.6 volumes of ice-cold isopropanol. Mix the contents and keep the tube at –20°C for 5–6 h or at –80°C for 1 h. This will result in DNA precipitation.
9. Spin the tubes at 10,000*g* for 2 min. Discard the supernatant.
10. Wash the precipitate with 1.0 mL of 70% (w/v) ethanol and dry the pellet by leaving the opened tube at room temperature. It takes approx 30 min to 60 min.
11. Dissolve the pellet in the minimum amount of TE (approx 100–200 μL).

3.11.2. Partial Digestion of Bacillus *DT7 genomic DNA*

1. Take 200 μL of genomic DNA in an Eppendorf tube. Add 22 μL of 10X *Bam*HI restriction buffer and mix gently. Add 30–40 units of *Bam*HI restriction enzyme (approx 3–4 μL) and mix the contents again.
2. Incubate the tube at 37°C and periodically (at about 10 min intervals) take out 15 μL of the sample in a separate Eppendorf tube. Immediately add 1 μL of 0.5 *M* EDTA, pH 8.0, to the tube and mix the contents. Inactivate the enzyme by incubating the tubes at 80°C for 15 min. Keep the tubes in ice until further use.
3. Take out 2 μL of the sample from each tube in a fresh Eppendorf tube. Add 13 μL of sterile distilled water and 5 μL of 5X loading dye. Mix the contents.
4. Transfer the above samples into the wells of 0.7% (w/v) agarose gel and run the gel at 70–90 V.
5. Take out the gel and stain it with ethidium bromide solution.
6. Observe DNA bands under UV light.
7. Look for the lane that shows *Bam*HI digested bands in the range of 1–10 kb.
8. Select the tube showing the aforementioned digested band pattern for further use.

3.11.3. Ligation Reaction

1. Add in equal amounts digested genomic DNA and pUC18 (*Bam*HI digested and dephosphorylated) in a fresh tube, keeping the total volume of the mixture 20 μL and concentration of vector and inserts the same.
2. Incubate the tube in a water bath (12°C) overnight.
3. Add 10 μL of the ligation mixture to 100 μL *E. coli* competent cells. Incubate in ice for 20 min.

4. Heat-shock the tube by immersing it in a 42°C heating block for 30 s.
5. Add 1 mL LB and incubate at 37°C for 1 h under stationary conditions.
6. Plate out 100 µL of the cells on LA-ampicillin (100 µg/mL) plates. You will need 10 such plates. Incubate plates at 37°C for 24 h.
7. The bacterial colonies appearing on the LA-ampicillin plates are the transformants. Some of the transformants will contain genes for pectinases.

3.11.4. Screening for Pectinase-Positive Clone

1. With the help of a sterile toothpick, pick a patch of each colony on two LA-ampicillin + 1 mM IPTG + 1% (w/v) pectin plates, and incubate at 37°C for 24 h.
2. Flood one set of plates with cetramide solution and look for the zone of clearance around bacterial colonies. These are the recombinant clones that express the pectinase-encoding genes of *Bacillus* sp. DT7.
3. Mark the pectinase-positive colonies on the other set of plates. Select them out and plate on fresh LA-ampicillin-pectin plates. Store these recombinant clones at 4°C or get them lyophilized for long-term storage.

3.12. Applications of Pectinases

3.12.1. Retting of Jute (Corchorus capsularis), Ramie (Boehmeria nivea), and Buel (Grewia optiva) Bast Fibers

1. Cut the stem of the source (jute, ramie, and buel) to a length of 3–4 cm.
2. Weigh the pieces and sterilize at 1 atmospheric pressure for 30 min.
3. Wash the sterile pieces with 1 L of sterile water aseptically.
4. Take 5 g of the washed stem in 250-mL conical flasks.
5. Add 20 mL sterile distilled water.
6. Inoculate the flasks with 2% (v/v) bacterial culture (grown as described under **Subheading 3.8.1.**) to achieve 2.0×10^2 cfu/mL in the final mixture (*6*).
7. Incubate the flasks at 37°C.
8. Every 24 h take a 2–3 mL sample of retting water and keep the initial volume constant by addition of 2–3 mL of sterile water.
9. Determine the galacturonic acid content (µmole/g dry stem).
10. A stage when no further increase in galacturonic acid content is recorded confirms the completion of retting, which is also evident from the separation of fibers from the stem.

3.12.2. Degumming of Ramie and Buel Bast Fibers

3.12.2.1. BACTERIAL TREATMENT

1. Inoculate 1 g fibers (autoclaved at 10 psi for 30 min) with 2% (v/v) bacterial culture (grown as described under **Subheading 3.8.1.**) to achieve 2.0×10^2 cfu/mL in the final mixture.

2. Adjust different moisture contents (50, 60, 71.4, 75, and 80% for ramie and 83, 91, 95, and 97% for buel fibers) in the fibers to optimize the required content for optimum degumming.
3. Withdraw samples periodically up to 96 h to assess amount of galacturonic acid released, final pH, and dry weight of the fibers. Galacturonic acid (GA) content (µmole/g dry fibers) may be estimated by dinitrosalicylic method.

3.12.2.2. ENZYMATIC TREATMENT

Enzymatic treatment of buel fibers can be carried out by the method of Baract-Pereira et al. *(18)*.

1. Treat 1 g of fibers with 10 mL of 0.01 *M* Tris-HCl, pH 8.0, containing different enzyme levels (100–500 U/mL).
2. Supplement the reaction mixture with 0.01% sodium azide (to prevent bacterial and fungal growth) and incubate under static conditions (45°C, 96 h).

3.12.2.3. CHEMICAL TREATMENT

Chemical treatment can be carried out by the method of Sharma *(19)*.

1. Incubate 2 g of decorticated fibers with 10 mL of 2% (w/v) NaOH solution under static conditions for 96 h in a water bath set at 90°C.
2. Withdraw samples periodically to assess degumming.

3.12.2.4. CHEMICAL AND SUBSEQUENT ENZYMATIC (COMBINED) TREATMENT

In combined treatment, first treat the fibers chemically (*see* **Subheading 3.11.1.3.**) and then with enzymes (*see* **Subheading 3.11.1.2.**). Prior to enzymatic treatment, wash the chemically treated fibers twice with distilled water to remove NaOH and resuspend in 0.01 *M* Tris-HCl, pH 8.0.

3.12.2.5. ASSESSMENT OF DEGUMMING OF FIBERS

1. Assess the extent of degumming by assaying the amount of GA released as a result of hydrolysis of pectic substances by dinitrosalicylic method.
2. Assess the reduction in dry weight of the fibers.
3. Prepare samples for scanning electron microscopy. Wash the degummed fibers twice with sterilized distilled water.
4. Fix in 2.5% glutaraldehyde solution made in 50 m*M* phosphate buffer, pH 7.2, for 2 h.
5. Wash the fibers twice with the same buffer and gradually dehydrate with acetone gradient 30–90%.
6. Suspend the fibers in 100% acetone.
7. Put the fibers on the stubs mounted with silver tape and allow them to dry at room temperature.
8. Sputter-coat the stubs with gold using fine-coat JEOL ion sputter, model JFC-1100.
9. Examine the gold-coated stubs at ×1000 and ×4000 magnifications under scanning electron microscope, model JSM 6100 SM JEOL.

3.12.3. Pretreatment of Pectic Wastewater from Fruit Juice Industry

3.12.3.1. SAMPLE COLLECTION

1. Collect wastewater samples from the fruit-processing plant in sterile containers and bring to the laboratory in ice containers.
2. Store the samples under refrigeration condition for further studies.

3.12.3.2. BACTERIAL PRETREATMENT

Pectic wastewater from the fruit juice plant can be treated by the methods described by Tanabe et al. *(20)*.

1. Take 25 mL of wastewater in a 250-mL Erlenmeyer flask and autoclave at 5 lb for 30 min.
2. Inoculate this water with 2% inoculum.
3. Incubate the flasks at 37°C for 36 h with constant shaking (150 rpm).
4. Withdraw the samples periodically during pretreatment period up to 36 h.

3.12.3.3. ENZYMATIC PRETREATMENT

For the pretreatment of pectic wastewater use crude pectinase produced in YEP medium.

1. To 25 mL of the wastewater (pH should be adjusted beforehand to 8.0 with 2 N NaOH), add different levels (100–500 U/mL of reaction mixture) of pectinase *(10)*.
2. Incubate the flasks at 45°C under shaking conditions (150 rpm).
3. Withdraw the samples periodically during the pretreatment period up to 36 h.

3.12.3.4. MEASUREMENT OF DEGRADATION OF PECTIC SUBSTANCES

1. Measure GA content (μg/mL of reaction mixture) by the method of Miller *(21)*. Take 100 μL suitably diluted cell-free supernatant, mix with 100 μL substrate (1% w/v polygatacturonic acid), and keep at 60°C for 5 min under static conditions.
2. Add 500 μL of DNSA, boil for 15 min, and bring the volume to 5 mL with distilled water.
3. Determine OD_{530} on spectrophotometer against a blank run in a similar manner but with distilled water. One unit (U) of enzyme is defined as the amount of enzyme that catalyzes the formation of 1 μmol of galacturonic acid per minute.
4. Measure COD level in the treated wastewater.
5. Assess molecular-weight distribution of pectic substances as described by Tanabe et al. *(20)*. Apply the treated wastewater onto a gel filtration column (1.5 × 35 cm) containing Sephadex G-25 equilibrated with 0.01 M Tris-HCl, pH 7.5, and elute with 20 mM NaCl at a flow rate of 15 mL/h.
6. Collect fractions of 1.5 mL volume.
7. Measure OD_{235} of the collected fractions (molecular extinction coefficient of GA at this wavelength is 4600/M/cm).
8. Estimate the GA content in each fraction.

9. As a standard for estimating molecular weight use α-D-galacturonic acid, D-cellobiose and pullulan (11,200 Da).

3.12.4. Clarification of Fruit Juices

1. For clarification of cold-pressed juices, add 300–600 U/mL of pectinase.
2. Incubate at 15–20°C for 12–18 h.
3. For hot clarification process, add 1800–4000 U/mL of pectinase to the juice.
4. Incubate the mixture at 45–50°C for 2–6 h.
5. The effectiveness of pectinases in clarifying fruit juices can be measured by the time required for the floc formation, rate of filtration after treatment for a specified time period, or decrease in OD_{530} of the filtered juice *(22)*.

4. Notes

1. Preferable sites for collection of samples for the isolation of microorganisms producing pectinases include soils around fruit- and vegetable-processing units, wastewater from fruit-juice industries, vermi-composting sites, rotten fruits and vegetables, etc.
2. Enrichment technique is carried out to increase the proportion of desirable microbes in the liquid culture medium by providing a specific carbon substrate that can be utilized only by desirable microbes.
3. The sample should be kept at 4°C until it is used in order to avoid the excessive growth of undesirable microbes.
4. For blank, take 200 μL of distilled water in place of 100 μL of enzyme and 100 μL of substrate. For enzyme control add 100 μL of distilled water instead of substrate. For substrate control add 100 μL of distilled water instead of enzyme.
5. 75% moisture level is found to be optimum for *Bacillus* sp. DT7.
6. In all the media 0.25% (w/v) pectin is added, which acts as an inducer as well as a carbon source.
7. Inoculum level can be varied depending on the time required for pectinase production.
8. Buffering can be done by making the substrate (PGA) in different buffers as well as by diluting the purified enzyme preparation with the buffers.
9. The alkali is added at the rate required to keep the mixture at pH 7.0 for 10 min or with weak enzyme for 30 min. The quantity of enzyme is adjusted so that the pH does not drift below 7.0 before it can be readjusted to 7.5.
10. If necessary, the brown-colored solution can be diluted with water to keep OD_{530} within 0–1.0.

References

1. Esquivel, J. C. C., Hours, R. A., Voget, C. E., and Mignone, C. F. (1999) *Aspergillus kawachii* produces an acidic pectin releasing enzyme activity. *Biosci. Bioeng.* **1,** 48–52.
2. Singh, S. A., Plattner, H., and Diekmann, H. (1999) Exopolygalacturonate lyase from a thermophilic *Bacillus* sp. *Enzyme Microb. Technol.* **25,** 420–425.
3. Kashyap, D. R., Vohra, P. K., Chopra, S., and Tewari, R. (2001) Applications of pectinases in the commercial sector: A review. *Bioresource Technol.* **77,** 215–227.

4. Kashyap, D. R., Chandra, S., Kaul, A., and Tewari, R. (2000) Production, purification and characterization of pectinase from a *Bacillus* sp. DT7. *World J. Micobiol. Biotechnol.* **16,** 277–282.

5. Kapoor, M., Beg, Q. K., Bhushan, B., Dadhich, K. S., and Hoondal, G. S. (2000) Production and partial purification and characterization of a thermo-alkali stable polygalacturonase from *Bacillus* sp. MG-cp-2. *Process Biochem.* **36,** 467–473.

6. Kobayashi, T., Koike, K., Yoshimatsu, T., Higaki, N., Suzumatsu, A., Ozawa, T., et al. (1999) Purification and properties of a low molecular weight, high alkaline pectate lyase from an alkalophilic strain of *Bacillus*. *Biosci. Biotech. Bioch.* **63,** 56–72.

7. Kashyap, D. R., Vohra, P. K., Soni, S. K, and Tewari, R. (2001) Degumming of buel *(Grewia optiva)* bast fibers by pectinolytic enzyme from *Bacillus* sp. DT7. *Biotechnol. Lett.* **23,** 1297–1301.

8. Henriksson, G., Akin, D. E., Slomczynski, D., and Eriksson, K. E. L. (1999) Production of highly efficient enzymes for flax retting by *Rhizomucor pusillus*. *J. Biotechnol.* **68,** 115–123.

9. Cao, J., Zheng, L., and Chen, S. (1992) Screening of pectinase producer from alkalophilic bacteria and study on its potential application in degumming of ramie. *Enz. Microb. Technol.* **14,** 1013–1016.

10. Kashyap, D. R., Soni, S. K., and Tewari, R. (2003) Pretreatment of pectinaceous wastewater from apple juice industry by *Bacillus* sp. DT7. *Bioresource Technol.,* to be published.

11. Tanabe, H. and Kobayashi, Y. (1987) Plant tissue maceration caused by pectinolytic enzymes from *Erwinia* spp. under alkaline conditions. *Agric. Biol. Chem.* **51,** 2845–2846.

12. Kashyap, D. R., Soni, S. K., and Tewari, R. (2003) Enhanced production of alkaline and thermotolerant pectinase by *Bacillus* sp. DT7 in solid state fermentation. *Bioresource Technol.* **80,** 251–254.

13. Dosanjh, N. S. and Hoondal, G. S. (1996) Production of constitutive, thermostable, hyperactive exopectinase from *Bacillus* GK-8. *Biotechnol. Lett.* **18,** 1435–1438.

14. Karbassi, A. and Vaughn, R. H. (1980) Purification and properties of poygalacturonic acid trans-eliminase from *Bacillus stearothermophilus*. *Can. J. Microbiol.* **26,** 377–384.

15. Sneath, P. H. A. (1994) Endospore forming Gram positive rods and cocci, in *Bergey's Manual of Systematic Bacteriology,* 9th ed. (Hensyl, W. M., ed.), Williams and Wilkins Publishers, Baltimore, MD, pp. 1104–1139.

16. Aguillar, G. and Huitron, C. (1990) Constitutive exo-pectinase produced by *Aspergillus* sp. CH-Y-1043 on different carbon sources. *Biotechnol. Lett.* **12,** 655–660.

17. Lowry, O, H., Rosebrough, M. J., Farr, A. L., and Randell, R. J. (1951) Protein estimation with Folin-phenol reagent. *J. Biol. Chem.* **193,** 265–275.

18. Baracat-Pereira, M. C., Vanetti, M. C. D., Aruojo, E. F. D., and Silva, D. O. (1993) Partial characterization of *Aspergillus fumigatus* polygalacturonase for the degumming of natural fibers. *J. Ind. Microbiol.* **11,** 139–142.

19. Sharma, H. S. S. (1987) Enzymatic degradation of residual non-cellulosic polysaccharides present on dew-retted flax fibers. *Appl. Microbiol. Biotechnol.* **26,** 358–362.

20. Tanabe, H., Kobayashi, Y., and Akamatsu, I. (1986) Pretreatment of pectic wastewater from orange canning by soft-rot *Erwinia carotovora. J. Ferment. Technol.* **64,** 265–268.

21. Miller, L. G. (1959) Use of dinitrosalicylic acid reagent for the determination of reducing sugars. *Anal. Chem.* **31,** 426–428.

22. Kilara, A. (1982) Enzymes and their uses in the processed apple industry: a review. *Process. Biochem.* **23,** 35–41.

13

Expression of Microbial Phytases in Yeast Systems and Characterization of the Recombinant Enzymes

Xin Gen Lei and Taewan Kim

Summary

Phytases are hexakisphosphate phosphorylases that initiate the hydrolysis of phosphate from phytate (*myo*-inositol hexakisphosphate), the major form of phosphorus in plant foods and feeds. These enzymes have been increasingly used worldwide for animal nutrition, environmental protection, and human health. In our laboratory, we have studied the expression of fungal and bacterial phytases in *Pichia pastoris* and *Saccharomyces cerevisiae*. Both sources of phytases can be expressed as functioning enzymes in the two yeast systems, and secreted into the media as extracellular proteins at relatively high yield and purity. This chapter describes detailed procedures for preparing phytase expression constructs, transforming the constructs, and fermenting the transformants. Partial purification, biochemical characterization, and in vitro function assay of the recombinant phytases are also discussed.

Key Words: *Aspergillus niger;* electroporation; environment; *Escherichia coli;* food; feed; gene expression; glycosylation; *Pichia pastoris,* phosphorus; phytase; phytic acid; purification; pH profile; *Saccharomyces cerevisiae;* thermostability; transformation.

1. Introduction

Phytase (*myo*-inositol hexakisphosphate phosphorylase, EC 3.1.3.8) catalyzes the release of phosphate from phytate (*myo*-inositol hexakisphosphate), which is the major form of phosphorus in feeds and foods of plant origin (*1*). Simple-stomached animals such as swine and poultry cannot digest phytate-phosphorus efficiently due to the lack of phytase in their gastrointestinal tracts. This results in the need for supplemental inorganic phosphorus, an expensive and nonrenewable resource, and environmental pollution by the unutilized phosphorus from animal waste. Supplemental microbial phytases effectively improve the utilization of phytate-phosphorus by swine, poultry, and fish, and reduce their phosphorus excretion by up to 50% (*2–4*). Therefore, phytase has become a tool in many parts of the world, either required by law or chosen voluntarily, for protecting the environment

From: *Methods in Biotechnology, Vol. 17: Microbial Enzymes and Biotransformations*
Edited by: J. L. Barredo © Humana Press Inc., Totowa, NJ

from phosphorus pollution of animal waste. In addition, phytase can be supplemented in human diets to improve bioavailability of micronutrients such as zinc and iron that are often poorly available owing to their chelation with phytate in plant foods.

Since the world market for the feed application of phytase is approx $500 million, manufacturing phytase has become a very competitive industry. Although different phytases are overexpressed in a number of hosts including animal, plant and microbe, yeast expression systems such as *Pichia pastoris* and *Saccharomyces cerevisiae* have attracted great attention for commercial phytase manufacturing. This is because these systems produce relatively high levels of heterologous phytases from both prokaryotic and eukaryotic origins *(5–7)*. The recombinant enzymes are secreted into the media as the predominant protein, requiring little downstream processing for product development. Their post-translational modifications such as glycosylation are functionally beneficial to phytase thermostability *(5,6,8)*. In addition, these organisms have been used for fermenting human foods or bioactive compounds. Thus, it is safe to use them for producing phytase as feed or feed supplements.

Aspergillus niger (PhyA) was the first characterized phytase *(9,10)* that contains 467 amino acid residues, with 10 potential glycosylation sites and two unique pH optima at 5.0 and 2.5 *(6)*. The enzyme is commercially available. *Escherichia coli* AppA2 was originally isolated from pig colon in our laboratory *(7)* that contains 433 amino acid residues with 3 putative *N*-glycosylation sites. This enzyme has an acidic pH optimum and has shown excellent effectiveness in animal diets *(4)*. Thus, we herein describe the detailed procedures for the expression of both PhyA and AppA2 in *P. pastoris* and *S. cerevisiae*. These procedures are simply divided into three major steps: construction of expression vectors, transformation of the vectors, and fermentation of the transformants. Meanwhile, we outline methods for partial purification, biochemical characterization, and in vitro function assay of the expressed phytases.

2. Materials

2.1. Construction of Expression Vectors

1. Phytase encoding gene *phyA* from *A. niger* (GeneBank accession no. M94550, provided by Dr. E. J. Mullaney of the U.S. Department of Agriculture) *(7)*.
2. Phytase encoding gene *appA2* from *E. coli* (GeneBank accession no. AY496073) *(9)*.
3. PCR machine (thermocycler) and *Taq*-polymerase.
4. Selected DNA restriction enzymes and T$_4$ DNA ligase.
5. TA-cloning vector.
6. Yeast expression vector (pYES, pGAP, or pPIC; Invitrogen, Carlsbad, CA).
7. DNA electrophoresis unit.
8. Gel extraction kit and UV-illuminator.

9. Electroporator or water bath (42°C).
10. Luria-Bertani (LB) medium: 1% tryptone, 0.5% yeast extract, and 0.5% sodium chloride.
11. Antibiotics in LB medium: zeocin (25 µg/mL) or ampicillin (100 µg/mL).
12. IPTG (isopropyl-β-D-thio-galactopyranoside).
13. X-gal (5-bromo-4-chloro-3-chloro-3-indoyl-β-D-galactose).

2.2. Transformation

2.2.1. Pichia pastoris System

1. *Pichia pastoris* X33 (Invitrogen).
2. pPICZαA or pGAPZαA (Invitrogen) ligated with phytase genes.
3. Electroporator.
4. 2-mm gap electroporation cuvet.
5. Ice-cold, sterile deionized H_2O.
6. Ice-cold, sterile 1 *M* sorbitol.
7. YPD media, pH 7: 2% peptone, 1% yeast extract, and 2% dextrose with zeocin (100 and 500 µg/mL).
8. BMGY (BMMY): 2% peptone, 1% yeast extract, 100 m*M* potassium phosphate, pH 6.0, 1.34% yeast nitrogen base w/o amino acids, 0.4 mg biotin/L, and 1% glycerol (or 0.5% methanol for BMMY expression medium).

2.2.2. Saccharomyces cerevisiae System

1. *Saccharomyces cerevisiae* INVSC1 (Invitrogen).
2. pYES2 (Invitrogen) ligated with phytase genes.
3. SCU⁻ (*S. cerevisiae* uracil deficient) minimal medium, pH 7.0: 0.08% CSM-URA, 0.17% yeast nitrogen base without amino acids, 0.5% ammonium sulfate, 1 *M* sorbitol, and 2% dextrose.
4. YPG medium: 2% peptone, 1% yeast extract, and 2% galactose, pH 7.0.

2.2.3. Chemicals for Chemical Transformation

1. 50% PEG (polyethylene glycol) 3350.
2. 1 *M* lithium acetate (LiAc).
3. 1 *M* lithium chloride (LiCl).
4. TE buffer: 10 m*M* Tris-HCl, pH 8.0, and 1 m*M* EDTA.
5. 0.2% single-stranded salmon sperm DNA (ssDNA) in TE buffer.

2.2.4. Plasmid Purification

1. Mini-prep kit.
2. Midi-prep kit.

2.3. Phytase Activity Assay

1. 0.2 *M* citrate buffer, pH 5.5 (for PhyA assay).
2. 1% sodium phytate in 0.2 *M* citrate buffer, pH 5.5 (pH adjusted by citric acid).

3. 0.2 *M* glycine-HCl buffer, pH 3.5 (for AppA2 assay).
4. 1% sodium phytate in 0.2 *M* glycine-HCl buffer, pH 3.5 (pH adjusted by HCl).
5. 15% TCA solution.
6. Reagent C (prepared fresh daily): Mix 3 volumes of 1 *M* sulfuric acid and 1 volume of 2.5% ammonium molybdate, then add 1 volume of 10% ascorbic acid and mix well.

2.4. Partial Purification

1. Ultra filtration unit with membrane, MW cut-off 30,000 (Millipore, Bedford, MA).
2. Spin column concentration unit, MW cut-off 30,000 (Millipore).
3. DEAE-cellulose resin (Sigma, St Louis, MO).
4. Sephadex G-100 (Sigma).
5. 10 m*M* Tris-HCl, pH 7.4.
6. 10 m*M* Tris-HCl with 0.3 *M* NaCl, pH 7.4.
7. 50 m*M* Tris-HCl with 0.15 *M* NaCl, pH 7.4.

2.5. Biochemical Characterization

2.5.1. Deglycosylation

1. Endo H_f (endoglycosylase, 0.1 IUB unit/μL).
2. 10X reaction buffer: 1 m*M* PMSF, 0.5 *M* sodium citrate, pH 5.5.
3. 10X glycoprotein denaturing buffer: 5% SDS, 10% β-mercaptoethanol.

2.5.2. pH Profile, Optimal Temperature, and Thermostability

1. 0.2 *M* glycine-HCl buffer for pH 2.0 to 3.5.
2. 0.2 *M* citrate buffer for pH 4.0 to 6.5.
3. 0.2 *M* Tris-HCl buffer for pH 7.0 to 8.5.

2.5.3. In Vitro Phytate-Phosphorus Hydrolysis

1. Ground soybean meal (SBM) or selected ingredients.
2. 0.2 *M* citrate buffer, pH 5.4, for final pH 5.5 after dissolving 1 g SBM in 10 mL buffer.
3. 0.2 *M* citrate buffer, pH 2.9, for a final pH 3.5.
4. 0.2 *M* glycine-HCl, pH 2.2, for a final pH 3.5.
5. 40 m*M* HCl for adjusting the mixture to a final pH 3.5.

3. Methods

The experimental protocol describes the expression, purification, and characterization of phytase in a sequential order. As mentioned above, two phytase genes (*phyA* and *appA2*), two yeast strains (*P. pastoris* and *S. cereivisiae*), and three expression vectors are involved in the expression *(7,9)*. The *S. cerevisiae* expression system bears a promoter *GAL1* inducible by galactose, and a signal peptide of PhyA for the secretion of the expressed phytase into the medium. The *P. pastoris* expression system uses either a constitutive promoter of glyceraldehyde-3-phosphate dehydrogenase gene *(GAP)* or an inducible promoter of alcohol oxidase gene *(AOX1)*. In both cases, α-factor is used as the signal peptide for the secretion of the expressed phytases *(see* **Note 1**). Despite all these differences in genes, vectors, and hosts, the basic phytase expression procedure, modified from the Invitrogen protocols *(11)*, remains similar.

3.1. Construction of Expression Vectors

DNA manipulations were performed by standard recombinant DNA methods *(12)* and are not described here in detail due to space limitations.

1. Design the expression vector (**Fig. 1**).
2. Prepare a set of primers matching the sequences at the 5′ and 3′ ends of the selected phytase coding region with appropriate restriction enzyme sites.
3. Amplify the coding region of phytase *(phyA* or *appA2)* by PCR.
4. Clone the PCR product into the TA-cloning vector (or ligate it directly with the expression vector after restriction digestion of the PCR product).
5. Transform the TA-cloning vector into *E. coli* competent cell by electroporation (2.5 kV, 129 ohms, 5.0 msec) or chemical activation (heat shock at 42°C for 45 s using CaCl$_2$-treated competent cells).
6. Select the white colonies from LB plates containing ampicillin, IPTG, and X-gal.
7. Extract plasmid DNA by Mini-prep or appropriate kits and digest it with restriction enzymes.
8. Separate the digested DNA by running an agarose gel (1%).
9. Visualize the gel on UV-illuminator, excise the band containing the target phytase DNA, and extract the DNA fragment using a gel extraction kit.
10. Ligate the DNA fragment with the selected expression vector cut with appropriate restriction enzymes.
11. Transform the ligated plasmid DNA into *E. coli* competent cells.
12. Culture the transformed cells in selection media (LB with 100 μg ampicillin/mL for pYES2 vector or LB with 25 μg zeocin/mL for pGAP or pPIC vector), and select the positive colonies.
13. Confirm the presence of the phytase DNA fragment by restriction digestion.

3.2. Transformation and Expression

Two methods, electroporation and chemical activation, are often used for phytase transformation into yeast cells *(see* **Note 2**).

3.2.1. Electroporation

3.2.1.1. PREPARATION OF SELECTED PLASMID FOR TRANSFORMATION

1. Culture the selected positive colony in 100 mL of LB medium containing ampicillin for *S. cerevisae* or zeocin for *P. pastoris* at 37°C overnight and extract plasmid DNA by Midi-prep kit.
2. Digest 5–10 μg of plasmid DNA overnight with 20 units of restriction enzymes (use *BspHI* for pGAP and *PmeI* for pPIC). Linearized vector integrates into the

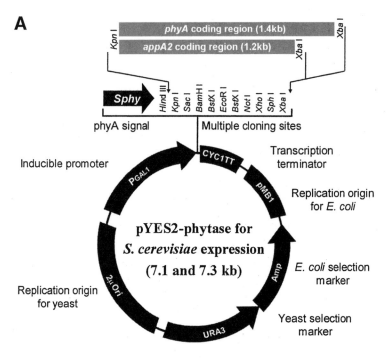

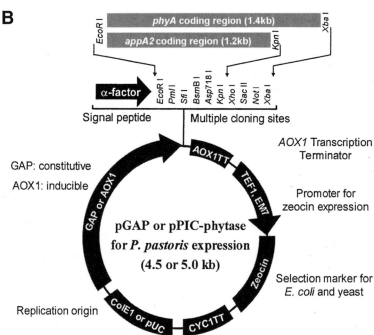

Fig. 1. Schematic drawing of phytase expression plasmids adapted from Invitrogen (Carlsbad, CA). (**A**) Phytase genes in the inducible pYES2 expression vector for *S. cerevisiae*. (**B**) Phytase genes in the constitutive pGAPZα expression vector or the inducible pPICZα expression vector for *P. pastoris*.

genomic DNA of *P. pastoris* with improved efficiency. The expression vector for *S. cerevisiae* is episomal, and does not need linearization.

3. Check the digestion with a small aliquot by an agarose gel (1%) electrophoresis.
4. Inactivate the restriction enzyme by heating at 65°C for 20 min.
5. Perform phenol extraction.
6. Precipitate DNA using 1/10 volume 3 *M* sodium acetate and 2.5 volumes of 100% ethanol and incubate at –80°C for 30 min.
7. Centrifuge the mixture to pellet DNA.
8. Wash the pellet with 80% ethanol twice, air-dry, and resuspend in 10 μL sterile ddH$_2$O.

3.2.1.2. PREPARATION OF COMPETENT CELLS

Although the following procedure deals with *P. pastoris* cells, all steps apply to *S. cerevisiae* except for preparing 200 mL of YPD for *S. cerevisiae* and growing cells to give an OD$_{600}$ around 4 to 5 (*see* **step 2** below).

1. Grow *P. pastoris* X33 strain in 5 mL of YPD broth in a 30-mL test tube overnight at 30°C with agitation (250 rpm) to give an OD$_{600}$ around 2 to 5.
2. Inoculate 0.1 to 0.5 mL of overnight culture to a 500-mL YPD medium in 2-L flask. Grow overnight again to give an OD$_{600}$ around 1.2 to 1.5 (1 × 10^7 cfu/mL). Doubling time of X33 at 30°C is approx 2 h.
3. Spin down cells at 1500g for 5 min at 4°C, and decelerate without brake.
4. Carefully decant supernatant and resuspend gently the cell pellet in 500 mL of cold deionized H$_2$O.
5. Spin down cells at 1500g for 5 min at 4°C and repeat the washing step.
6. Decant supernatant and resuspend gently the cell pellet in 50 mL of ice-cold 1 *M* sorbitol.
7. Spin down cells at 1500g for 5 min at 4°C and repeat the washing step.
8. Carefully decant supernatant and resuspend cell pellet in 1 *M* sorbitol using a pipet for a final volume of 2 mL to give an OD$_{600}$ around 200 to 300.
9. Dispense 80 μL of competent cells to each prechilled Eppendorf tube on ice.

3.2.1.3. ELECTROPORATION

1. Keep the Eppendorf tubes containing competent cells on ice.
2. For *P. pastoris*, add 10 μL of linearized plasmid DNA to each tube to give 5–10 μg of DNA (up to 30 μg). For *S. cerevisiae*, add 0.1 to 1 μg of DNA to each of the competent cell tubes.

3. Pipet up and down to mix and keep on ice until use.
4. Transfer the competent cell mixture to the electroporation cuvet kept on ice.
5. Pulse the cells at 1.5 kV (7.5 kV/cm) and 129 ohms (pulse length: 4.9 msec).
6. Add 1 mL of ice-cold 1 *M* sorbitol to the cuvet and transfer the cuvet contents to a sterile tube.
7. Incubate at 30°C for 1–2 h without shaking.
8. Spread 150–300 μL of the mixture on YPDZ agar plates containing 1 *M* sorbitol (100 and 500 μg zeocin/mL) for *P. pastoris* or SCU⁻ agar plate for *S. cerevisiae*.
9. Incubate the plates for 2–3 d at 30°C until colonies appear.
10. Isolate 10 to 20 colonies and inoculate each into YPD with zeocin (or SCU⁻ broth).
11. Grow overnight at 30°C with shaking. The culture can be used for the preparation of glycerol stock and protein expression.

3.2.2. Chemical Method

3.2.2.1. PREPARATION OF COMPETENT CELLS

1. Inoculate the yeast strain into 100 mL of liquid YPD.
2. Incubate the culture at 30°C with shaking until it reaches OD_{600} of 0.5 to 0.8.
3. Harvest the culture in sterile 50-mL centrifuge tubes at 1500g for 5 min at RT.
4. Pour off the medium, resuspend the cells in 25 mL of sterile water, and centrifuge again.
5. Pour off the water, resuspend the cells in 1 mL of 100 m*M* LiAc solution for *S. cerevisiae* (or 100 m*M* LiCl solution for *P. pastoris*), and transfer the suspension to a 1.5-mL microfuge tube.
6. Pellet the cells (at top speed) for 15 s and remove the supernatant with a micropipet.
7. Resuspend the cells to a final volume of 500 μL in 100 m*M* LiAc solution (or LiCl).

3.2.2.1. TRANSFORMATION

1. Vortex the cell suspension and pipet 50 μL of samples into microfuge tubes. Pellet the cells at 5000g for 15 s at RT and remove the LiAc solution with a micropipet.
2. Add the following solutions in the order listed:
   - 240 mL of 50% PEG.
   - 36 mL of 1 *M* LiAc (or LiCl).
   - 25 mL of SS-DNA (2 mg/mL): boiled for 5 min and chilled on ice.
   - 50 mL of plasmid DNA (5–10 mg).
3. Vortex each tube vigorously until the cell pellet is completely mixed.
4. Incubate at 30°C for 30 min.
5. Heat-shock in a water bath at 42°C for 20 min.
6. Centrifuge at 6000–8000g for 15 s and remove the supernatant with a micropipet.
7. Pipette 0.5 mL of sterile water into each tube and gently resuspend the pellet.
8. Inoculate 100–300 μL of the suspension on the selection agar plates.

3.2.3. Screening Colonies for Phytase Activity

1. *P. pastoris* transformants are selected based on zeocin resistance, and high copies of transformants can be obtained by using high zeocin concentrations (500 or 1000 µg/mL). The *S. cerevisiae* transformants are selected on SCU⁻ agar plates.
2. Selected transformants should be tested for phytase activity expression (see below).
3. Insoluble (turbid) calcium phytate may be overlayered on agar plates as the main source of medium phosphorus. Extracellular phytase produced by the transformants may degrade calcium phytate, forming clear zones in the agar media.

3.2.4. Phytase Expression

3.2.4.1. Constitutive Expression in *P. pastoris* (pGAPZαA Vector)

1. Inoculate *P. pastoris* transformants into a 15-mL culture tube with YPD medium, and incubate overnight at 30°C with shaking (250–300 rpm). Measure OD_{600} after 14–16 h and maintain the absorbance between 2 and 6.
2. Use 0.1 mL of the overnight culture to inoculate 50 mL of YPD in a 250-mL flask.
3. Incubate the culture for 2–3 d at 30°C with shaking (at 250–300 rpm) and take samples every 12 h for studying the time-course of phytase expression.
4. Record OD_{600} before spinning down the cells and store the supernatant at –80°C until analysis.

3.2.4.2. Inducible Expression in *P. pastoris* (pPICZαA Vector)

1. Inoculate *P. pastoris* transformants into a 15-mL culture tube containing 8 mL of BMGY medium, and grow the culture overnight at 30°C until culture reaches an OD_{600} around 2 to 6.
2. Spin down cells at 1500g for 5 min at RT. Decant the supernatant and resuspend cells in BMMY (expression media with methanol) to give an OD_{600} around 1.
3. Grow cells at 30°C with shaking (200–250 rpm). Induce phytase expression by adding methanol (maintain at 0.5% in the culture) every 24 h.
4. Grow cells for 7–8 d and take samples every 24 h for the time-course study and phytase expression.

3.2.4.3. Inducible Expression in *S. cerevisiae* (pYES2 Vector)

1. Inoculate isolated *S. cerevisiae* transformants into 8 mL of SC-U minimal medium containing 2% glucose or raffinose (or YPD), and incubate overnight with agitation (250 rpm) at 30°C.
2. Measure OD_{600} of the overnight culture and calculate the amount of the culture needed for an initial OD_{600} of 0.4 in 50 mL of induction medium (YPG).
3. Remove the amount of overnight culture needed and pellet the cells at 1500g for 5 min at 4°C.
4. Resuspend the cells in 1–2 mL of induction medium and mix with 50 mL of induction medium (SCU⁻ with 2% gal or YPG). Grow at 30°C with shaking.

5. Collect samples at 6, 12, 24, 36, 48, and 72 h. Spin down the samples at 1500*g* for 5 min at 4°C and keep the supernatant for phytase assay.

3.3. Phytase Activity Assay

Phytase activity is often measured based on the enzyme's ability to hydrolyze sodium phytate at certain pH, temperature, and buffer conditions. Thus, activities of different phytases may not be compared directly. For PhyA, 1 unit is defined as the amount of activity that releases 1 µmole of inorganic phosphorus from sodium phytate per minute at pH 5.5 and 37°C. For AppA2, the enzyme activity is often measured at pH 3.5 and 37°C (*see* **Note 3**).

1. Dilute medium supernatant samples in 0.2 *M* citrate buffer (pH 5.5) for PhyA and in 0.2 *M* glycerin-HCl (pH 3.5) for AppA2 to give 15–240 mU/mL of solution (1000–5000-fold dilution for *Pichia* culture and 10–20-fold dilutions for *Saccharomyces* culture).
2. Pipet 0.5 mL of diluted sample into two test tubes each and keep in a water bath at 37°C for 5 min.
3. Add 0.5 mL of sodium phytate to each tube and vortex (substrate prewarmed for 10 min at 37°C).
4. Incubate the tubes at 37°C for 15 min.
5. Add 1 mL of 15% TCA to each tube to stop the reaction. Mix and cool to room temperature. Spin the tubes for 10 min at 2000*g,* if precipitate is present.
6. Pipet 0.2 mL of reaction mixture from each tube to a new test tube.
7. Add 1.8 mL of deionized water to each tube.
8. Add 2.0 mL of fresh reagent C and mix.
9. Incubate at 50°C for 15 min and cool to room temperature.
10. Measure OD_{820} against that of sample blank.
11. Blank test: Add 1 mL of TCA solution to the reaction tube before adding 0.5 mL of substrate, and proceed as with the sample tubes.
12. Phosphorus standards:
    a. Use 9.0 m*M* phosphate stock solution to prepare different concentrations of phosphate as shown below:

9.0 m*M* P-stock	1:1600	1:800	1:400	1:200	1:100
Phytase activity (mU/mL)	15	30	60	120	240

    b. Pipet 2 mL of each solution into new test tubes.
    c. Add 2 mL of reagent C to each tube and proceed as with the sample tubes.

3.4. Partial Purification

All procedures are conducted at 4°C; *see* **Note 4.**

1. Prepare 200 mL of yeast culture and centrifuge at 15,000*g* for 10 min to remove yeast cells.
2. Concentrate the supernatant by ultrafiltration up to 10–20 mL, using a membrane with an apparent molecular cutoff of 30 kDa.

3. Apply the concentrate onto the DEAE-cellulose column (2.5 × 40 cm) equilibrated with 10 mM Tris-HCl, pH 7.4, and elute at 0.1–0.15 mL/min with a linear gradient of 0–0.3 M NaCl after washing with 2 column volumes of the same buffer.
4. Identify the fractions of phytase by measuring OD_{280} and phytase activity.
5. Concentrate the fractions with high activities using the spin column concentrator to 0.5–1 mL.
6. Apply the concentrated fraction onto the Sephadex G-100 column (2.5 × 100 cm) equilibrated with 50 mM Tris-HCl, pH 7.4, containing 150 mM NaCl and elute the protein fractions at 2 mL/h.
7. Identify the fractions of phytase by measuring OD_{280} and phytase activity.
8. Calculate the specific phytase activity of each fraction. Normally, the purity of phytase protein after this step is approx 90%.

3.5. Biochemical Characterization

It is important to determine the molecular mass of the expressed proteins and their levels of glycosylation because their yeast hosts have a strong ability to glycosylate proteins and glycosylation of phytases affects their properties. As a feed or food supplement, phytase needs to have effective pH and temperature ranges that are similar to the gastrointestinal tract conditions of animals. The enzyme is also desirable to be thermostable for processing and storage. More important, the ability of the expressed phytase to hydrolyze phytate-phosphorus in feeds or food such soybean meal in vitro is a prerequisite and a good indicator for its effectiveness in the digestive systems of animals and humans.

3.5.1. Deglycosylation

1. Prepare 100 µL of phytase enzyme solution (1 mg/mL).
2. Add 11 µL of 10X glycoprotein denaturing buffer and incubate at 100°C for 10 min.
3. Add 0. 3 IUB units of Endo H_f and 12 µL of 10X reaction buffer to the solution.
4. Incubate for 2 h at 37°C.
5. Run an SDS-PAGE to check the change in size of the expressed protein by deglycosylation.

3.5.2. SDS-PAGE and Western Blotting (**Fig. 2**)

1. Prepare the phytase sample and boil it in the sample buffer for 2 min. For Western blot, use 0.1–1 µg purified phytase (0.01–0.1 U), and for SDS-PAGE (Coomassie blue staining), use 1–10 µg purified phytase (0.1–1 U).
2. Prepare 10% polyacrylamide gel and load sample on the gel.
3. Run the gel at 100 V until the dye reaches the bottom of the gel.
4. Stain the gel with Coomassie blue solution or transfer the gel to the nitrocellulose membrane for Western blotting.

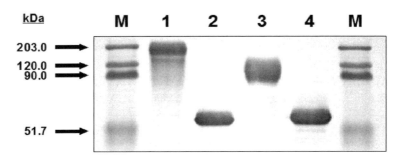

Fig. 2. Western blot analysis of *A. niger* PhyA phytase before and after deglycosylation by Endo H$_f$. M: pre-stained molecular marker; lane 1: PhyA expressed in *S. cerevisiae;* lane 2: deglycosylated PhyA expressed in *S. cerevisiae;* lane 3: PhyA expressed
in *P. pastoris;* and lane 4: deglycosylated PhyA expressed in *P. pastoris.*

3.5.3. pH Profile

Prepare buffers and substrate with different pH: 0.2 *M* glycine-HCl buffer for
pH 2.0–3.5, 0.2 *M* citrate buffer for pH 4.0–6.5, and 0.2 *M* Tris-HCl buffer for
pH 7.0–8.5. Sodium phytate is used as a substrate and prepared in the same
buffer for each designated pH.

1. Dilute the purified enzymes with each buffer to give 0.2 units for PhyA or 0.1 units
   for AppA2 at pH 5.5.
2. Assay the enzyme activity at different pH.
3. Plot the phytase activity against the reaction pH **(Fig. 3)**.

3.5.4. Optimal Temperature

1. Dilute PhyA with 0.2 *M* citrate buffer, pH 5.5, and AppA2 with 0.2 *M* glycine-HCl
   to give 0.05 U/mL. Store them on ice until use.
2. Incubate the diluted enzyme and substrate for 5 min at different temperatures (37,
   45, 55, 65, 75, 85, and 95°C).
3. Start hydrolysis reaction by adding prewarmed substrate and measure the enzyme
   activity according to the standard phytase assay method.
4. Plot the activity against different temperatures.

3.5.5. Thermostability

1. Dilute PhyA with 0.2 *M* citrate buffer, pH 5.5, and AppA2 with 0.2 *M* glycine-HCl
   to give 10 μg/mL. Store them on ice until use.
2. Incubate the sample tubes in the water bath with different temperatures for 15 min
   (4, 37, 45, 55, 65, 75, 85, and 95°C).
3. Chill on ice for 30 min and proceed the phytase activity determination (dilute the
   samples to give 0.2 U/mL).
4. Plot the residual activity against different heating temperatures **(Fig. 4)**.

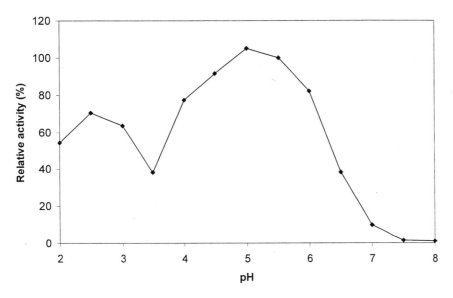

Fig. 3. The pH profile of *A. niger* PhyA phytase expressed in *S. cerevisiae*. Enzyme was diluted to give 0.2 units/mL with buffer. Phytase activity was determined using sodium phytate dissolved in the assay buffer for each selected pH.

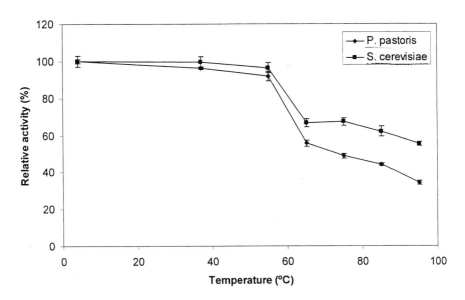

Fig. 4. Thermostability of *A. niger* PhyA phytase expressed in *S. cerevisiae* and *P. pastoris*. Each purified enzyme was diluted with 0.2 *M* citrate buffer, pH 5.5, to give 10 μg of phytase/mL, incubated at different temperatures for 15 min, and chilled on ice for 30 min. The residual activity was measured according to the standard phytase assay method.

3.5.6. In Vitro Hydrolysis Test (see **Note 5**)

1. Dilute PhyA sample with 0.2 *M* citrate buffer, pH 5.5, to give an enzyme activity of 1 U/mL (AppA with 0.2 *M* glycine-HCl buffer, pH 3.5).
2. Weigh 1 g of ground soybean meal and transfer to a 50-mL flask.
3. Dispense 10 mL of 0.2 *M* citrate buffer, pH 5.5, or 0.2 *M* glycine-HCl buffer, pH 3.5, to each flask and keep at 37°C for 20 min with agitation.
4. Add 0.25 units of enzyme to each tube and mix well. Incubate at 37°C for 1 h with agitation.
5. Prepare blank by adding distilled water instead of enzyme solution.
6. Add 10 mL of 15% TCA solution to stop the reaction.
7. Centrifuge at 13,000*g* for 5 min to remove denatured protein and oily layer.
8. Take 50 μL of reaction and transfer to a new test tube to measure phosphorus released from soybean meal according to the standard phytase assay.
9. Calculate the amount of released inorganic phosphorus from 1 g soybean meal (per hour).

4. Notes

1. Compared with the inducible expression system in *Pichia,* the constitutive system requires much less expensive media and fermentation equipment to produce phytase, and also reaches the expression plateau within a shorter period of time (48 h vs 7 d). However, the inducible system gives a much higher phytase yield at the expression peak. Compared with the *Pichia* systems, the *Saccharomyces* system produces phytases at a lower yield, but with greater glycosylation that improves their thermostability.
2. Because the *Pichia* expression vector is inserted into the genomic DNA by homologous integration, phytase expression in individual transformants depends on the copy number of the integrated vector. Thus, the electroporation method is preferable for producing transformants with multicopy insert. Factors such as purity and amount of the plasmid vector, competent cell, and transformation condition (pulse length and voltage) should be considered for efficient transformation. Up to 30 μg more DNA for each transformation appears to give better chances for multiple-copy colonies. Sorbitol is usually added to the selection media for cell stabilization but has no effect on colony numbers. For *Saccharomyces,* both electroporation and chemical methods give satisfactory results using 0.1–1.0 μg of pure DNA. For chemical transformation, it is very important to add PEG solution first because it protects the cells from the detrimental effects of the high concentration of LiAc.
3. Activities of different phytases may be assayed in different conditions upon their enzymatic characteristics such as optimal pH and temperature and substrate specificity. Thus, phytases of different sources cannot be simply compared without considering the specific definition of each enzyme activity unit. Because the stomach pH of animals is around 3.5, we have chosen to measure the activity of AppA2 using 0.2 *M* glycine-HCl, pH 3.5. Phytase samples are usually diluted with the assay buffer. Since sodium phytate has some buffering capacity, the pH

of the substrate solution should be adjusted after dissolving the substrate in the selected buffer.

4. Recombinant phytase expressed in yeast is highly glycosylated and may not be easily precipitated by ammonium sulfate. Ultrafiltration or freezer drying may be used as an alternative concentration. Phytases expressed in *Pichia* usually represent the predominant protein band on SDS-PAGE using Coomassie blue staining. However, several steps including ultrafiltration, ion-exchange chromatography, and gel-exclusion chromatography are required to get pure protein fractions. Ultrafiltration using the membrane with a molecular cut-off size of 30 kDa gives fast concentration with 95% recovery. Phytase can be eluted as the first few fractions from the DEAE cellulose resin equilibrated with 10 m*M* Tris-HCl, pH 7.4, with a gradient of NaCl from 0–0.3 *M,* and then separated as a single-speak fraction by gel exclusion chromatography.

5. For hydrolysis of phytate-phosphorus in soybean meal, the actual pH of the mixture should be adjusted to the target pH because soybean meal may affect the mixture pH. According to our experience, mixing soybean meal with 0.2 *M* citrate buffer, pH 5.4, yields a final pH of 5.5, while mixing soybean meal with 0.2 *M* glycine-HCl, pH 2.2, results in a final pH of 3.5. Ingredients such as wheat products contain high levels of intrinsic phytase that may be confounded with the effects of extrinsic phytases.

Acknowledgments

The research at the authors' laboratory was developed under the auspices of the Cornell University Center for Biotechnology, a NYSTAR Designated Center for Advanced Technology supported by New York State and industrial partners.

References

1. Pandey, A., Szakacs, G., Soccol, C. R., Rodriguez-Leon, J. A., and Soccol, V. T. (2001) Production, purification and properties of microbial phytases. *Bioresource Technol.* **77,** 203–214.
2. Lei, X. G., Ku, P. K., Miller, E. R., and Yokoyama, M. T. (1993) Supplementing corn-soybean meal diets with microbial phytase linearly improves phytate phosphorus utilization by weanling pigs. *J. Anim. Sci.* **71,** 3359–3367.
3. Lei, X. G. and Stahl, C. H. (2001) Biotechnological development of effective phytases for mineral nutrition and environmental protection. *Appl. Microbiol. Biotechnol.* **57,** 474–481.
4. Augspurger, N. R., Webel, D. M., Lei, X. G., and Baker, D. H. (2003) Efficacy of an *E. coli* phytase expressed in yeast for releasing phytate-bound phosphorus in young chicks and pigs. *J. Anim. Sci.* **81,** 474–483.
5. Han Y. M., Wilson, D. B., and Lei, X. G. (1999) Expression of an *Aspergillus niger* phytase gene (phyA) in *Saccharomyces cerevisiae. Appl. Environ. Microbiol.* **65,** 1915–1918.
6. Han Y. M. and Lei, X. G. (1999) Role of Glycosylation in the functional expression of and *Aspergillus niger* phytase (phyA) in *Pichia pastoris. Arch. Biochem. Biophys.* **364,** 1, 83–90.

7. Rodriguez, E., Han, Y. M., and Lei, X.G. (1999) Cloning, sequencing, and expression of an *Escherichia coli* acid phosphatase/phytase gene (*app*A2) isolated from pig colon. *Biochem. Biophys. Res. Commun.* **257,** 117–123.

8. Rodiguez, E., Wood, Z. A., Karplus, A. P., and Lei, X. G. (2000) Site-directed mutagenesis improves catalytic efficiency and thermostability of *Escherichia coli* pH 2.5 acid phosphatase/phytase expressed in *Pichia pastoris. Arch. Biochem. Biophys.* **382,** 105–112.

9. Mullaney, E. J., Gibson, D. M., and Ullah, A. H. J. (1991) Positive identification of a lambda gtl1 clone containing a region of fungal phytase gene by immunoprobe and sequence verification. *J. Appl. Microbiol. Biotechnol.* **35,** 611–614.

10. Ullah, A. H. J. and Dischunger, H. C. Jr. (1993) *Aspergillus ficuum* phytase: complete primary structure elucidation by chemical sequencing. *Biochem. Biophys. Res. Commun.* **192,** 747–753.

11. Invitrogen catalog no. V825–20, V195–20 and V200–20, Carlsbad, CA.

12. Sambrook, J. and Russell D. W. (2001) *Molecular Cloning, A Laboratary Manual,* Third ed. Cold Spring Harbor Laboratory Press, Cold Spring Harbor, NY.

14

Glucose Dehydrogenase for the Regeneration of NADPH and NADH

Andrea Weckbecker and Werner Hummel

Summary

Glucose dehydrogenases (GDHs) occur in several organisms such as *Bacillus megaterium* and *Bacillus subtilis*. They accept both NAD$^+$ and NADP$^+$ as cofactor and can be used for the regeneration of NADH and NADPH. In order to demonstrate their applicability we coupled an NADP$^+$-dependent, *(R)*-specific alcohol dehydrogenase (ADH) from *Lactobacillus kefir* with the glucose dehydrogenase from *B. subtilis*. The ADH reduces prochiral ketones stereoselectively to chiral alcohols. The reduction requires NADPH, which was regenerated by the glucose dehydrogenase. Glucose dehydrogenase from *B. subtilis* (EC 1.1.1.47) is a tetramer with a molecular weight of 126,000. The enzyme shows a pH optimum at 8.0 and a broad temperature optimum at 45–50°C.

We investigated the conversion of acetophenone in a cell-free system with purified ADH and GDH. Furthermore, we constructed two plasmids containing the genes encoding ADH and GDH by inserting them one after the other. These two plasmids differ from each other in the order of the genes. Because of the low solubility of the compounds, we examined the reaction in a water/organic solvent two-phase system.

Key Words: Glucose dehydrogenase; *Bacillus subtilis;* alcohol dehydrogenase; *Lactobacillus kefir;* coexpression; whole-cell biotransformation; two-phase system.

1. Introduction

Glucose dehydrogenase (GDH, E.C. 1.1.1.47) catalyzes the oxidation of β-D-glucose to β-D-glucono-1,5-lactone with simultaneous reduction of the cofactor NADP$^+$ to NADPH or, to a lesser extent, NAD$^+$ to NADH. The enzyme occurs in a variety of organisms such as *Bacillus megaterium (1–4)*, *Bacillus subtilis (5–7)*, *Gluconobacter suboxydans (8)*, *Halobacterium mediterranei (9)*, *Thermoplasma acidophilum (10,11)*, and *Sulfolobus solfataricus (12)*.

From: *Methods in Biotechnology, Vol. 17: Microbial Enzymes and Biotransformations*
Edited by: J. L. Barredo © Humana Press Inc., Totowa, NJ

Glucose dehydrogenase accepts $NADP^+$ as well as NAD^+ as a cofactor. Therefore, this enzyme is a good candidate for the regeneration of NADPH or NADH. This is of interest in asymmetric reductions of prochiral ketone compounds to produce chiral hydroxy or amino acids or alcohols, which are valuable building blocks and starting materials for the synthesis of important drugs. The majority of the currently applied oxidoreductases depend on nicotinamide coenzymes. For NADH, the regeneration based on formate/formate dehydrogenase (FDH) has reached the most advanced stage of development *(13)*. Nevertheless, the only moderate stability and low specific activity of FDH are severe limitations, especially for industrial applications. For the regeneration of NADPH, there is no established method available.

One area of application of GDH is the regeneration of NADH. A few examples are known in literature. In the synthesis and conversion of 2-keto-6-hydroxyhexanoic acid to L-6-hydroxynorleucine by reductive amination using beef liver glutamate dehydrogenase, the GDH from *Bacillus* sp. regenerates NADH *(14)*. The same purpose serves the GDH from *Bacillus* sp. during the production of L-carnitine from 3-dehydrocarnitine by L-carnitine dehydrogenase from *Pseudomonas putida* IAM12014 *(15)*.

A further application of GDH is the regeneration of NADPH. For the asymmetric reduction of ethyl 4-chloro-3-oxobutanoate, an aldehyde reductase from *Sporobolomyces salmonicolor* was coupled and coexpressed with the GDH from *Bacillus megaterium* for use as an NADPH regenerator *(16)*. GDH from *Gluconobacter scleroides* KY3613 was used for cofactor recycling in the production of L-leucovorin *(17)*.

Glucose-6-phosphate dehydrogenase can also regenerate NADPH. Commercially available glucose-6-phosphate dehydrogenase was used to regenerate NADPH *(18)*, as well as glucose-6-phosphate dehydrogenase from *Leuconostoc mesenteroides (19)*.

We coupled the glucose dehydrogenase from *Bacillus subtilis* with an $NADP^+$-dependent *(R)*-specific alcohol dehydrogenase from *Lactobacillus kefir* overexpressed in *E. coli* for the stereoselective production of alcohols. Since *E. coli* cells do not have a sufficient cofactor regeneration system we developed a coexpression system to express both genes in *E. coli* and to perform whole-cell biotransformations.

2. Materials

1. pET-21a(+) expression system (Novagen, Madison, WI).
2. *E. coli* strains Tuner(DE3), BL21(DE3), Origami(DE3) (Novagen).
3. Oligonucleotide primers.
4. Restriction enzymes, T7 DNA polymerase, T4 DNA ligase.
5. Agarose.

6. Luria-Bertani (LB) medium: 10 g/L tryptone, 5 g/L yeast extract, 10 g/L sodium chloride, pH 7.5.
7. Ampicillin.
8. IPTG (isopropyl-β-D-thio-galactopyranoside).
9. Ultrasonic disintegration.
10. Disintegration buffer: 100 mM Tris-HCl, pH 7.2, and 1 mM MgCl$_2$.
11. Standard reaction buffer: 100 mM TEA, pH 7.5, and 1 mM MgCl$_2$.
12. Assay for ADH: 11 mM acetophenone in 50 mM TEA, pH 7.0, and 0.25 mM NADPH, monitoring the decrease in OD$_{340}$ at 30°C spectophotometrically for one minute.
13. Assay for GDH: 100 mM D-glucose, 75 mM Tris-HCl, pH 8.0, 2 mM NADP$^+$, and 1 mM MgCl$_2$, monitoring the increase in OD$_{340}$ at 30°C in a spectrophotometer for one minute.
14. SDS-PAGE (sodium dodecyl sulfate-polyacrylamide gel electrophoresis) equipment (Invitrogen, Karlsruhe, Germany).
15. Chromatography equipment.
16. Phenyl-650C (Tosoh Bioscience, Stuttgart, Germany).
17. Ultrafiltration tools (Amicon, Millipore, Bedford, MA).
18. Gas chromatography equipment.

3. Methods

3.1. Regeneration of NADPH by GDH

We isolated an NADP$^+$-dependent *(R)*-specific alcohol dehydrogenase from *Lactobacillus kefir*. This enzyme is a valuable catalyst for the asymmetric reduction of keto compounds. It has a broad substrate spectrum and reduces a wide range of aliphatic, aromatic and cyclic ketones *(20–23)*. In order to realize high efficiency under ecological and economical concerns the enzyme should be used as a whole-cell catalyst coupled with an intracellular enzymatic regenerating step.

For the regeneration of NADPH a few enzymes were taken into account as described above. We decided to work with the glucose dehydrogenase from *Bacillus subtilis*. Glucose is an inexpensive substrate, the specific activity of this enzyme is similar to that of the ADH from *L. kefir*, and the sequence of this enzyme is known *(6)*. The principle of the reaction scheme we tried to perform is shown in **Fig. 1.**

3.2. Cloning and Expression of the GDH Gene

Chromosomal DNA of *B. subtilis* strain 168 was prepared. To amplify the GDH gene from this chromosomal DNA by PCR, upstream and downstream primers were designed on the basis of the known GDH nucleotide sequence. The PCR-generated DNA fragment was ligated into pET-21a(+) cleaved with *Sac*I and *Hind*III and then transformed into *E. coli* BL21. After ampicillin

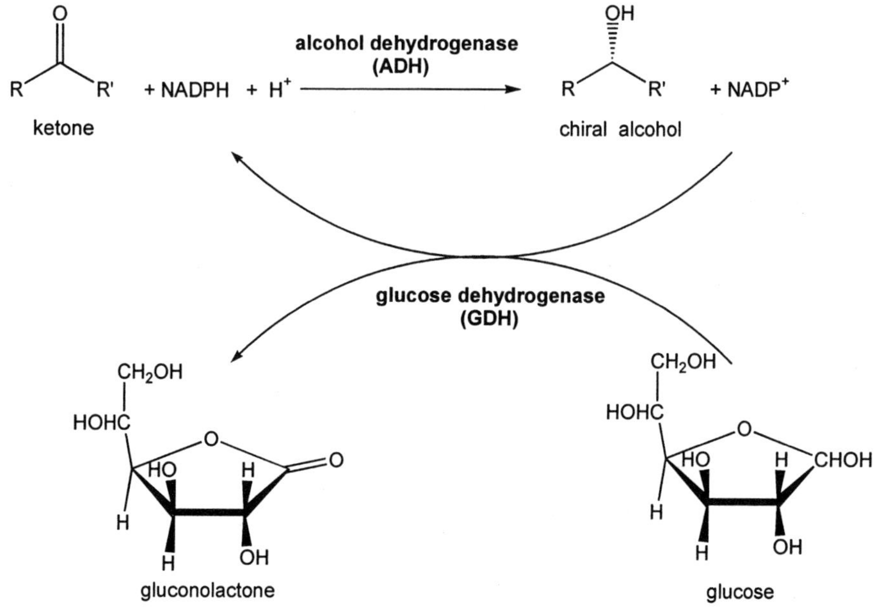

Fig. 1. Coupling of ADH and GDH for the reduction of prochiral ketones and the simultaneous NADPH regeneration.

selection, several clones were picked and the nucleotide sequence of the DNA was compared with the published sequence. The alignment showed 100% identity; GDH consists of 786 bp encoding a 261-amino acid protein.

3.3. Characterization of GDH

After cloning, we characterized the recombinant GDH biochemically. Purification of the enzyme was performed by hydrophobic interaction chromatography on phenyl-650C:

1. *E. coli* culture was grown at 37°C in LB medium containing ampicillin for 5 h and harvested by centrifugation.
2. Cells were disrupted by sonication.
3. The debris was removed by centrifugation at 11,000g for 10 min at 4°C.
4. The supernatant was loaded onto a 30 mL phenyl-650C column equilibrated with 50 mM triethanolamine buffer, pH 7.0, 1.5 M (NH$_4$)$_2$SO$_4$ and 1 mM MgCl$_2$. The protein was eluted with a total 300-mL gradient of 1.5 to 0 M (NH$_4$)$_2$SO$_4$.

GDH enzymatic activity was measured as follows:

1. The reaction mixture contained 75 mM Tris-HCl, pH 8.0, 2 mM NADP$^+$, and 0.1 M D-glucose.

Table 1
Kinetic Constants and Properties of Some Glucose Dehydrogenases

Organism	K_M value NADP+ [mM]	K_M value NAD+ [mM]	K_M value β-D-glucose (NADP+) [mM]	pH optimum	Optimum temperature [°C]	Reference
Bacillus subtilis	0.26	0.23	42.9	8.0	–	*5*
Gluconobacter suboxydans	0.01	– (no activity with NAD+)	5	8.5–9.0	50	*8*
Sulfolobus solfataricus	0.03	1.2	0.44	8.0	77	*12*

2. The enzyme activity was measured by following the increase in absorbance of NADPH at 340 nm.
3. One unit of GDH activity was defined as the amount of enzyme catalyzing the formation of 1 μmol NADPH per min at 30°C.

After ultrafiltration, the enzyme was purified about 15-fold and the specific activity was 255 U/mg. The protein was characterized biochemically. Properties of some GDHs published in literature are shown in **Table 1.**

GDHs from various organisms show different properties. In contrast to the enzymes from *B. subtilis* and *S. solfataricus,* which accept both NADP+ and NAD+ as coenzyme, GDH from *G. suboxydans* is dependent on NADP+ only. The pH optima for the listed enzymes are similar to each other while the K_m values vary. The K_m values for NADP+ and NAD+ for the enzyme from *B. subtilis* are similar to each other while the *S. solfataricus* protein shows a much lower value for NADP+. Particular variations are noticed for the K_m values for β-D-glucose with NADP+ as coenzyme. The value for the enzyme from *B. subtilis* is about 100-fold higher than that from *S. solfataricus.*

Furthermore we investigated the GDH concerning properties that are important for the performance of coupled experiments. No cross reactions with educt and product of the standard ADH reaction, the reduction of acetophenone to *(R)*-phenylethanol, were observed.

Because of the instability of glucono-1,5-lactone, the reverse reaction of the enzyme, the reduction of D-glucono-1,5-lactone to β-D-glucose was not further investigated.

3.4. Reduction of Acetophenone in a Cell-Free System

2 U purified recombinant ADH and 4 U purified recombinant GDH were added to a reaction mixture containing 10 mM acetophenone, 100 mM glucose,

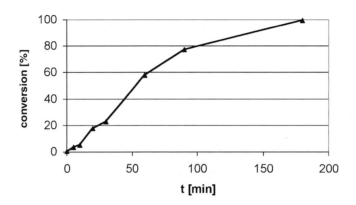

Fig. 2. Conversion of acetophenone by cell-free system containing ADH and GDH after incubation at 37°C.

and 1 μ*M* NADP⁺. This mixture was incubated at 37°C. After specified times, samples were taken and analyzed by gas chromatography (**Fig. 2**).

After an incubation time of 3 h, acetophenone was completely reduced to *(R)*-phenylethanol.

3.5. Coexpression of ADH and GDH

To perform whole-cell biotransformations with high efficiencies we cloned GDH and ADH encoding genes in pET-21a(+). The ADH encoding gene was located upstream of the gene coding for GDH (pAW-3). To compare the expression rates we constructed a second plasmid (pAW-4) where GDH encoding gene was placed upstream (**Fig. 3**).

Both plasmids were transformed in *E. coli* Tuner(DE3), BL21(DE3), and Origami(DE3) cells. The activities of both enzymes were investigated and the expression rates of the two coexpression systems were compared (**Table 2**).

Cells carrying pAW-3 and pAW-4 showed both ADH and GDH activities. The ADH activity was measured as follows:

1. The reaction mixture consists of 50 m*M* triethanolamine, pH 7.0, 0.25 m*M* NADPH, 1 m*M* MgCl₂, and 11 m*M* acetophenone.
2. The enzyme activity was measured by following the decrease in absorbance of NADPH at 340 nm.
3. One unit of ADH activity was defined as the amount of enzyme catalyzing the oxidation of 1 μmol NADPH per min at 30°C.

ADH activity was threefold higher in the pAW-3 system than in pAW-4, while the ratios were inverted in pAW-4 (*see* **Note 1**). Activities with the

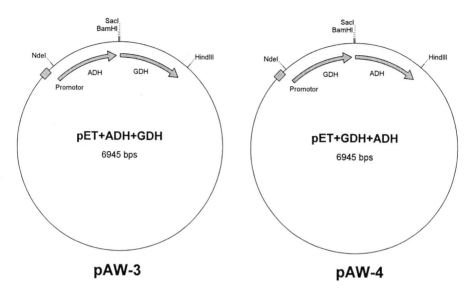

Fig. 3. Structure of the expression plasmids pAW-3 and pAW-4 carrying ADH and GDH encoding genes.

Table 2
Comparison of Enzyme Activities and Expression Ratios of Plasmids pAW-3 and pAW-4

E. coli expression strain	pAW-3 (ADH-GDH)			pAW-4 (GDH-ADH)		
	Spec. act. ADH [U/mg]	Spec. act. GDH [U/mg]	GDH/ADH [%]	Spec. act. ADH [U/mg]	Spec. act. GDH [U/mg]	ADH/GDH [%]
Tuner	40.62	11.19	27.5	9.14	32.90	27.7
BL21	41.60	12.93	31.1	13.77	58.01	23.7
Origami	67.68	12.40	32.9	5.29	17.28	30.6

E. coli BL21 cells were in both cases higher than with *E. coli* Tuner and Origami.

Additionally, we compared both plasmids concerning the reduction of acetophenone. To a mixture of 20 m*M* acetophenone and 200 m*M* glucose in TEA 50 m*M*, pH 7.5, we added *E. coli* BL21 pAW-3 cells. The reaction mixture was incubated at 30°C and gently shaken. After certain times we took samples and analyzed them by gas chromatography (**Fig. 4**).

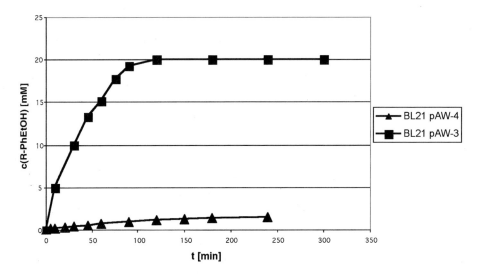

Fig. 4. Comparison of pAW-3 and pAW-4 concerning the reduction of acetophenone to *(R)*-phenylethanol.

Surprisingly, the conversion rate by *E. coli* BL21 pAW-4 was only 10% of the conversion rate reached by pAW-3. The reaction seems to be substantially limited by the activity of the ADH, and an excess of ADH must be available whereas an excess of GDH did not support the reduction. Therefore, the following experiments were carried out with cells carrying pAW-3.

A problem in carrying out whole-cell biotransformations is the limited permeability of the membrane. Educts, products, cofactors, and other compounds are not able to pass the membrane in many cases so that the reaction process becomes aggravating (*see* **Note 2**).

3.6. Conversion of Acetophenone with E. coli Cells Coexpressing ADH and GDH Genes in a Two-Phase System

Because of the low solubility of acetophenone and *(R)*-phenylethanol in aqueous systems, we decided to introduce a water/organic solvent two-phase system to the acetophenone reduction. For this purpose we investigated several organic solvents (**Fig. 5**).

1. A reaction mixture consists of 50 m*M* triethanolamine buffer, pH 7.5, 1 m*M* MgCl$_2$, 500 m*M* D-glucose, 20 m*M* acetophenone, and 0.1 g (wet weight) cells in a final volume of 100 mL, was mixed with 50 mL of several organic solvents

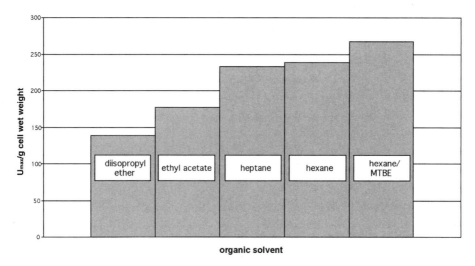

Fig. 5. Rate of conversion of acetophenone in various organic solvents used in two-phase systems (MTBE = methyl-*t*-butyl ether).

(diisopropyl ether, ethyl acetate, heptane, hexane, and hexane/MTBE (methyl-*t*-butyl ether) (1:1), respectively in a 500-mL reaction vessel.

2. The mixture was stirred at 30°C and 120 rpm.
3. The concentrations of acetophenone and *(R)*-phenylethanol after several times were determined by gas chromatography.

The solvent with the highest reaction rate was the 1:1 mixture of hexane/MTBE. Applying this system, we reached high solubilities of educt and product. For long-term applications, however, this solvent system has the drawback of a significantly reduced stability of the cells. Monitoring OD_{660} of the suspension as a measure of integrity of the cells over a period of 3 d, we observed that the cells suspended in the water/organic solvent two-phase system were quite more sensitive against solvents and lysis began after only a short time **(Fig. 6)**.

This figure also demonstrates that the activity to reduce acetophenone is significantly decreased in the presence of organic solvents. In conclusion, we reached higher solubilities of the compounds by introduction of the water/organic solvent two-phase system, and the recovery and isolation of the compounds can be reached more easily. On the other hand, the stability of the *E. coli* cells was reduced severely in the presence of organic solvents (*see* **Note 3**).

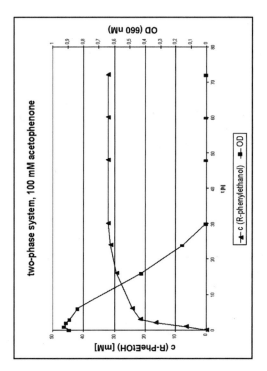

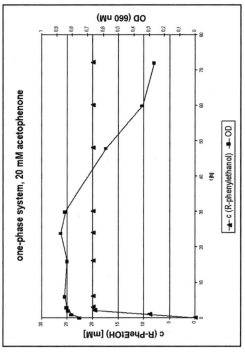

Fig. 6. Comparison of the long-time stability of the cells in one- and two-phase system.

4. Notes

1. To perform and study whole-cell biotransformations with high efficiencies we coupled the NADP⁺-dependent, *(R)*-specific ADH from *L. kefir* that reduces prochiral ketones stereoselective to enantiomerically pure alcohols with the GDH from *B. subtilis* for the regeneration of NADPH. We constructed two different plasmids. Each one contained both genes in reverse order (pAW-3 and pAW-4). Both systems showed ADH as well as GDH activity. Cells containing pAW-3 that contains the ADH as first gene showed much higher conversion rates than pAW-4 with the GDH as first gene. Although the development of a whole-cell system bearing both enzyme activities led to a suitable catalyst, some problems arose from the detailed characterization studies. A problem with coexpression systems is the different expression rate of the genes cloned one after the other in one single vector. The gene in the immediate vicinity of the promotor is getting expressed with much higher rates than the genes that have a longer distance to the promoter. A solution to overcome this problem may be the use of one single plasmid with two promoters so that each gene has its own promoter. Furthermore, it is possible to use two different but compatible plasmids to adjust the amounts of enzyme activities *(16)*. Alternatively, each of the enzymes can be produced in a different *E. coli* strain so that the strains can be mixed in different amounts according to the different enzyme activities *(24)*.

2. A difficulty that occurs with carrying out whole-cell biotransformations is the limited permeability of the membrane so that educts, products, cofactors, and other compounds are not able to pass the membrane. To solve this problem the permeability of the membrane can be increased by using toluene or other agents *(24)*.

3. Another serious question to be solved is the poor solubility of many organic compounds in aqueous systems. The introduction of a water/organic solvent two-phase system can provide a solution. The solubility of the appropriate compounds is increased clearly and the recovery of the product is much easier. However, a limited stability of *E. coli* cells in organic solvents was observed in our studies. To overcome this problem strains that are more stable in organic medium should be used as the host strain for the coexpression system.

Acknowledgments

Part of this work was financially supported by the Deutsche Bundesstiftung Umwelt (Verbundprojekt Biokatalyse).

5. References

1. Pauly, H. E. and Pfleiderer, G. (1975) D-glucose dehydrogenase from *Bacillus megaterium* M 1286: purification, properties and structure. *Hoppe Seylers Z Physiol. Chem.* **356,** 1613–1623.

2. Jany, K. D., Ulmer, W., Froschle, M., and Pfleiderer, G. (1984) Complete amino acid sequence of glucose dehydrogenase from *Bacillus megaterium. FEBS Lett.* **165,** 6–10.

3. Heilmann, H. J., Magert, H. J., and Gassen, H. G. (1988) Identification and isolation of glucose dehydrogenase genes of *Bacillus megaterium* M1286 and their expression in *Escherichia coli*. *Eur. J. Biochem.* **174,** 485–490.
4. Mitamura, T., Urabe, I., and Okada, H. (1989) Enzymatic properties of isozymes and variants of glucose dehydrogenase from *Bacillus megaterium*. *Eur. J. Biochem.* **186,** 389–393.
5. Fujita, Y., Ramaley, R., and Freese, E. (1977) Location and properties of glucose dehydrogenase in sporulating cells and spores of *Bacillus subtilis*. *J. Bacteriol.* **132,** 282–293.
6. Lampel, K. A., Uratani, B., Chaudhry, G. R., Ramaley, R. F., and Rudikoff, S. (1986) Characterization of the developmentally regulated *Bacillus subtilis* glucose dehydrogenase gene. *J. Bacteriol.* **166,** 238–243.
7. Hilt, W., Pfleiderer, G., and Fortnagel, P. (1991) Glucose dehydrogenase from *Bacillus subtilis* expressed in *Escherichia coli*. I: Purification, characterization and comparison with glucose dehydrogenase from *Bacillus megaterium*. *Biochim. Biophys. Acta* **1076,** 298–304.
8. Adachi, O., Kazunobu, M., Shinagawa, E., and Ameyama, M. (1980) Crystallization and characterization of NADP-dependent D-glucose dehydrogenase from *Gluconobacter suboxydans*. *Agric. Biol. Chem.* **44,** 301–308.
9. Bonete, M. J., Pire, C., Llorca, F. I., and Camacho, M. L. (1996) Glucose dehydrogenase from the halophilic archaeon *Haloferax mediterranei*: enzyme purification, characterisation and N-terminal sequence. *FEBS Lett.* **383,** 227–229.
10. Smith, L. D., Budgen, N., Bungard, S. J., Danson, M. J., and Hough, D. W. (1989) Purification and characterization of glucose dehydrogenase from the thermoacidophilic archaebacterium *Thermoplasma acidophilum*. *Biochem. J.* **261,** 973–977.
11. Bright, J. R., Byrom, D., Danson, M. J., Hough, D. W., and Towner, P. (1993) Cloning, sequencing and expression of the gene encoding glucose dehydrogenase from the thermophilic archaeon *Thermoplasma acidophilum*. *Eur. J. Biochem.* **211,** 549–554.
12. Giardina, P., de Biasi, M. G., de Rosa, M., Gambacorta, A., and Buonocore, V. (1986) Glucose dehydrogenase from the thermoacidophilic archaebacterium *Sulfolobus solfataricus*. *Biochem. J.* **239,** 517–522.
13. Hummel, W. (1999) Large-scale applications of NAD(P)-dependent oxidoreductases: recent developments. *Trends Biotechnol.* **17,** 487–492.
14. Hanson, R. L., Schwinden, M. D., Banerjee, A., Brzozowski, D. B., Chen, B. C., Patel, B. P., et al. (1999) Enzymatic synthesis of L-6-hydroxynorleucine. *Bioorg. Med. Chem.* **7,** 2247–2252.
15. Lin, S.-S., Miyawaki, O., and Nakamura, K. (1999) Continuous production of L-carnitine with NADH regeneration by a nanofiltration membrane reactor with coimmobilized L-carnitine dehydrogenase and glucose dehydrogenase. *J. Biosci. Bioeng.* **87,** 361–364.
16. Kataoka, M., Yamamoto, K., Kawabata, H., Wada, M., Kita, K., Yanase, H., and Shimizu, S. (1999) Stereoselective reduction of ethyl 4-chloro-3-oxobutanoate by *Escherichia coli* transformant cells coexpressing the aldehyde reductase and glucose dehydrogenase genes. *Appl. Microbiol. Biotechnol.* **51,** 486–490.

17. Eguchi, T., Kuge, Y., Inoue, K., Yoshikawa, N., Mochida, K., and Uwajima, T. (1992) NADPH regeneration by glucose dehydrogenase from *Gluconobacter scleroides* for l-leucovorin synthesis. *Biosci. Biotechnol. Biochem.* **56,** 701–703.
18. Hummel, W. (1990) Enzyme-catalyzed synthesis of optically pure R(+)-phenylethanol. *Biotechnol. Lett.* **12,** 403–408.
19. Wong, C.-H. and Whitesides, G. M. (1981) Enzyme-catalyzed organic synthesis: NAD(P)H cofactor regeneration by using glucose 6-phosphate and the glucose-6-phosphate dehydrogenase from *Leuconostoc mesenteroides. J. Am. Chem. Soc.* **103,** 4890–4899.
20. Hummel, W., Boermann, F., and Kula, M.-R. (1989) Purification and characterization of an acetoin dehydrogenase from *Lactobacillus kefir* suitable for the production of (+)-acetoin. *Biocatalysis* **2,** 293–308.
21. Bradshaw, C. W., Hummel, W., and Wong, C.-H. (1992) *Lactobacillus kefir* alcohol dehydrogenase: a useful catalyst for synthesis. *J. Org. Chem.* **57,** 1532–1536.
22. Hummel, W. and Riebel, B. (1996) Chiral alcohols by enantioselective enzymatic oxidation. *Enzyme Engineering* **13,** 713–716.
23. Kruse, W., Hummel, W., and Kragl, U. (1996) Alcohol-dehydrogenase-catalyzed production of chiral hydrophobic alcohols. A new approach leading to a nearly waste-free process. *Recueil des Travaux Chimiques des Pays-Bas* **115,** 239–243.
24. Wilms, B., Wiese, A., Syldatk, C., Mattes, R., and Altenbuchner, J. (2001) Development of an *Escherichia coli* whole cell biocatalyst for the production of L-amino acids. *J. Biotechnol.* **86,** 19–30.

15

Acetate Kinase From *Methanosarcina thermophila,* a Key Enzyme for Methanogenesis

Prabha Iyer and James G. Ferry

Summary

Materials and procedures are described for the overproduction and purification of acetate kinase from *Methanosarcina thermophila.* Methods are detailed for large-scale preparation of the unaltered enzyme and for overproduction and one-step purification of the histidine-tagged recombinant acetate kinase. Also included is a method for routine assay of the enzyme, and enzyme-linked assay procedures for obtaining kinetic constants in the forward and reverse reaction directions. Finally, a method is included for the overproduction and purification of a variant of the acetate kinase with an altered K_m for acetate. A method is presented utilizing the variant for determination of acetate levels in biological samples.

Key Words: Acetate kinase; methanogenesis; overexpression; purification; acetate quantification.

1. Introduction

Methanogenesis is the term applied to the decomposition of complex organic matter to methane in oxygen-free (anoxic) environments, which is a significant component of the global carbon cycle *(1)*. The process involves a food chain of at least three metabolic groups of microbes in which the fermentative and acetogenic groups degrade the organic matter to hydrogen gas, formate, carbon dioxide, and acetate. The methanogenic group converts these products to methane, completing the process. At least two thirds of the methane is derived from acetate. The enzyme acetate kinase is found in all three groups, where it catalyzes the terminal step in acetate production (the forward reaction in equation [1]) for the fermentative and

$$CH_3CO_2PO_3^{2-} + ADP \rightleftharpoons CH_3COO^- + ATP \qquad (1)$$

acetogenic groups, and the first step (reverse direction in equation 1) in the acetate-utilizing genus *Methanosarcina* of the methanogenic group. Thus,

From: *Methods in Biotechnology, Vol. 17: Microbial Enzymes and Biotransformations*
Edited by: J. L. Barredo © Humana Press Inc., Totowa, NJ

acetate kinase is a key enzyme required for production and consumption of acetate, the major intermediate in methanogenesis. The enzyme from *Methanosarcina thermophila* has emerged as the model for physiological, biochemical, and biotechnological investigations. This chapter describes methods for the overexpression and purification of the enzyme.

The measurement of acetate in biological fluids has clinical and industrial applications. Traditional methods for the determination of acetate are based on gas chromatography. While this method has a low detection limit, it requires both expensive equipment and specially trained personnel. Assays that employ acetate kinase are simpler, require less time per sample, and utilize inexpensive equipment; however, the cost of commercially available acetate kinases are prohibitive and have reported K_m values as high as 120–300 mM acetate, which decreases the sensitivity of detection. Here is reported the use of a variant of acetate kinase from *M. thermophila* for measurement of acetate in biological samples that represents an improvement in existing methods.

2. Materials

1. Spectrophotometer capable of continuous monitoring in the visible light range.
2. 2.0 M hydroxylamine.
3. 2.0 M potassium acetate.
4. 200 mM MgCl$_2$.
5. 100 mM ATP.
6. 145 mM Tris-HCl, pH 7.5.
7. 10% trichloroacetic acid.
8. 2.5% FeCl$_3$ in 2 N NCl.
9. 200 mM Tricine-KOH buffer, pH 8.2.
10. 50 mM phosphoenolpyruvate.
11. 10 mM NADH.
12. 10 mM dithiothreitol.
13. Pyruvate kinase.
14. Lactate dehydrogenase.
15. Hexokinase.
16. Glucose-6-phosphate dehydrogenase in 100 mM Tris-HCl, pH 7.4.
17. 20 mM acetyl phosphate.
18. *Escherichia coli* strain BL21 (DE3) *(2)*.
19. Plasmids pMTL107 *(2)*, pET*ack (5)*, and pET*ack*E79D *(5)*.
21. Luria-Bertani (LB) broth: 10 g/L bacto-tryptone, 5 g/L bacto-yeast extract, 10 g/L NaCl, and 20 g/L agar. Adjust to pH 7.0 with 5 N NaOH. Autoclave at 121°C for 15 min.
22. IPTG (isopropyl-β-D-thio-galactopyranoside).
23. Lactose.
24. Ampicillin.
25. Ni-nitrilotriacetic acid silica spin kit (Qiagen, Valencia, CA).
26. Low-speed centrifuge and large-capacity rotor.

27. French press and cell.
28. Streptomycin sulfate.
29. High-performance liquid chromatography system (HPLC).
30. Q-Sepharose 5/10 Fast Flow column (Pharmacia, Piscataway, NJ).
31. Phenyl Sepharose 26/10 column (Pharmacia).
32. Mono-Q 10/10 column (Pharmacia).
33. Buffer A: 25 mM Bis Tris-HCl, pH 6.5.
34. Buffer B: 50 mM Tris-HCl, pH 7.2.
35. Buffer C: 25 mM Tris-HCl, pH 7.6.
36. 1.0 M KCl.
37. 0.9 M ammonium sulfate.
38. 50 mM NaH$_2$PO$_4$, pH 8.0, and 10 mM imidazole.
39. 50 mM NaH$_2$PO$_4$, pH 8.0, and 20 mM imidazole.
40. 50 mM NaH$_2$PO$_4$, pH 7.0, 300 mM NaCl, and 250 mM imidazole.
41. Apple and orange juice (Minute Maid Co., Houston, TX).
42. Sheep blood serum (Penn State sheep facility, University Park, PA).

3. Methods

3.1. Overproduction and Purification of Acetate Kinase

3.1.1. Enzyme Assays

The basis of the hydroxamate assay (*see* **Note 1**) in the reverse direction of acetyl phosphate synthesis (equation [1]) is shown in equations (2) and (3).

$$\text{acetyl phosphate} + \text{hydroxylamine} \rightarrow \text{acetyl hydroxamate} \qquad (2)$$
$$\text{acetyl hydroxamate} + \text{FeCl}_3 \rightarrow \text{ferric acetylhydroxamate} \qquad (3)$$

The colored ferric acetylhydroxamate complex (reaction [3]) is measured spectrophotometrically at 540 nm.

1. Prepare the reaction mixture (333 μL) containing (in final concentrations): 705 mM hydroxylamine, 200 mM potassium acetate, 10 mM MgCl$_2$, and 10 mM ATP in 145 mM Tris-HCl, pH 7.5.
2. Prewarm the mixture at 37°C prior to initiating the reaction with 50 μL of enzyme solution.
3. After incubation for 10 min, stop the reaction by the addition of 333 μL of 10% trichloroacetic acid (*see* **Note 2**).
4. Initiate the color reaction by the addition of 333 μL of 2.5% FeCl$_3$ in 2 N HCl.
5. Allow 15 min for color development before recording the OD$_{540}$.

The enzyme-linked assay in the direction of acetyl-phosphate synthesis couples ADP formation to the oxidation of NADH ($\varepsilon_{340} = 6.22$ per mM per cm) through pyruvate kinase (equation [4]) and lactic dehydrogenase (equation [5]).

$$\text{phosphoenolpyruvate} + \text{ADP} \rightarrow \text{pyruvate} + \text{ATP} \qquad (4)$$
$$\text{pyruvate} + \text{NADH} \rightarrow \text{lactate} + \text{NAD} \qquad (5)$$

1. Perform the assay at 37°C with an assay mixture containing (final concentrations): 100 mM Tricine-KOH, pH 8.2, 200 mM potassium acetate, 1.5 mM ATP, 2 mM MgCl$_2$, 2 mM phosphoenolpyruvate, 0.4 mM NADH, 2 mM dithiothreitol, 9 units of pyruvate kinase, and 26 units of lactic dehydrogenase. The combined activities of the linked enzymes are routinely assayed to determine that they are not limiting.
2. Initiate assays by addition of acetate kinase to the assay mixture.
3. Monitor OD$_{340}$ with time to determine the rate.

Acetate kinase activity in the direction of acetate synthesis links ATP formation to NADP reduction through hexokinase and glucose-6-phosphate dehydrogenase.

1. The 0.5 mL assay mixture contains (in final concentrations): 0.2 mM dithiothreitol, 10 mM MgCl$_2$, 4.4 mM glucose, 1 mM NADP, 10 units of hexokinase, and 10 units of glucose 6-phosphate dehydrogenase in 100 mM Tris-HCl, pH 7.4.
2. Add the acetate kinase to the mixture prewarmed to 37°C.
3. Initiate the reaction with 20 mM acetyl phosphate.
4. Monitor the reduction of NADP at 340 nm.

3.1.2. Overexpression and Purification of Acetate Kinase

Overexpression of the unmodified enzyme (*see* **Note 3**) is accomplished with plasmid pMTL107 *(2)*, which is based on the T7 expression system *(3,4)*.

1. Transform competent *E. coli* strain BL21 (DE3) cells and plate overnight.
2. Culture cells in standard LB broth supplemented with 100 µg/mL of ampicillin.
3. Induce six 1-L cultures at an OD$_{660}$ of 1.6 with 1% (final concentration w/v) Bacto-lactose or 0.4 mM (final concentration) IPTG.
4. Harvest the cells by centrifugation at 13,000g for 10 min at 4°C. The cell pellets can be stored for several months at −80°C without loss of enzyme activity.
5. Resuspend approximately 45 g of cell pellet to a final volume of 130 mL in buffer A.
6. Lyse cells by one passage through a French pressure cell (20,000 psi).
7. Centrifuge the lysate at 78,000g for 25 min at 4°C.
8. Add streptomycin sulfate (1% final concentration, w/v) to the supernatant solution and stir with a glass rod until dissolved.
9. Centrifuge the solution as in **step 7.**
10. Load the supernatant solution onto a 5/10 Q-Sepharose Fast Flow column equilibrated with buffer A.
11. Develop the column with a 1.8-L linear gradient of between 0.0 and 1.0 M KCl at a rate of 6 mL per min.
12. Collect fractions (12 mL) and assay using the hydroxamate assay (**Subheading 3.1.1.**). The peak fraction of activity elutes in approximately 250 mL KCl.
13. Pool the fractions with the highest activities, to which an equal volume of buffer A is added.
14. Load the enzyme solution (100 mL) onto a phenyl Sepharose 26/10 column equilibrated with buffer B containing 1.8 M ammonium sulfate.

15. Elute the enzyme with a decreasing linear gradient of 0.9 to 0.0 M ammonium sulfate at 3 mL per min and collect 6-mL fractions. The enzyme elutes in approximately 720 mM ammonium sulfate.
16. Pool the fractions with the highest activity, dilute 20-fold with buffer C, and load on a Mono-Q 10/10 column equilibrated with buffer C. The purified acetate kinase elutes at 230 mM KCl using a 200-mL 0.0 to 1.0 M KCl gradient at 2 mL per min.

For overexpression of the histidine-tagged acetate kinase, a T7-based expression vector pET*ack (5)* is used. In this plasmid, a 60-nucleotide leader sequence with 6 tandem histidine codons are fused in-frame to the 5′ end of the wild-type *ack* gene.

1. Transform *E. coli* BL21 (DE3) with the expression vector.
2. Inoculate into 50 mL of LB medium containing 100 µg/mL of ampicillin, and grow at 37°C.
3. At an OD_{600} of 0.7, add IPTG to a final concentration of 1 mM.
4. After 1.5–2.0 h, harvest the cells by centrifugation at 13,000g for 10 min at 4°C. The cell pellet can be stored at –80°C for several months without loss of enzyme activity. The recombinant acetate kinase is purified using the Ni-nitrilotriacetic acid silica spin kit.
5. Lyse cells by one passage through a French pressure cell (20,000 psi).
6. Apply lysate to the spin column (5 mg protein per mL of resin) and wash the column with 5 bed volumes of 50 mM phosphate buffer, pH 8.0, containing (final concentration) 10 mM imidazole, followed by a second wash with 5 bed volumes of the same buffer containing (final concentration) 20 mM imidazole.
7. Elute the enzyme with 3 bed volumes of 50 mM phosphate buffer, pH 7.0, containing (final concentrations) 300 mM NaCl and 250 mM imidazole.

3.2. Measurement of Acetate in Biological Samples

The histidine-tagged E97D variant acetate kinase from *M. thermophila* is overproduced from plasmid pET*ack*E97D *(5)* and purified as described in **Subheading 3.1.2.** All experiments utilize the hydroxamate assay (**Subheading 3.1.1.**). The volume of the test sample added to the assay mixture is between 90 and 125 µL.

Previous studies *(5)* showed that acetate kinase variant E97D from *M. thermophila* has nearly an eightfold lower K_m for acetate (*see* **Note 4**) than the wild-type enzyme (**Table 1**). The K_m value is severalfold lower than those reported for other commercially available acetate kinases from *Bacillus stearothermophillus* (120 mM) *(6)* or *E. coli* (7–300 mM) *(6,7)*; thus, the variant enzyme is evaluated in a method to determine acetate concentrations in biological samples. The method is evaluated for detection of acetate in serum, orange juice (*see* **Note 5**), and apple juice. The sensitivity and accuracy of the method are assessed by estimating the recovery of acetate from samples to which known amounts of acetate have been added (**Fig. 1**). Recovery is nearly quantitative

Table 1
**Kinetic Constants of Wild-Type and E97D Acetate Kinases
from *M. thermophila***

Enzyme	K_m acetate (mM)	k_{cat} (s^{-1})	k_{cat}/K_m (s^{-1}/mM)
Wild-type	22	1042	47
E97D	2.8	190	68

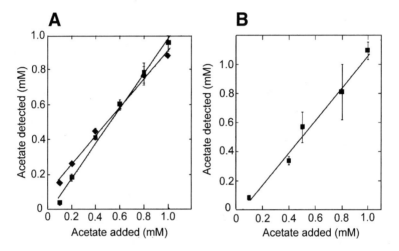

Fig. 1. Estimation of acetate concentrations in biological samples. The indicated concentrations of potassium acetate are added to orange juice (■) and apple juice (◆) (**A**) and serum (**B**). Acetate is estimated using the method described in **Subheading 3.1.1.**

between 0.1 and 1.0 m*M* acetate. The method is sufficiently sensitive to consistently detect acetate levels as low as 0.1 m*M* and is linear up to 1 m*M* without dilution of the sample. Although the enzyme displays a K_m of 24 m*M* propionate and a k_{cat}/K_m of only 3 (data not shown), propionate concentrations between 0.1 and 1 m*M* produce an absorbance between 40 and 50% of that observed for acetate (data not shown). Thus, the method is not appropriate for samples in which propionate is suspected in concentrations that would interfere with the detection of acetate. The specificity of the enzyme was further examined with several other compounds with the potential to serve as substrates (**Table 2**), none of which substituted for acetate.

This method for determining acetate in biological samples has several advantages over those currently used. This enzyme can be overproduced in *E. coli* in

Table 2
Reactivity of Acetate Kinase Toward Selected
Compounds Compared to Acetate

Compound	Reactivity (%)
Acetate	100
Butyrate	2
Ethanol	1
Formate	2
Glycerol	1
Glycine	1
Glycolic acid	4

The enzyme reaction mixture contains 5 mM of each compound. The assay is performed as described in **Subheading 3.1.1.** and the OD$_{540}$ is compared.

sufficient quantities so as to be economical and useful on a large scale. The K_m for acetate of this variant acetate kinase is an order of magnitude lower than those of enzymes used in current commercial assays; thus, the method reported here is likely to be severalfold more sensitive. The only instrumentation required is a basic spectrophotometer capable of measuring absorbance in the visible range of the spectrum. The entire procedure takes about 30 min and no specialized training of personnel is required. Extraction of acetate from the sample is not necessary and the sample requires no extensive preparation. No pre-precipitation of serum proteins is required, as the proteins precipitate on addition of trichloroacetic acid during the assay and are removed by centrifugation before measuring the absorbance. This method should also be easily adapted to determine acetate in solid foods such as cheese or mayonnaise. The only modification required would be to suspend the foodstuffs in the assay buffer and filter the supernatant, which can then directly be used in the assay. Most of the reagents can be stored at 4°C and are stable for several months. Only acetate kinase and ATP require storage below 0°C and these can be stored for long periods without compromising the effectiveness of the assay.

4. Notes

1. Two assays are described for activity in the direction of acetyl-phosphate synthesis. The hydroxamate assay first described in 1954 *(8)* is the simplest and most convenient; however, it has the potential to inhibit the enzyme *(9)* and is best suited for routine activity measurements and not for determining kinetic parameters. The *M. thermophila* enzyme is inhibited noncompetitively (J. G. Ferry, unpublished results). The enzyme-linked assay in the direction of acetyl-phosphate synthesis, albeit more

cumbersome, is more accurate and has the advantage of continuous monitoring to conveniently determine rates.

2. Use caution when using trichloroacetic acid. Wear protective gloves and eyewear.
3. Methods for overexpression and purification of the unmodified and histidine-tagged acetate kinase are presented. Yields of up to 65 mg are typical for the unmodified enzyme. The facile purification of the histidine-tagged enzyme is especially useful for purification of site-specific amino acid replacement variants, although lower in yield; nonetheless, scale-up of this method is possible.
4. The variant also exhibits a lower k_{cat} than does the wild-type so that the catalytic efficiency (k_{cat}/K_m) is similar to that of the wild-type. Since the enzyme is from a thermophilic organism, no activity is lost on storage at 21°C for up to 6 h. The variant enzyme can be stored at –20°C for at least a year without loss in activity. It can also be lyophilized and retains full activity upon reconstitution.
5. Pulp particles from orange juice are removed by filtration through a 0.2-μm syringe filter.

Acknowledgments

This work was supported by Department of Energy grant DE-FG02-95ER20198 and National Institutes of Health grant GM44661.

References

1. Ferry, J. G. (1997) Methane: small molecule, big impact. *Science* **278,** 1413–1414.
2. Latimer, M. T. and Ferry, J. G. (1993) Cloning, sequence analysis, and hyperexpression of the genes encoding phosphotransacetylase and acetate kinase from *Methanosarcina thermophila. J. Bacteriol.* **175,** 6822–6829.
3. Tabor, S. and Richardson, C. C. (1985) A bacteriophage T7 RNA polymerase/promoter system for controlled exclusive expression of specific genes. *Proc. Natl. Acad. Sci. USA* **82,** 1074–1078.
4. Studier, F. W. and Moffatt, B. A. (1986) Use of bacteriophage T7 RNA polymerase to direct selective high-level expression of cloned genes. *J. Mol. Biol.* **189,** 113–130.
5. Singh-Wissmann, K., Ingram-Smith, C., Miles, R. D., and Ferry, J. G. (1998) Identification of essential glutamates in the acetate kinase from *Methanosarcina thermophila. J. Bacteriol.* **180,** 1129–1134.
6. Nakajima, H., Suzuki, K., and Imahori, K. (1978) Purification and properties of acetate kinase from *Bacillus stearothermophilus. J. Biochem.* **84,** 193–203.
7. Fox, D. K. and Roseman, S. (1986) Isolation and characterization of homogeneous acetate kinase from *Salmonella typhimurium* and *Escherichia coli. J. Biol. Chem.* **261,** 13,487–13,497.
8. Rose, I. A., Grunberg-Manago, M., Korey, S. R., and Ochoa, S. (1954) Enzymatic phosphorylation of acetate. *J. Biol. Chem.* **211,** 737–756.
9. Webb, B. C., Todhunter, J. A., and Purich, D. L. (1976) Studies on the kinetic mechanism and the phosphoryl-enzyme compound of the *Escherichia coli* acetate kinase reaction. *Arch. Biochm. Biophys.* **173,** 282–292.

16

Immobilization of Enzymes by Covalent Attachment

Scott J. Novick and J. David Rozzell

Summary

Enzymes are finding increasing use for the production of agrochemicals, pharmaceuticals, and fine chemicals. They are almost always used in the immobilized form in order to simplify their removal from the product stream. In addition, immobilization often enhances the stability of the enzyme. Immobilization can be performed in a number of ways. This chapter discusses various methods, properties, and uses of covalently immobilized enzymes.

Key Words: Immobilized enzymes; covalent immobilization.

1. Introduction

1.1. Historical Perspective

An immobilized enzyme is generally defined as "the imprisonment of an enzyme molecule in a distinct phase that allows exchange with, but is separated from, the bulk phase in which substrate effector or inhibitor molecules are dispersed and monitored" *(1)*. Immobilized enzyme technology dates back to the 1910s to the 1930s, when proteins were physically adsorbed onto surfaces such as charcoal, kaolinite, cellulose, and glass beads *(2–4)*. But it was not until the 1950s and '60s with the work of Katchalski-Katzir, and Chibata and coworkers that real advancements were beginning to be made in the development and applications of immobilized enzyme materials *(5)*. This early work culminated in the First Enzyme Engineering Conference in 1971. The first industrial use of immobilized enzymes was for amino acid production. Chibata and coworkers at Tanabe Seiyaku (Japan) in 1969 used an immobilized L-aminoacylase in a packed bed reactor to resolve various DL-amino acids into their enantiomerically pure forms. Since that time, immobilized enzymes have become increasingly important for the production of many important chiral compounds (i.e., amines and alcohols) for the pharmaceutical and fine chemical industries.

From: *Methods in Biotechnology, Vol. 17: Microbial Enzymes and Biotransformations*
Edited by: J. L. Barredo © Humana Press Inc., Totowa, NJ

Table 1
Stabilization Effects Immobilization Imparts to Enzymes *(8)*

1. Prevention of either proteolysis or aggregation by spatial fixation of enzyme molecules to the support.
2. Unfolding of the enzyme is reduced due to multipoint covalent or adsorptive attachment to the support, and/or intramolecular crosslinking of the enzyme.
3. Multimeric enzymes would have a lower likelihood to dissociate if all subunits are attached to the support.
4. Denaturing agents (e.g., chemical inactivators) can be excluded from the enzyme by the support or inactivated by the support before reaching the enzyme (e.g., decomposition of hydrogen peroxide, produced during the oxidation of glucose by glucose oxidase, catalyzed by activated carbon).
5. Shifting by a charged support of the local pH, thus preventing pH inactivation of the enzyme.
6. Exclusion by the support (e.g., an encapsulation membrane) of proteases from the enzyme's environment.
7. Increased thermal stability due to multipoint attachment of enzyme to support.

1.2. Reasons for Enzyme Immobilization

The principal advantage of immobilizing enzymes is to retain the catalyst in the reactor. This can greatly improve the economics of a process. For a continuous process, a soluble enzyme would be washed out of the reactor along with the product stream. A process like this would not be economically feasible if the biocatalyst is very expensive (as is often the case) and cannot be reused. Although an ultrafiltration setup could be used to retain the enzyme, it is often too costly, both in capital and operation, on a large scale. Also, having a soluble enzyme in the product would not be desirable if the biocatalyst can cause the product to undergo side reactions or if there are toxicity effects associated with the catalyst, as will often be the case if the product is an intravenous drug *(6)*. Another advantage of immobilizing enzymes is to increase enzyme activity or stability especially under denaturing conditions *(7,8)*. Thermal stability can often be improved by many orders of magnitude compared to the soluble enzyme *(9–11)*. Activity of an enzyme in nonaqueous media can also be significantly higher than the native enzyme *(12–18)*. Another important advantage is the ability to control the microenvironment of the immobilized enzyme. For example, by immobilizing an enzyme on an acidic support (such as poly [acrylic acid]), the catalyst can be used at higher pHs, where the substrate may be more soluble, while the pH of the microenvironment surrounding the enzyme could be much closer to the enzyme's optimum pH. These and other stabilizing effects of immobilization are listed in **Table 1.**

There are also limitations to immobilizing enzymes. Some inherent catalytic activity is almost always lost during the immobilization procedure. Enzymes possess highly defined, yet relatively fragile three-dimensional structures that must come in contact and interact with the rigid support. These binding forces, such as covalent bonds or adsorptive interactions, are often more powerful than the secondary forces, such as hydrogen bonding and hydrophobic and ionic interactions, which hold proteins in their proper configuration for enzymatic activity. In addition, no covalent immobilization method is able to bind only the nonessential elements of every enzyme (if they even exist) to the support, and all supports create asymmetric force fields and change the water activity around the biocatalyst *(6)*. In addition, apparent activity can be decreased by mass transfer limitations. However, the increase in stability and ease of removal from the product stream and reuse often more than make up for any decrease in activity.

1.3. Enzyme Immobilization Methods

In general five techniques have been described for immobilization of enzymes. It is important to point out that there is no one universal immobilization system; instead, a range of methodologies must be evaluated depending on the enzyme to be immobilized and the overall process in which the immobilized enzyme is to be used. Also, most immobilization methods, although conceptually distinct, often overlap to a certain extent, and in some cases, multiple immobilization methods are employed.

One of the simplest and most economical immobilization methods is adsorbing an enzyme onto a support. The enzyme is bound to the support via ionic or nonionic interactions. Supports often include carbohydrate-based or synthetic polymer ion-exchange resins or uncharged supports such as polymers, glasses, and ceramics. The main drawback of this method is the leaching of enzyme from the support.

Cross-linking enzyme molecules with themselves, or more often with an inert protein such as gelatin or bovine serum albumin, results in an insoluble active enzyme preparation that can be readily handled or manipulated in a continuous reactor. Glutaraldehyde, adipimate esters, and diisocyanates are often used as the cross-linking agent. Significant inactivation of the enzyme may result during the cross-linking step and is the major drawback of this method.

Entrapment of an enzyme within a polymeric matrix is another method used for enzyme immobilization. This is often done by mixing the enzyme with a monomer and a cross-linker, and polymerizing the monomer around the enzyme. Leaching of the enzyme out of the matrix and mass transfer limitations of substrate diffusing into the matrix can limit the use of this technique.

Encapsulating or confining an enzyme within a membrane is another method for enzyme immobilization. Ultrafiltration membranes or hollow fibers made of polyethersulfone, cellulose nitrate or acetate, or nylon are often used. The pore size must be properly chosen to allow substrate and product to enter and exit the membrane while still retaining the enzyme. Since the enzyme exists in its soluble form, activity is usually high. Membrane fouling and reduced flow rates are drawback of this technique.

The fifth immobilization method, covalent attachment of enzymes to a support, will be the subject of the rest of this chapter.

1.4. Covalent Enzyme Immobilization

Covalent attachment of enzymes to an insoluble support is an often-used method of enzyme immobilization. It is especially useful when leaching of enzyme from the support is a concern. The enzyme is usually anchored via multiple points and this generally imparts greater thermal, pH, ionic strength, and organic solvent stability onto the enzyme since it is more rigid and less susceptible to denaturation. Covalently immobilized enzymes are also often more resistant to degradation by proteolysis.

There are, however, some drawbacks to covalent enzyme immobilization. Typically it is more expensive and complex to covalently immobilize an enzyme compared to the other methods due to the higher costs of the support. The support often needs to be activated prior to immobilization. The increased stability and typically minimal enzyme leaching often more than make up for these shortcomings.

Enzymes contain a number of functional groups capable of covalently binding to supports. **Table 2** lists these groups along with their relative frequency in a typical protein *(19–21)*. Of the functional groups of enzymes listed, $-NH_2$, $-CO_2H$, and $-SH$ are most frequently involved in covalent immobilization. Amines and sulfhydryls are good nucleophiles, while the ability to activate carboxylates so they are reactive toward nucleophiles makes these groups important as well. The phenolic ring of tyrosine is also extremely reactive in diazo-coupling reactions, and its hydroxyl group can be an excellent nucleophile at basic pH. Aldehydes can react with the guanidino group of arginine and, although histidine displays a lower nucleophilicity, it can react in some cases with supports activated with tosylates, tresylates, or other good leaving groups.

The supports to which the enzymes are attached to can vary greatly. They can be either natural polymers, such as modified cellulose, starch, dextran, agal polysaccharides, collagen, and gelatin; or they can be synthetic, such as polystyrene, polyacrylamide, polyacrylates, hydroxyalkyl methacrylates, and polyamides. Inorganic supports can also be used, such as porous glass, metal oxides, metals, sand, charcoal, and porous ceramics. The variety of chemistries available for covalent attachment allows the conditions of immobilization to be tailored to

Table 2
Reactive Functional Groups in Enzymes and Their Average Occurrence in a Typical Protein (19–21)

Structure of reactive group	Reactive group	Occurrence in average protein
$-NH_2$	ε-Amino of lysine and N-terminus	5.9
$-CO_2H$	Carboxylate of glutamic acid, aspartic acid, and C-terminus	6.3 (Glu), 5.3 (Asp)
$-SH$	Thiol of cysteine	1.9
—OH	Phenolic of tyrosine	3.2
	Guanidino of arginine	5.1
	Imidazole of histidine	2.3
$-S-S-$	Disulfide of cystine	—
	Indole of tryptophan	1.4
$-CH_3-S-$	Thioether of methionine	2.2
$-CH_2OH$	Hydroxyl of serine and threonine	6.8 (Ser), 5.9 (Thr)

each enzyme system. This also allows the microenvironment of the enzyme to be tailored by appropriate modification of the support surface; hydrophobic moieties or ionically charged groups may be used to alter the support to enhance the enzyme-catalyzed reaction of interest. Some supports, such as those containing

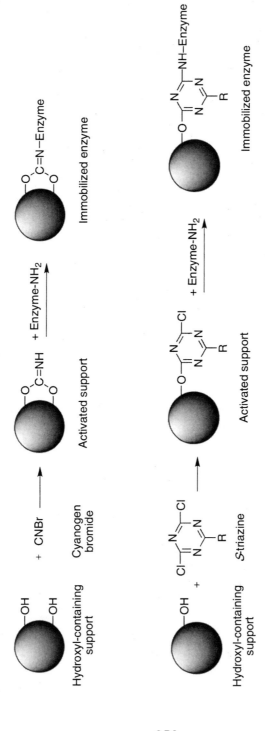

Fig. 1. Enzyme immobilization onto hydroxyl containing supports via activation with cyanogen bromide (top) or *S*-triazine derivatives (bottom).

epoxide groups can be used directly for enzyme binding. However, most supports require preactivation before enzymes are able to bind to it. The following sections describe some typical covalent attachment methodologies.

1.4.1. Covalent Attachment Onto Polyhydroxyl Supports

Polyhydroxyl supports, such as porous glass, and especially polysaccharides are among the most commonly used matrices for enzyme immobilization. Because hydroxyl groups are poor leaving groups they must first be activated. This is typically done with cyanogen bromide *(22)*. Other activating agents such as S-triazine derivatives have also been used. Once the support is activated it is able to covalently couple to an enzyme usually through the ε-amino group of lysine or through the amino terminus. The mechanism of derivatization polyhydroxyl supports with the above two derivatizing agents and the subsequent enzyme immobilization is shown in **Fig. 1.**

Supports that have been preactivated with cyanogen bromide can be stored for periods of up to 1 yr at freezer temperatures. Preactivated supports are also available commercially. Once the support is activated, coupling of the enzyme requires no more than exposing the enzyme to the activated support in an aqueous solution for a few hours, followed by extensive washing to remove any protein that is not covalently bound.

This method is extremely popular in the lab scale; however, it has not been widely used in large-scale applications. The activating agent, cyanogen bromide, is extremely toxic, and most carbohydrate supports, such as cellulose, agarose, and dextran, have poor mechanical stability compared to other support materials. Also, since the supports are natural polysaccharides, microbial contamination and degradation are a concern. Finally, the bond between the enzyme and the support is potentially susceptible to hydrolytic cleavage, which would cause leaching of the enzyme from the support over time.

1.4.2. Covalent Attachment onto Carboxylic Acid-Bearing Supports

Carboxylic acid-containing supports, such as copolymers of (meth)acrylic acids with (meth)acrylic esters have also been used as an immobilization support. These must also be activated and this is usually done with a carbodiimide reagent. Under slightly acidic conditions (pH 4.75–5.0) carbodiimides react with carboxylic acid groups to give the highly reactive *O*-acylisourea derivatives. The supports are then washed to remove excess reagent and the enzyme is coupled to the activated support at neutral pH to give stable amide, thioester, or ester linkages, depending on the residue reacting with the support. The most widely used water-soluble carbodiimides are 1-ethyl-3-(3-dimethylamino propyl)-carbodiimide (EDC) and 1-cyclohexyl-3-(2-morpholino-ethyl)-carbodiimide (CMC), both of which are available commercially. The reaction

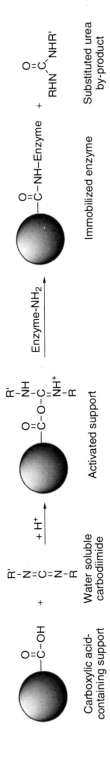

Fig. 2. Activation of a carboxylic acid containing support with a carbodiimide followed by enzyme coupling.

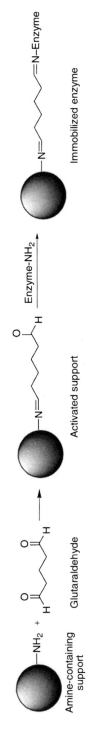

Fig. 3. Activation of an amine-bearing support with glutaraldehyde followed by enzyme coupling.

254

scheme for of activating a carboxylic acid-containing support and subsequent enzyme coupling is shown in **Fig. 2.**

1.4.3. Covalent Attachment Onto Amine-Bearing Supports

Amine-bearing supports are among the most used and the most useful supports for covalent enzyme immobilization. These supports can either be organic or inorganic supports bearing amine functionality. The most frequent technique for introducing amine groups on inorganic supports is via aminosilane attachment *(23–25)*. For example, 3-aminopropyltriethoxysilane can be coupled to porous glass to give pendent amine groups *(26)*. This silane has been developed through the pioneering work at Corning Glass Works *(23)*.

Another common amine-bearing support is polyethyleneimine-coated particles. Polyethyleneimine is a common polyamine derived from the polymerization of ethyleneimine to give highly branched polymers containing approximately 25% primary amines, 25% tertiary amines, and 50% secondary amines. This polymer can be coated onto various supports including alumina *(27)*, carbon *(28)*, diatomaceous earth *(29)*, and polyvinyl chloride-silica composites *(30,31)*.

The coupling of an enzyme to amine-bearing supports can be done in a number of ways. The most common way is through the use of difunctional reagents, such as diimidate esters, diisocyanates, and dialdehydes. Glutaraldehyde is often used, as it is one of the least expensive difunctional reagents available in bulk. This reagent reacts in a complex fashion to form Schiff bases with amine groups on the support and produces pendent aldehydes and α,β-unsaturated carbonyl functionalities through which enzymes may attach. Enzyme attachment is accomplished simply by mixing the enzyme with the activated support. A simplified example of this is shown in **Fig. 3.** The acid-labile Schiff bases can be reduced to more stable secondary amine bonds with sodium borohydride to increase the stability of the enzyme-support linkage.

Crump and coworkers *(32)* have described the immobilization of an L-amino acid transaminase onto a polyethyleneimine coated PVC-silica support matrix that was activated with glutaraldehyde. Very high binding efficiency and residual activity were obtained. After washing, 93% of the enzyme offered was bound to the support (total loading was about 10%) and the enzyme retained approximately 89% of the soluble activity. Both these values are unusually high for immobilized enzymes, but not necessarily atypical for this type of support and immobilization chemistry.

Enzymes can also be covalently bonded directly to amine-bearing supports via the enzyme's carboxyl groups. These must first be activated with a carbodiimide or similar reagent prior to immobilization. The activation step can cause enzyme inactivation and thus this method is not used as often.

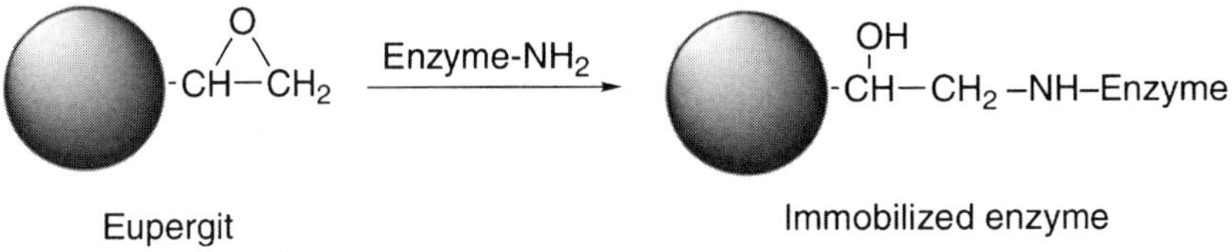

Fig. 4. Enzyme immobilization to Eupergit via free amino groups.

Diisocyanates have also been used as a coupling agent between amine-bearing supports and enzymes *(33)*. If alkaline conditions are used a substituted urea bond is formed between an amine on the enzyme and the isocyanate. If moderately acidic conditions are employed, the isocyanate will react with a hydroxyl group on the enzyme and form a urethane bond. Isothiocyanates have also been used successfully *(23)*.

Another amine-bearing support, developed by Leuta and coworkers *(34)*, is mineral or carbon particles coated with chitosan. Chitosan is deacylated chitin, a polymer of glucosamine, and contains an available amino group for chemical activation and enzyme binding using methods similar to those described for the other amine-bearing supports.

1.4.4. Covalent Attachment to Reactive Polymer Supports

Due to the preactivated nature of epoxy-containing supports, these materials have gained considerable attention as commercially useful support matricies for enzyme immobilization. A commercial epoxy-containing support is available from Röhm Pharma Polymers (Piscataway, NJ) under the trade name Eupergit. The material is a crosslinked copolymer of methacrylamide and oxirane containing monomers and consists of spherical beads of about 200 μm in diameter. Eupergit is available in two varieties, Eupergit C and Eupergit C 250 L, with their differences being their oxirane content and pore size. Eupergit C has average pore radius of 10 nm and an oxirane content of 600 μmol/g, while Eupergit C 250 L has a pore size and oxirane content of 100 nm and 300 μmol/g, respectively *(35)*. Eupergit C 250 L is targeted for the immobilization of large molecular weight enzymes (>100 kDa). Immobilization of enzymes to Eupergit is relatively simple. The enzyme solution is brought in contact with the Eupergit beads either quiescently or with slight mixing (magnetic stirbars should be avoided to prevent fractionation of the beads) for 24–96 h. This can be done either at room temperature, or if the enzyme is unstable, 4°C will also work. Various pHs can be used for the binding. Under neutral and alkaline conditions the amino groups on the enzyme are principally responsible for binding to the support (**Fig. 4**). Under acidic and neutral conditions sulfhydryl and carboxyl groups take part in binding.

Immobilization to Eupergit does not change the charged state of the enzyme. Typically, it is best to bind the enzyme to the support at the pH at which activity is optimum for the enzyme. The various parameters mentioned above—mixing type, immobilization time, temperature, pH, and also ionic strength (0.5–1 M buffer or neutral salt is often optimal)—can be varied to optimize the amount of enzyme immobilized and the residual activity. Once the enzyme is bound to the support, the binding is stable over the long term and it is stable over a wide pH range, from 1.0 to 12.0. Also, because Eupergit is electrically neutral, pH changes do not effect the swelling of the gel.

After the enzyme has been bound, typically only about 1% of the available epoxy group are involved in enzyme immobilization. The remaining groups will slowly hydrolyze into diols or they can be quenched with a variety of compounds that can effect the microenvironment around the immobilized enzyme by making it more hydrophilic, hydrophobic, or charged. This in turn can effect the stability or activity of the bound enzyme. Bovine serum albumin, dithiothreitol, Tris-buffer, mercaptoethanol, various amino acids (i.e., lysine or glycine), and ethanolamine are among some of the quenching reagents that have been used, and in many cases activity of the immobilized enzyme can be altered depending on the quenching reagent.

There have been two extensive reviews recently published concerning the immobilization of enzymes to Eupergit *(35,36)*. In these reviews, the details of the immobilization of nearly two dozen different enzymes are presented. In addition to Eupergit, other epoxy-containing polymers have been investigated for the covalent attachment of enzymes *(37–42)*.

Polyacrolein beads is another useful reactive-polymer carrier for covalent enzyme immobilization. Margel *(43)* synthesized such beads and encapsulated them into agarose prior to enzyme binding. Because these supports are polyaldehydes, enzymes are bound in a similar way as with glutaraldehyde activated supports. Various oligomers such as poly(lysine) and poly(glycine) have been attached to the polyacrolein beads to act as spacers between the particles and the enzyme. In both cases the poly(amino acids) are attached to the support through their terminal amino groups, or ε-amino groups in the case of poly(lysine), via Schiff bases (which can then be reduced). The enzyme is attached to the poly(lysine)-derivatized polyacrolein via the lysine ε-amino groups using glutaraldehyde as a linker. For the poly(glycine)-derivatized polyacrolein support, the terminal carboxyl group is activated with a water-soluble carbodiimide followed by enzyme binding. In some cases the use of these spacers has shown a significant increase in activity, especially for large-molecular-weight substrates. Covalent enzyme immobilization to paramagnetic polyacrolein beads has also been investigated *(44)*. Binding of enzymes to unmodified polyacrolein is shown in **Fig. 5.**

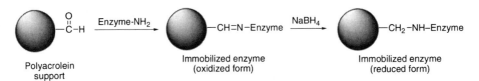

Fig. 5. Enzyme immobilization to unmodified polyacrolein via free amino groups, followed by reduction of the Schiff base with sodium borohydride.

1.5. Assaying the Properties of Immobilized Enzymes

There are three important properties of immobilized enzymes that are often evaluated: activity, enzyme loading, and stability. Prior to the measurement of these properties, the immobilized enzyme materials should be washed extensively to remove any unbound enzyme that may be entrapped in the pores of the particles or loosely bound through noncovalent interactions.

1.5.1. Activity Assay

There are two basic methods to measure activity—batch and continuous. In the batch method the immobilized enzyme is added to a flask or vial and the substrate solution is then added to initiate the reaction. At various time points, an aliquot of the mixture is removed and filtered (this is most easily done through a syringe filter) to remove any of the immobilized enzyme particles and to quench the reaction. This aliquot can then be analyzed using the appropriate analytical method, such as liquid chromatography, gas chromatography or spectrophototometry. If product continues to be produced in this aliquot after filtration, it is a good indication that there may be significant leaching of soluble enzyme off the support. This can occur if the support is not washed extensively enough after immobilization or if the binding is labile under the assay conditions. To get more accurate activity measurements the supports should be rewashed.

There are two basic methods for performing a continuous activity assay. In the packed-bed plug-flow tubular reactor (PFTR) method, the immobilized enzyme is packed into a column and substrate is pumped though the column and the substrate and/or product concentration is measured in the effluent. In the continuous stirred tank reactor, the solution and the immobilized enzyme are well mixed so there are minimal concentration gradients. To prevent the loss of immobilized enzyme out of the exit, a filter is added at the effluent or a tube is added at the exit that is long enough such that at the given flow rate gravity prevents the particles from leaving the reactor. Modeling a batch or continuous immobilized enzyme reactor can be found in many reactor engineering or bioprocess engineering textbooks.

1.5.2. Determining Protein Loading in Covalently Immobilized Enzyme Materials

It is often useful to have information on how much enzyme is bound to the support after an immobilization was performed. This information is needed when optimizing the immobilization conditions or when calculating the residual enzyme activity. Measurement can be done either indirectly or directly on the particles themselves.

In the indirect method, the initial amount of protein offered to the support is determined using any of the variety of protein assays available. After the immobilization is complete and the particles are washed, the same protein assay is done on the supernatant and wash solutions. The difference in the mass of enzyme offered and the amount in the immobilization supernatant and wash solutions will give the amount of enzyme bound to the support.

In the direct method, the amount of enzyme actually bound to the support is determined. A number of methods to determine this have been published. In one method, the bicinchoninic acid protein assay (often referred to as the BCA assay) is used *(45)*. In this assay, the immobilized enzyme is incubated in the BCA assay solution. The enzyme bound to the support reacts with the BCA solution in the same manner a soluble enzyme would, by reducing Cu^{2+} to Cu^{1+} in the presence of peptide bonds, which complex with the bicinchoninic acid to form a aqueous purple-colored solution. The absorbance of this solution will be proportional to the amount of enzyme immobilized. Bovine serum albumin (BSA) is often used as protein standards to quantify the amount of enzyme bound to the support.

Coomassie-based protein dyes have also been used to directly quantify the amount of enzyme bound to a support. In one method, the dye is allowed to bind to the enzyme attached to the support, after which the residual unbound dye is removed from the particles *(46)*. The bound dye is then eluted from the particles by adding sodium dodecylsulfate and sodium bicarbonate. The absorbance of this solution can then be read at 595 nm and the amount of protein bound to the support can be determined by comparing to a BSA standard curve.

In another method, the Bradford Protein Assay solution (a Coomassie-based protein assay) is used *(47)*. With no protein present, the Bradford solution is brown with a λ_{max} of 465 nm (with protein the color is blue with a λ_{max} of 595 nm). When the Bradford solution is mixed with the immobilized enzyme particles, some of the dye will bind to the protein on the beads (turning them blue) and the absorbance at 465 nm will decrease. This decrease at 465 nm can be correlated to the amount of bound protein by comparing to a standard curve of BSA or other suitable protein. The Bradford solution is acidic, so any enzyme that is immobilized through acid-labile links cannot be used with this method. Also, if the supernatant turns blue it is a good indication that significant leaching of the enzyme off the support is occurring.

Other methods also exist, including photometric, fluorometric, radiochemical, and ELISA-based methods. Elemental analysis on nitrogen or sulfur can also be used as long as the support material does not contain these elements. Amino acid analysis after acid hydrolysis can be used as well to determine protein content. All of these have differing sensitivity, work-up, and costs, all of which should be considered *(48)*.

1.5.3. Determining Stability of Covalently Immobilized Enzymes

Stability of immobilized enzymes can be measured in a number of ways. The simplest way is to pack the immobilized enzyme into a continuous reactor such as a column (plugged flow reactor) or a vessel with mixing (continuous stirred tank reactor). Substrate is then pumped through the reactor and the effluent is analyzed for product and/or substrate concentration. Depending on the stability of the enzyme, this is allowed to run for days or even months and the decrease in product concentration or the increase in outlet substrate concentration is monitored to determine the working lifetime of the immobilized enzyme. It is important to choose the proper amount of immobilized enzyme and flow rate such that less than 100% conversion is obtained. If 100% conversion is achieved, then it is unknown whether the entire immobilized enzyme takes part in the reaction. For example, if only half the enzyme present in the reactor is needed for 100% conversion under a given set of reaction conditions, significant inactivation (up to 50% of the enzyme) could occur with no observable change in conversion. The continuous reactor can also be run with various cosolvents or at various pHs or temperatures to determine the stability of the immobilized enzymes under these conditions.

The stability of immobilized enzymes can also be determined batchwise. In this technique, measured amounts of the immobilized enzyme are placed into separate vials along with the solvent of interest. The vials are allowed to incubate at a given temperature for a given amount of time. One or more vials are then sacrificed and the activity of the immobilized enzyme is measured (alternatively, the immobilized enzyme particles can be washed of substrate and product and returned to the initial incubation conditions). This is repeated over time to determine how the activity changes. Another batchwise method of measuring stability is to add the substrate solution to the immobilized enzyme and measure the reaction time course. The immobilized enzyme particles are then washed and this is repeated. The time it takes to reach the required conversion, the conversion at a set time, or the reaction time course can be plotted versus the number of recycles to give an indication of long-term stability/usability. Reactors capable of performing this automatically are commercially available *(36)*.

1.6. Industrial Uses of Covalently Immobilized Enzymes

A large number of enzymes are used in industry for the synthesis of a wide range of compounds. Most of these applications can be placed in either the food industry (both for human and animal consumption) or the pharmaceutical/fine chemical industry. Many of these applications involve immobilized enzymes with some of them covalently immobilized. A few examples of the latter will be discussed in **Subheadings 1.6.1., 1.6.2.,** and **1.6.3.**

1.6.1. High-Fructose Corn Syrup With Immobilized Glucose Isomerase

The largest use of immobilized enzymes is for the isomerization of glucose, from corn, to the much sweeter fructose. The resultant high-fructose corn syrup (HFCS) is used as a sweetener in a variety of foodstuffs, especially sweetened beverages and baked goods. Current US production of HFCS exceeds 9.3 million short tons (dry weight) per year (http://www.ers.usda. gov/briefing/sugar/ Data/Table27.xls). Glucose isomerase (also called xylose isomerase due the high affinity of the commercially available catalysts toward xylose) is used industrially to perform this reaction. Many of the commercial catalysts are immobilized whole cells; however, Miles Kali-Chemie (Germany) developed a glucose isomerase from *Streptomyces rubiginosus* covalently immobilized onto silanized ceramics and sold under the trade name Optisweet *(49,50)*. Typically this reaction is run at 50–60°C to limit microbial contamination. The reactor is a continuous packed-bed reactor with a residence time of 0.17–0.33 h. The half-life of the enzyme is over 100 d; however, it is replaced after about 12.5% activity loss to maintain the necessary activity for the reactor *(49)*.

1.6.2. Semi-Synthetic Penicillins with Immobilized Penicillin Amidase

Another major use of immobilized enzymes is in the synthesis of semi-synthetic penicillins. Worldwide production of these semi-synthetic β-lactam antibiotics is more than 20 thousand tons per year *(49)*. The starting material for these compounds is 6-aminopenicillanic acid (6-APA). It is too expensive to make 6-APA synthetically or by fermentation. Instead, benzyl penicillin (also called penicillin G), which can be made relatively cheaply via fermentation, is hydrolyzed by penicillin amidase (also called penicillin acylase) into 6-APA and phenylacetic acid. This is accomplished industrially by using penicillin amidase from *Escherichia coli* or *Bacillus megaterium* immobilized onto Eupergit C. The reaction is usually carried out in a stirred-tank batch reactor. The immobilized enzyme can be reused nearly 1000 times before the half-life is reached. About 1700 kg of penicillin G is able to be hydrolyzed with 20 g of protein immobilized onto 300 g of dry polymer *(51)*. Once the 6-APA is formed and purified, it is chemically acylated with various side groups to make the semisynthetic antibiotics such as

Fig. 6. Synthesis of L-aspartate from fumaric acid and ammonia catalyzed by L-aspartase.

ampicillin (D-phenylglycine side chain), amoxicillin (D-*p*-hydroxyphenylglycine side chain), and penicillin V (phenoxymethyl side chain). The preparation of a new immobilized penicillin G acylase yielding derivatives thermoestable and resistant to organic solvents is shown in **Chapter 17**.

1.6.3. L-Amino Acids with Immobilized L-Aspartase

L-amino acids are produced in large quantities for human consumption in the form of supplements, ingredients in artificial sweeteners, intermediates in the synthesis of pharmaceuticals, and as additives in animal feed. Covalently immobilized enzymes have been used to produce L-aspartase, a main ingredient in the artificial sweetener aspartame. This amino acid can be synthesized from two inexpensive starting materials, fumaric acid and ammonia (**Fig. 6**). The enzyme that catalyzes this reaction is L-aspartate-ammonia lyase, also called L-aspartase. This enzyme, from *E. coli,* has been covalently immobilized onto PVC-silica supports that have been coated with a polyamine and activated with glutaraldehyde. The process is relatively simple. The two starting materials are passed through a packed bed containing the immobilized enzyme. The effluent is acidified to pH 2.8 and chilled, causing the L-aspartate to precipitate, and it can then be recovered by filtration. Conversion, optical purity, and chemical purity of this reaction are all over 99%. The product concentration is 200 g/L with a space time yield of 3 kg/L·h. The reactor volume was 75 L, therefore producing more than 5 metric tons of product per day. The half-life of the immobilized enzyme was more than 6 mo (*49,52*).

1.7. Conclusions

Covalent immobilization of enzymes represents a robust method for the attachment of enzymes to insoluble supports. A variety of supports are available including synthetic and natural organic polymers and inorganic minerals, metal oxides, and glasses. The chemistry and functional groups used to bind enzymes to the supports can vary greatly and can be tailored depending on the specific application. In addition to the ease of enzyme recovery, stability of the immobilized enzyme is usually much higher than the soluble enzyme and thus can be reused multiple times. Applications of covalently immobilized enzymes have been demonstrated from the lab scale to the industrial production of multiton quantities. Increasing numbers of enzyme are becoming commercially available in immobilized form, including lipases, proteases, nitrilases, amino acid dehydrogenases, oxidoreductases, and others *(53)*. This trend will continue as enzymes find more applications, particularly for chiral chemical production. This is a vibrant field that continues to evolve to this day.

2. Materials

2.1. Covalent Enzyme Immobilization

2.1.1. Covalent Attachment Onto Polyhydroxyl Supports

1. Polysaccharide support material, such as Sepharose 4B, Sephadex, agarose, cellulose, or dextran.
2. 2 *M* sodium carbonate.
3. 1 g/mL cyanogen bromide (CNBr) dissolved in DMF.

2.1.2. Covalent Attachment Onto Carboxylic Acid-bearing Supports

1. Carboxylic acid-bearing support such as Bio-Rex 70 Resin (BioRad, Hercules, CA), IRC50 (Supelco, St. Louis, MO), carboxymethylcellulose, acrylic acid copolymers, or succinylated glass beads.
2. 0.1 *M* sodium acetate buffer, pH 4.5.
3. 1-Cyclohexyl-3-(2-propyl)carbodiimide (EDC).
4. 0.1 *M* sodium phosphate buffer, pH 7.0.

2.1.3. Covalent Attachment Onto Amine-Bearing Supports

1. Amine-containing support, such as aminopropyl-glass, aminoethyl-cellulose, PEI-coated materials (e.g., silica or alumina) or other similar material.
2. 0.05 *M* phosphate buffer, pH 7.0.
3. 25% glutaraldehyde (GA).

2.1.4. Covalent Attachment to Reactive Polymer Supports

1. Eupergit C (Röhm Pharma Polymers, Piscataway, NJ).
2. 0.05 *M* phosphate buffer, pH 7.0.

2.2. Determining Protein Loading in Covalently Immobilized Enzyme Materials

2.2.1. Indirect Method

1. Immobilized enzyme.
2. Bradford reagent (Sigma Chemical Co., St. Louis, MO).
3. Bovine serum albumin (BSA) protein standards.
4. 0.05 *M* phosphate buffer, pH 7.0.

2.2.2. Direct Method (BCA)

1. Immobilized enzyme.
2. BCA protein assay kit (contains Reagent A and Reagent B).
3. BSA protein standard.

3. Methods

3.1. Covalent Enzyme Immobilization

3.1.1. Covalent Attachment Onto Polyhydroxyl Supports

1. Wash the support material with distilled water and remove residual water using suction filtration to form a packed cake.
2. Add 10 g of washed support material to a flask and add 2 M sodium carbonate until the total volume is about 1.2 times the settled bed volume (*see* **Note 1**).
3. Cool slurry to 0°C.
4. Add 1 mL of 1 g/mL of CNBr dissolved in DMF and mix vigorously for 2 min (*see* **Note 2**).
5. Wash the now-activated support material with at least 5 volumes of cold distilled water.
6. Add the activated support material to a solution of the enzyme in the appropriate buffer. Let incubate at 4°C for 12 to 24 h (*see* **Notes 3–5**).
7. Wash the immobilized enzyme material with the buffer used in **step 6.**

3.1.2. Covalent Attachment Onto Carboxylic Acid-Bearing Supports

1. Add 5 mL of the hydrated support to 15 mL of 0.1 M sodium acetate buffer, pH 4.5.
2. Add 200 mg EDC to the support slurry. Mix for 20 min at room temperature (*see* **Note 6** and **Note 7**).
3. Wash the beads with 500 mL of cold 0.1 M sodium phosphate buffer, pH 7.0.
4. Add the activated beads to 15 mL of the enzyme solution and gently agitate for 24–48 h at 4°C (*see* **Notes 8** and **9**).
5. Wash the immobilized enzyme beads with at least 500 mL of 0.1 M sodium phosphate buffer, pH 7.0.

3.1.3. Covalent Attachment Onto Amine-Bearing Supports

1. Suspend 10 g of the moist amine-bearing support in 100 mL of 0.05 M phosphate buffer, pH 7.0.
2. Add 10 mL of 25% glutaraldehyde (GA) and mix well for 1–2 h (*see* **Notes 10** and **11**).
3. Wash and filter the GA-activated support material with 500 mL of the above buffer at least five times to remove any residual GA (*see* **Notes 12** and **13**).
4. Suspend 10 g of the moist GA-activated support in 30 mL of the enzyme solution in the above buffer and mix well at room temperature or 4°C depending on enzyme stability (*see* **Note 14**).
5. After 12–24 h filter the immobilized enzyme material and wash well with the above buffer (*see* **Note 15**).

3.1.4. Covalent Attachment to Reactive Polymer Supports

1. Add 1.0 g of dry Eupergit C to 6.0 mL of the enzyme solution in 0.05 M phosphate buffer, pH 7.0 (*see* **Notes 16–19**).
2. Gently mix the suspension at room temperature or 4°C depending on the stability of the enzyme.
3. After 24–48 h wash the immobilized enzyme with 50 mL of the buffer in **step 1** followed by suction filtration. Repeat the washing three to five times (*see* **Note 20**).

3.2. Determining Protein Loading in Covalently Immobilized Enzyme Materials

3.2.1. Indirect Method

1. Before adding the enzyme solution to the support for immobilization, record the volume of the enzyme solution and the dry weight of the support.
2. Determine the protein concentration in the enzyme solution (prior to adding it to the support material) using the Bradford protein assay method. This is done by adding 0.1 mL of the enzyme solution to 0.9 mL of Bradford reagent and mixing. OD_{595} is recorded after it has stabilized (usually 5–10 min). Protein concentration is calculated by comparing to a standard curve using a suitable protein standard such as BSA. The linear range for this method is approximately 0–0.5 mg/mL protein, therefore, if necessary, the enzyme solution should be diluted in the phosphate buffer (*see* **Note 21**). After the immobilization is complete, pour off the supernatant into a separate container.
3. Wash the immobilized enzyme as necessary and add the washings to the supernatant in **step 2.**
4. Record the combined volume of the supernatant and washings (from **step 3**) and measure the protein concentration as described in **step 2** (*see* **Notes 22–24**).
5. Use the following equation to calculate the amount of protein bound:

$$\frac{\text{mg enzyme bound}}{\text{g dry wt of support}} = \frac{\left(\begin{array}{cc}\text{enzyme conc.} & \text{volume of} \\ \text{in initial} & \text{initial} \\ \text{solution} & \times & \text{solution} \\ \text{(mg/mL)} & \text{(mL)}\end{array}\right) - \left(\begin{array}{cc}\text{enzyme conc.} & \text{volume} \\ \text{in final} & \text{of final} \\ \text{solution} & \times & \text{solution} \\ \text{(mg/mL)} & \text{(mL)}\end{array}\right)}{\text{g dry weight of support}}$$

3.2.2. Direct Method

1. Make up 20 mL of the BCA working solution by mixing 19.6 mL of reagent A with 0.4 mL of reagent B.
2. Make up 1.0 mL BSA stock solution of 25.2 mg/mL.
3. In 3–5 mL (total volume) capped tubes make up the following solutions (add the working solution last) (*see* **Notes 25** and **26**):

Sample	BSA standard, 25.2 mg/mL (μL)	Water (μL)	Final BSA conc (mg/mL)	Immobilized enzyme (mg dry)	Working solution (mL)
Std-1	0	100	0	0	2.0
Std-2	8.3	91.7	0.1	0	2.0
Std-3	25	75	0.3	0	2.0
Std-4	50	50	0.6	0	2.0
Std-5	75	25	0.9	0	2.0
Std-6	100	0	1.2	0	2.0
Immobilized enzyme	0	100	N/A	10–20	2.0

4. Shake the standards solutions and immobilized enzyme suspensions well at 37°C for 30 min, then cool the tubes to at or below room temperature.
5. Remove particles by filtration or centrifugation and measure OD_{562} of the supernatant (*see* **Notes 27–29**).
6. Construct a calibration curve using the BSA standards, using the 0 mg/mL BSA sample as a blank (*see* **Notes 30–32**).
7. Based on the calibration curve calculate the protein concentration in the immobilized enzyme samples.
8. Use the following equation to determine the enzyme loading:

$$\frac{\text{mg enzyme bound}}{\text{g dry weight}} = \frac{\text{concentration of protein in immob. enzyme sample (mg/mL)} \times 2.1 \text{ mL}}{\text{amount of immob. enzyme used in assay (g dry weight)}}$$

4. Notes

1. A procedure similar to this can be performed where, instead of using concentrated buffer, the pH is maintained at 11.0 by titrating with 2 or 4 N NaOH. This method often results in the doubling of the binding capacity compared to the buffer method.
2. CNBr is highly toxic and proper safety precautions should be employed when handling it.
3. Although the activated support materials should be used soon after activation, it can be stored at –20°C under airtight conditions with a loss of 10% or less per month of its original binding capacity.
4. Ideally the enzyme should be in a buffer at an alkaline pH to reduce the amount of protonated amino groups. However, it is more important to have the enzyme at a pH it is most stable, so this should be chosen if the enzyme is not stable at high pH.
5. The binding of protein to the support can be monitored by performing protein assays on the supernatant (e.g., Bradford or BCA total protein assay) and comparing to the initial protein concentration. For some enzymes, incubations times longer than 24 h may be necessary to achieve maximal enzyme binding.

6. 1-Cyclohexyl-3-(2-morpholino-ethyl)-carbodiimide metho-*p*-toluene-sulfonate (CMC) may also be used to activate carboxylic acid-containing supports.

7. The EDC concentration is about 100-fold molar excess to the carboxylate groups on the support. The activated support should be washed well before adding enzyme.

8. EDC concentration, activation time, coupling time, coupling pH, and wash conditions (i.e., washing with NaCl and/or urea to remove any unbound enzyme) should all be optimized for a given support-enzyme system.

9. An alternative method is to add the support, EDC, and enzyme all at once such that support activation and enzyme immobilization occurs simultaneously. In this case, a 1:1 molar ratio of EDC to support carboxyl groups should be used. Allow this to react in 0.05 *M* sodium phosphate buffer, pH 7.0, at 4°C for 1 h before washing. Longer times may be necessary depending on the enzyme and support. The EDC may also react with the carboxyl groups on the enzyme causing inactivation. If this occurs, the other method should be used.

10. A higher concentration of GA can be used and this may increase the binding capacity of the support material. However, since GA is often detrimental to the enzyme, more extensive washing should be performed.

11. If PEI coated materials are used, they will turn to a pinkish color after activation with GA.

12. A saturated solution of 2,4-dinitrophenylhydrazine in 0.2 *N* HCl can be used to detect residual GA in the washing solutions. Add approx 0.2 mL of the solution used to wash the GA-activated support to 0.5 mL of saturated 2,4-dinitrophenylhydrazine solution. The formation of a yellow precipitate indicates the presence of residual GA and the support material should be further washed.

13. The GA-activated support can be stored in a moist form at 4°C for at least 1 yr without significant loss in binding capacity.

14. Shaking as opposed to magnetic stirring should be used if the support is friable.

15. Enzyme loading in this type of immobilization can be as high as 30% dry weight and higher.

16. The dry Eupergit C will absorb about 3 times its weight in water (1 g dry will have a mass of about 4 g when wet).

17. The Eupergit should be kept dry until use as the epoxy groups can hydrolyze.

18. Approximately 5–10% enzyme loading on a dry basis is typical, however this is dependent on the enzyme and higher or lower loading levels are possible.

19. The ionic strength and the pH of the enzyme solution can significantly affect the loading amount and the residual activity. Often a high ionic strength (1 *M* sodium chloride) gives better binding, but this is dependent on the enzyme and a range of ionic strengths and pH values should be evaluated.

20. Post-treatment of the beads after immobilization to quench the remaining epoxy groups can effect the immobilized enzyme activity. *See* **Subheading 1.4.4.** for details on this.

21. Other protein assays may be used besides the Bradford assay, i.e., BCA, Lowry, absorbance at 280 nm.

22. If any enzyme precipitates during the immobilization, this could overestimate the amount of enzyme bound.
23. If the amount of enzyme bound to the support is very small, the inherent variability in the protein assay may not give accurate protein loading results.
24. If the wash solution contains compounds that interfere with the protein assay, unreliable data may result.
25. This method is a general method and can be modified as necessary depending the enzyme loading. The method above assumes a 1–10% enzyme loading based on dry weight.
26. The BCA protein assay has a working range of 0.02–2 mg/mL protein.
27. For low loading or small sample sizes the "enhanced protocol" can be used (incubation at 60°C for 30 min).
28. This should be done as soon as possible after cooling as the standards will continue to react with the BCA reagent while the immobilized enzyme will not if removed or centrifuged.
29. If the absorbance of the immobilized enzyme samples falls outside the range of the calibration curve, then the procedure should be repeated with a change in either the standards or immobilized enzyme concentration.
30. For more accurate results, the standards and the immobilized enzyme samples should be done in at least duplicate.
31. The support without any enzyme bound should also be tested to see if the BCA shows a response toward it. If it does, this should be taken into account and subtracted from the results of the immobilized enzyme supports.
32. The standards should be used every time this procedure is performed as the assay is highly dependent on the temperature and time of incubation, which may be difficult to replicate every time.

References

1. Trevan, M. D. (ed.) (1980) *Immobilized Enzymes,* John Wiley & Sons, Ltd., New York.
2. Langmuir, I. and Schaefer, V. J. (1938) Activities of urease and pepsin monolayers. *J. Am. Chem. Soc.* **60,** 1351–1360.
3. Nelson, J. M. and Griffin, E. G. (1916) Adsorption of invertase. *J. Am. Chem. Soc.* **38,** 1109–1115.
4. Nelson, J. M. and Hitchcocks, D. I. (1921) Activity of adsorbed invertase. *J. Am. Chem. Soc.* **43,** 1956–1961.
5. Goldstein, L. and Katchalski-Katzir, E. (1976) Immobilized enzymes—a survey, in *Immobilized Enzyme Principles* (Goldstein, L., ed.), Vol. 1, Academic Press, Inc., New York.
6. Rosevear, A., Kennedy, J. F., and Cabral, J. M. S. (eds.) (1987) *Immobilised Enzymes and Cells,* Adam Hilger, Bristol.
7. Clark, D. S. (1994) Can immobilization be exploited to modify enzyme activity? *Trends Biotechnol.* **12,** 439–443.
8. Klibanov, A. M. (1979) Enzyme stabilization by immobilization. *Anal. Biochem.* **93,** 1–23.

9. Kawamura, Y., Nakanishi, K., Matsuno, R., and Kamikubo, T. (1981) Stability of immobilized a-chymotrypsin. *Biotechnol. Bioeng.* **23,** 1219–1236.
10. Mozhaev, V. V., Siksnis, V. A., Torchilin, V. P., and Martinek, K. (1983) Operational stability of copolymerized enzymes at elevated temperatures. *Biotechnol. Bioeng.* **25,** 1937–1945.
11. Mozhaev, V. V. (1993) Mechanism-based strategies for protein thermostabilization. *Trends Biotechnol.* **11,** 88–95.
12. Adlercreutz, P. (1991) On the importance of the support material for enzymatic synthesis in organic media. Support effects at controlled water activity. *Eur. J. Biochem.* **199,** 609–614.
13. Ingalls, R. G., Squires, R. G., and Butler, L. G. (1975) Reversal of enzymatic hydrolysis: rate and extent of ester synthesis as catalyzed by chymotrypsin and subtilisin carlsberg at low water concentratons. *Biotechnol. Bioeng.* **17,** 1627–1637.
14. Reslow, M., Adlercreutz, P., and Mattiasson, B. (1987) On the importance of the support material for bioorganic synthesis: influence of water partition between solvent, enzyme and solid support in water-poor reaction media. *Eur. J. Biochem.* **172,** 573–578.
15. Tanaka, A. and Kawamoto, T. (1991) Immobilized enzymes in organic solvents. *Bioprocess Technol.* **14,** 183–208.
16. Wang, P., Sergeeva, M. V., Lim, L., and Dordick, J. S. (1997) Biocatalytic plastics as active and stable materials for biotransformation. *Nat. Biotechnol.* **15,** 789–793.
17. Dordick, J. S. (1989) Enzymatic catalysis in monophasic organic solvents. *Enzyme Microb. Tech.* **11,** 194–211.
18. Mozhaev, V. V., Sergeeva, M. V., Belova, A. B., and Khmelnitsky, Y. L. (1990) Multipoint attachment to a support protects enzyme from inactivation by organic solvents: α-chymotrypsin in aqueous solutions of alcohols and diols. *Biotechnol. Bioeng.* **35,** 653–659.
19. Srere, P. A. and Uyeda, K. (1976) Functional groups on enzymes suitable for binding to matrices. *Methods Enzymol.* **44,** 11–19.
20. Kennedy, J. F., Melo, E. H. M., and Jumel, K. (1990) Immobilized enzymes and cells. *Chem. Eng. Prog.* **86,** 81–89.
21. Fasman, G. D. (ed.) (1989) *Predictions of Protein Structure and the Principles of Protein Conformation,* Plenum Press, New York.
22. Mosbach, K. (1976) Immobilized Enzymes. Vol. 44, in *Methods in Enzymology* (Kaplan, N. O., ed.), Academic Press, New York.
23. Weetall, H. H. (1969) Trypsin and papain covalently coupled to porous glass: preparation and characterization. *Science* **166,** 615–617.
24. Weetall, H. H. and Filbert, A. M. (1974) Porous glass for affinity chromatography applications. Vol. 34, in *Methods in Enzymology* (Wilchek, M., ed.), Academic Press, New York.
25. Messing, R. A. and Weetall, H. H. (1970) Chemically coupled enzymes. U.S. Pat. 3,519,538.
26. Kadima, T. A. and Pickard, M. A. (1990) Immobilization of chloroperoxidase on aminopropyl-glass. *Appl. Environ. Microb.* **56,** 3473–3477.

27. Rohrback, R. P. (1985) Support matrix and immobilized enzyme system. U.S. Pat. 4,525,456.
28. Lantero, O. J. (1984) Immobilization of biocatalysts on granular carbon. U.S. Pat. 4,438,196.
29. Chiang, J. P. and Lantero, O. J. (1987) Immobilization of biocatalysts on granular diatomaceous earth. U.S. Pat. 4,713,333.
30. Goldberg, B. S. (1978) Method of immobilizing proteinaceous substances. U.S. Pat. 4,169,014.
31. Goldberg, B. S. (1978) Immobilized proteins. U.S. Pat. 4,102,746.
32. Crump, S. P. and Rozzell, J. D. (1992) Biocatalytic production of amino acids by transaminases, in *Biocatalytic Production of Amino Acids and Derivatives* (Rozzell, J. D. and Wagner, F., eds.), Vol. 1, Oxford University Press, New York, pp. 44–58,
33. Messing, R. A. and Yaverbaum, S. (1978) Immobilization of proteins on inorganic support materials. U.S. Pat. 4,071,409.
34. Leuba, J.-L., Renker, A., and Flaschel, E. (1990) Enzyme immobilization on mineral particles coated with chitosan. U.S. Pat. 4,918,016.
35. Boller, T., Meier, C., and Menzler, S. (2002) Eupergit oxirane acrylic beads: How to make enzymes fit for biocatalysis. *Org. Process. Res. Dev.* **6,** 509–519.
36. Katchalski-Katzir, E. and Kraemer, D. M. (2000) Eupergit C, a carrier for immobilization of enzymes of industrial potential. *J. Mol. Catal. B-Enzym.* **10,** 157–176.
37. Hosaka, S., Murao, Y., Masuko, S., and Miura, K. (1983) Preparation of microspheres of poly(glycidyl methacrylate) and its derivatives as carriers for immobilized proteins. *Immunol. Commun.* **12,** 509–517.
38. Malmsten, M. and Larsson, A. (2000) Immobilization of trypsin on porous glycidyl methacrylate beads: effects of polymer hydrophilization. *Colloid Surface B* **18,** 277–284.
39. Mujawar, S. K., Kotha, A., Rajan, C. R., Ponrathnam, S., and Shewale, J. G. (1999) Development of tailor-made glycidyl methacrylate-divinyl benzene copolymer for immobilization of D-amino acid oxidase from *Aspergillus* species strain 020 and its application in the bioconversion of cephalosporin C. *J. Biotechnol.* **75,** 11–22.
40. Turkova, J., Blaha, K., Malanikova, M., Vancurova, D., Svec, F., and Kalal, J. (1978) Methacrylate gels with epoxide groups as supports for immobilization of enzymes in pH range 3–12. *Biochim. Biophys. Acta.* **524,** 162–169.
41. Drobnik, J., Saudek, V., Svec, F., Kalal, J., Vojtisek, V., and Barta, M. (1979) Enzyme immobilization techniques on poly(glycidyl methacrylate-co-ethylene dimethacrylate) carrier with penicillin amidase as model. *Biotechnol. Bioeng.* **21,** 1317–1332.
42. Kotha, A., Raman, R. C., Ponrathnam, S., Kumar, K. K., and Shewale, J. G. (1998) Beaded reactive polymers 3: Effect of triacrylates as crosslinkers on the physical properties of glycidyl methacrylate copolymers and immoblization of penicillin G acylase. *Appl. Biochem. Biotech.* **74,** 191–203.
43. Margel, S. (1982) Agarose polyacrolein microsphere beads. New effective immunoabsorbents. *FEBS Lett.* **145,** 341–344.

44. Varlan, A. R., Sansen, W., Loey, A. V., and Hendrickx, M. (1996) Covalent enzyme immobilization on paramagnetic polyacrolein beads. *Biosens. Bioelectron.* **11,** 443–448.

45. Stich, T. (1990) Determination of protein covalently bound to agarose supports using bicinchiconic acid. *Anal. Biochem.* **191,** 343–346.

46. Lewis, W. S. and Schuster, S. M. (1990) Quantitation of immobilized proteins. *J. Biochem. Bioph. Meth.* **21,** 129–144.

47. Bonde, M., Pontoppidan, H., and Pepper, D. S. (1992) Direct dye binding – a quantitative assay for solid-phase immobilized proteins. *Anal. Biochem.* **200,** 195–198.

48. Orschel, M., Katerkamp, A., Meusel, M., and Cammann, K. (1998) Evaluation of several methods to quantify immobilized proteins on gold and silica surfaces. *Colloid Surface B* **10,** 273–279.

49. Liese, A., Seelbach, K., and Wandrey, C. (eds.) (2000) *Industrial Biotransformations,* Wiley-VCH, Weinheim, Germany.

50. Hartmeier, W. (1988) *Immobilized Biocatalysts,* Springer-Verlag, New York.

51. Katchalski-Katzir, E. (1993) Immobilized enzymes—learning from past successes and failures. *Trends Biotechnol.* **11,** 471–478.

52. Rozzell, D. (2003) Personal communication.

53. Catalog of Enzyme Products 2003, BioCatalytics, Inc., Pasadena, CA.

17

Preparation of an Industrial Biocatalyst of Penicillin G Acylase on Sepabeads

Improving the Design of Penicillin G Hydrolysis

Cesar Mateo, Valeria Grazu, Olga Abian, Manuel Fuentes, Gloria Fernández-Lorente, José M. Palomo, Roberto Fernández-Lafuente, and José M. Guisán

Summary

The enzyme penicillin G acylase (PGA) is currently employed at an industrial scale in the hydrolysis of penicillin and cephalosporin G. Here, we describe the preparation of a new immobilized preparation of the enzyme that yields derivatives that are very thermostable and resistant in the presence of organic solvents. The stabilization is obtained via a double strategy. First, using a new epoxy support (Sepabeads), a high degree of multipoint covalent attachment is achieved. A three-step methodology is proposed: (1) immobilization on the support at high ionic strength, (2) promotion of an intense multipoint covalent attachment by incubation for a long time at alkaline pH, and (3) blocking hydrophylization of the support surface. These derivatives are more rigid and therefore more resistant to inactivation by both temperature or organic solvents. Second, the contact of the enzyme with the organic solvent molecules is reduced by generating a highly hydrophilic microenvironment using polymers (polyethylenimine, aldehyde-dextran, and sulfate-dextran). The exact protocol for the production of this environment has been found to be critical to achieve the desired results (final derivatives remain active in 90% of organic solvents). This derivative could be used in the hydrolysis of antibiotics even in media fully saturated with organic solvents, making simpler the current processes of production of 6-aminopenicillanic acid (6-APA) or 7-aminodeacetoxy-cephalosporanic acid (7-ADCA).

Key Words: Multipoint covalent attachment; stabilization of proteins; hydrolysis of penicillin G; enzymes in organic solvents.

1. Introduction

The industrial use of enzymes usually requires their previous immobilization. Thus, the biocatalyst may be recovered after the reaction, and also we can avoid some mechanisms of inactivation of the enzyme related to its interaction

From: *Methods in Biotechnology, Vol. 17: Microbial Enzymes and Biotransformations*
Edited by: J. L. Barredo © Humana Press Inc., Totowa, NJ

with other molecules or interfaces (e.g., preventing aggregation, proteolysis, or interaction with gas bubbles). Moreover, immobilized enzymes permit improvement of both the reactor design and its performance. Bearing in mind that immobilization is necessary, it should be convenient to couple this step to the improvement of enzymes' properties. Although activity and selectivity may be altered by immobilization, stability is the most commonly accepted property that may be improved via immobilization. Applicability to most enzymes is usually reduced by the low stability of many enzymes.

Penicillin G acylase is the enzyme used at an industrial scale to hydrolyze penicillin G or cephalosporin G, producing 6-aminopenicillanic acid (6-APA) or 7-aminodeacetoxycephalosporanic acid (7-ADCA). The stability of the enzyme is quite high, which has permitted the rapid implementation of this process. Certainly, the productivity of the biocatalyst may be enhanced by improving the enzyme stability. However, the enzyme is readily inactivated by the organic solvent's action. This is a problem when it is used in synthetic routes, but it is also a problem in hydrolysis. To understand this, we must pay attention to the current scheme of 6-APA production (**Fig. 1**).

Penicillin G is extracted from the fermentation broth using an organic solvent, e.g., methyl isobutyl ketone (MIBK) or butyl acetate, at low pH *(1)* (**Fig. 1**). Then the penicillin G is back-extracted to an aqueous phase at higher pH and further crystallized to yield its potassium salt (KPen G). Subsequently, KPen G is dissolved in aqueous medium and hydrolyzed by penicillin G acylase (PGA) at pH 7.0–8.0 to obtain phenylacetic acid (PAA) and 6-APA *(2)*. From this scheme, elimination of the crystallization step is an evident way of simplifying the 6-APA production process. The main problem for this strategy is that the biocatalyst needs to be stable and active under these conditions *(3)*. For this reason, the stabilization of penicillin G acylase against the action of organic solvents may greatly improve the final performance of the biocatalyst.

Random immobilization might not promote any conformational stabilization of immobilized enzymes. In fact, there are reports of immobilization procedures with no effect or even negative effects on the stability of some enzymes *(4–8)*. However, it is generally accepted that such stabilization should be achieved if the immobilization of each enzyme molecule occurs through several residues, mainly if the reactive groups in the support are secluded from its surface through very short spacer arms. In this way, all the residues of the enzyme molecule involved in immobilization have to preserve their relative positions almost completely unaffected during any conformational change promoted by heat, organic solvents, or any other distorting agents *(9–11)*. Thus, such multipoint covalently immobilized enzymes should become much more stable than their soluble counterparts

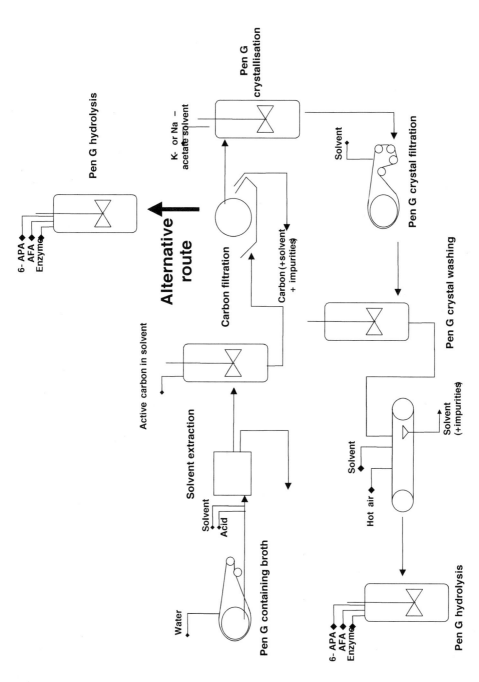

Fig. 1. Schematic overview of the conventional enzymatic process of penicillin G purification and 6-APA production. An alternative route is proposed.

or than randomly immobilized preparations against any distorting reagent (temperature, drastic pH values, organic solvent).

1.1. Stabilization by Multipoint Covalent Attachment on Epoxy-Sepabeads

There are many protocols for protein immobilization described in the literature, but most of them may be difficult to perform at industrial scale, where an extended handling of the support may be necessary, and the use of toxic or contaminant substances should be avoided. These problems are not considered at laboratory scale, where the experiment is performed by specialized staff under very controlled conditions *(12–20)*.

In this context, epoxy-activated supports are almost ideal matrices to perform very easy immobilization of proteins and enzymes at both laboratory and industrial scales *(21–23)*. These activated supports are very stable during storage and also suspended in neutral aqueous media. Hence, they can be handled easily and long-term before and during immobilization procedures. Furthermore, epoxy-activated supports are able to directly form very stable covalent linkages with different protein groups (amino, thiol, phenolic) under very mild experimental conditions (e.g., pH 7.0). The immobilization follows a two-step mechanism: first a rapid mild physical adsorption between the protein and the support is produced, then the covalent reaction between adsorbed protein and epoxide groups occurs *(21,24–27)* (**Fig. 2**). For this requirement, commercial epoxy supports utilized to immobilize proteins are fairly hydrophobic in order to adsorb proteins when they are incubated at high ionic strength in the presence of the support, while hydrophilic supports (e.g., agarose) are not recommended for this purpose because this preliminary physical hydrophobic adsorption is not possible. After the protein immobilization, remaining epoxy support may be easily blocked in order to stop any kind of undesired covalent support-protein reaction *(22,28)*.

Eupergit C and Sepabeads may be used to improve the enzyme stability via multipoint covalent attachment, but using more drastic experimental conditions than those necessary to covalently immobilize the proteins (perhaps via a single, or only a few, very reactive residue/s) in order to enhance the reactivity of nucleophilic groups of the enzyme surface: higher pH values (around 9.0–10.0), long reaction time, moderately high temperature, and so on. *(21,22)*. However, Eupergit C behaves as a support with an internal geometry that does not offer large plane surfaces to the proteins for the enzyme multi-interaction, yielding significant stabilization factors (around 100) but not as important as with the use of other supports (e.g., glyoxyl agarose, Sepabeads). The internal morphology of the support's surface is a key point to achieve a very intense multipoint covalent attachment because it determines the possibilities of enzyme support multi-interaction.

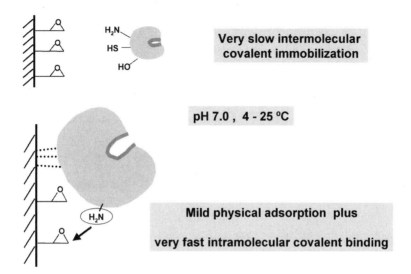

Fig. 2. Immobilization of enzymes on epoxy supports. The immobilization follows a two-step mechanism: first, a rapid mild physical adsorption between the protein and the support is produced; second, the covalent reaction between adsorbed protein and epoxide groups occurs.

Here, we propose the use of Sepabeads-EP (a highly activated epoxy-support) not only to immobilize penicillin G acylase, but also as a very adequate method to achieve multipoint covalent attachment between the immobilized enzyme and the support. Sepabeads-EP has some excellent properties to give stabilizing multipoint immobilizations of proteins and adequate mechanical properties to the final biocatalyst. Sepabeads is produced via a very intense crosslinking in the presence of porogenic agents (Technical information, Resindion, Milan, Italy). This makes Sepabeads very rigid supports that may be used in stirred tanks or bed reactors. The support does not swell when hydrated, so it can be changed from aqueous media to anhydrous media without exhibiting significant changes in volume or geometry. Finally, Sepabeads-EP internal geometry offers large internal plane surfaces where the enzyme may undergo intense interactions with the support (the internal pores are surrounded by convex surfaces to give cylindrical pores). Also, they have in their surface a dense monolayer of reactive and stable epoxy groups (around 100 μmol/g of wet support), equivalent to that of Eupergit C.

In this way, a very intense intramolecular attachment might be achieved between the already immobilized enzyme and these highly activated supports. Here, we describe the protocol to use epoxy-Sepabeads to achieve the stabilization of penicillin G acylase via covalent multipoint attachment, following a strategy that yields stabilization factors as high as 10,000 when it is compared with the one-point covalent attached penicillin G acylase.

1.2. Stabilization of Enzymes Against Inactivation by Organic Solvents: Generation of Artificial Microenvironment

An immobilized enzyme (therefore fully dispersed and immobilized inside a porous support) usually undergoes very rapid inactivation when it is exposed to the presence of medium-high concentrations of water-soluble organic solvents or organic solvents in anhydrous systems, even at neutral pH values and moderate temperatures. However, the exact mechanism of these dramatic inactivations should be quite simple. Under these mild conditions, immobilized enzymes hardly undergo chemical modifications (the co-solvents are chemically inert, and pH and temperature are very mild) and aggregation is prevented by the dispersion on the support surface. For all these reasons we can assume that, under these conditions, the unique cause of these dramatic inactivations of enzyme molecules is their interaction with organic co-solvents and the main inactivating effect is the promotion of conformational changes on the enzyme structure *(4)*.

On the basis of this simple mechanism of dramatic inactivation, two clear strategies for enzyme stabilization against organic co-solvents can be proposed *(29–31)*: (1) The prevention of the cause of inactivation: promotion of a drastic diminution of the co-solvent concentration in the immediate surroundings of the enzyme molecules. (2) The prevention of inactivating effects: a drastic "rigidification" of enzyme molecules in order to reduce the intensity of conformational changes induced by co-solvents.

The second strategy may be achieved by following protocols for covalent multipoint attachment, as the described in **Subheading 1.1.** Here, we have focused in the first phenomenon: when an enzyme has to work in the presence of organic co-solvents in the bulk solution, its effective stability may be significantly increased if we are able to greatly reduce the actual concentration of co-solvent existing in the microenvironment of the enzyme molecule (e.g., 15–30%) *(29–32)*. This goal could be achieved via generation of a hydrophilic shell surrounding every enzyme molecule. Obviously, this hydrophilic shield cannot be constituted by a very close structure not allowing the transfer of substrates and products from or to the bulk solution to or from the enzyme active site. On the contrary, the hydrophilic shell must have a very open structure and the only limitation to the access of co-solvent molecules has to be achieved through "partition effects."

By using polyethylenimine (PEI), aldehyde-dextran, and sulfate-dextran, we have been able to significantly improve the stability against the effects of organic co-solvents of penicillin G acylase. The optimization in the preparation of the hydrophilic microenvironment has played a critical role. In order to get such very strong partition effects between co-solvents and the enzyme environment, the open hydrophilic shell surrounding every enzyme molecule has to be an extremely hydrophilic one. To achieve this objective, the use of some "random coil" polyfunctional polymers is here proposed. The following polymers were utilized: (1) Polyethylenimine is commercially available and there are preparations with very different molecular weights (Mw). This polymer contains a very high concentration of primary, secondary, and tertiary amino groups. Most of them are ionized at neutral pH value; hence, this polymer is able to become adsorbed on proteins and other polyanionic polymers. In addition to that, the presence of nonionized primary amino groups also allows the covalent attachment of this polymer on aldehyde-supports and aldehyde-polymers. (2) Aldehyde-dextran can be obtained via periodate oxidation of commercial dextrans with different Mw *(33)*. These polyfunctional polymers may covalently react with primary amino groups placed in enzymes, support, or other polymers (e.g., PEI). (3) Sulfate-dextran is also commercially available in different Mw preparations. This is a very hydrophilic polymer able to become adsorbed on polyethylenimine with formation of polymeric salts (very similar to a polymeric ammonium sulfate salt).

By using different combinations of these polymers, we have optimized the preparation of artificial hyper-hydrophilic microenvironments formed by polymeric salts and completely surrounding multipoint covalent immobilized ones. This strategy has permitted to prepare a penicillin G acylase immobilized preparation not only useful to work in the aqueous media fully saturated of organic solvent, but even in the presence of 90% organic solvent or anhydrous media.

2. Materials

1. Penicillin G acylase (Antibióticos S. A., León, Spain).
2. 25 mM NaOH.
3. 1 M sodium phosphate buffer, pH 7.0.
4. 4 M sodium phosphate buffer, pH 7.0.
5. 100 mM sodium phosphate buffer, pH 8.0.
6. 100 mM sodium phosphate buffer, pH 9.0.
7. 5 mM sodium phosphate buffer, pH 7.0.
8. 3 M sodium bicarbonate buffer, pH 10.5.
9. 500 mM sodium phosphate buffer, pH 7.0.
10. 50 mM sodium phosphate buffer, pH 7.0.
11. pHstat (Mettler Toledo, Madrid, Spain).
12. Multipoint covalent attachment buffer: 100 mM phosphate buffer, 100 mM phenylacetic acid, and 20% glycerol, pH 8.5.
13. Blocking solution: 3 M glycine and 100 mM phenylacetic acid, pH 8.5.
14. Sepabeads (Resindion srl, Milan, Italy).
15. Aldehyde-dextran prepared as described by Abian et al. *(30)*.
16. Sodium bicarbonate buffer: 100 mM sodium bicarbonate buffer, 100 mM phenylacetic acid, and 25% (v/v) glycerol, pH 10.05.
17. Polyethylenimine, dextran, and sulfate-dextran (Sigma, St. Louis, MO).

18. Water-MIBK biphasic system: 25 mM phosphate buffer and 2% MIBK, then adjust at pH 8.0.
19. Water-butanone biphasic system: 25 mM phosphate buffer and 28% butanone, pH 8.0.

3. Methods

3.1. Activity Assay of Penicillin G Acylase

1. The reaction mixture contains 30 mM penicillin G in 100 mM phosphate buffer, pH 8.0, pre-incubated at 25°C for 15 min.
2. Penicillin G acylase activity was measured using a pHstat, by titration of the phenylacetic acid released during the hydrolysis of penicillin G at 25°C. Twenty-five mM NaOH were used as a titrating agent.

3.2. Immobilization of Enzymes on Epoxy Supports

1. Wash 10 g of support 10 times with 3 volumes of distilled water.
2. Mix the enzyme solution (1 mg/mL in 1 M phosphate buffer, pH 7.0) and 10 g of washed support. Leave under mild stirring at 20°C.
3. Follow the immobilization of the enzyme by measuring the activity of the supernatant and suspension, compared with a reference solution where the support is substituted for the equivalent water volume.
4. To take supernatant samples, use some pipet-filter or short centrifugations.
5. To check the covalent immobilization, incubate 1 mL of the preparations during 30 min in 10 mL of 5 mM phosphate buffer, pH 7.0. and measure the activity in the supernatant and suspension as described in **step 3.1.**

3.3. Enzyme-Support Multipoint Covalent Attachment

1. After the immobilization process, incubate 1 mL of the preparations in 10 mL of multipoint covalent attachment buffer for 4 d.
2. Wash the preparations with an excess of water to remove the enzyme that might be only adsorbed to the matrix.
3. To end the process and hydrophilize the support surface, block the remaining epoxy groups by addition of 10 mL of blocking solution. This step has a very relevant effect on enzyme stability (**Fig. 3**).
4. Stir the suspension over 24 h, and then wash with an excess of water.

3.4. Optimal Protocol for Protein Immobilization-Stabilization on Sepabeads

An optimal protocol for immobilization-stabilization of PGA is proposed:

1. Immobilization in 1 M phosphate buffer, pH 7.0, for 24 h on Sepabeads-EP.
2. Incubation of the immobilized enzyme at pH 9.0 in 100 mM phosphate buffer for 72 h.
3. Hydrophilization of the support surface by incubating the immobilized preparation for 24 h at pH 8.5 in presence of 3 M glycine (or other amino acids).

This immobilized preparation shows a remarkable stabilization regarding the soluble enzyme (around 10,000-fold more stable) or the enzyme immobilized

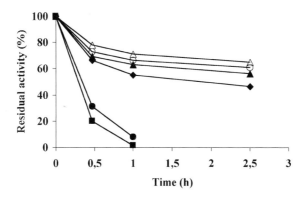

Fig. 3. Effect of blocking with different substances on the stability of PGA immobilized on standard Sepabeads. Immobilization was performed in 1 *M* sodium phosphate, pH 7.0, and blocking was performed as described in **Subheading 3.3.** after 24 h. Solid Circles: no-blocked derivative. Squares: blocking with mercaptoethanol. Rhombus: blocking with ethylendiamine. Solid triangles: blocking with Lys. Empty circles: blocking with Glu. Empty triangles: blocking with Gly.

and stabilized in Eupergit C support following the same optimized protocol (around 100-fold) (**Fig. 4**).

3.5. Co-Immobilization of Polyethyleneimine and the Enzyme

1. Add 20 g of polyethyleneimine (PEI; MW from 700 to 1,000,000 Da) to 50 mL of an aqueous solution of 200 m*M* phenylacetic acid. Adjust to pH 9.0 with 3 *M* sodium bicarbonate buffer, pH 10.5, and bring final volume to 100 mL.
2. Add 7 g of PGA-Sepabeads preparation (without blocking the epoxy groups), and leave at 20°C under mild stirring.
3. After 16 h reaction, wash the enzyme PEI co-immobilized with an excess of 0.5 *M* phosphate buffer, pH 7.0, at 25°C to eliminate the excess of PEI in the preparation.
4. After this process, block the remaining epoxy groups of the support as described in **Subheading 3.3** (*22*) (*see* **Note 1**).

This treatment does not have a significant effect on thermostability, but promotes a noteworthy increment on enzyme stability in the presence of organic solvents (**Fig. 5**).

3.6. Modification with Aldehyde-Dextran

1. Suspend 7 g of wet PGA-PEI-Sepabeads preparation in 145 mL of aldehyde-dextran (6000 Da) prepared at pH 7.0 (*32*).
2. Maintain the reaction mixture at 25°C under gentle stirring for 12 h.
3. Wash with an excess of water to eliminate the excess of aldehyde-dextran (*see* **Note 2**).

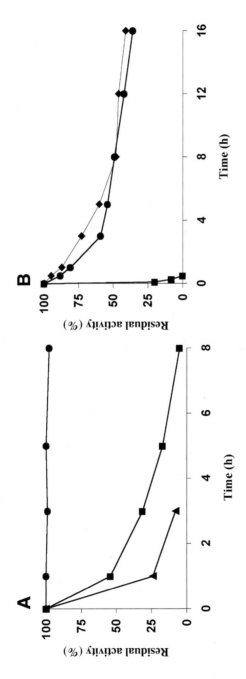

Fig. 4. Thermal inactivation of different PGA derivatives at pH 8.0 and 45°C (**A**) or 55°C (**B**). Experiments were carried out as described in **Subheading 3.9.** Triangles: PGA immobilized under standard conditions on Eupergit C. Squares: PGA immobilized, incubated, and blocked under optimal conditions on Eupergit C. Circles: PGA immobilized, incubated, and blocked under optimal conditions on Sepabeads. Rhombus: PGA immobilized on glyoxyl agarose.

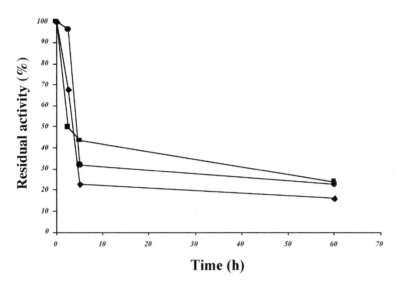

Fig. 5. Effect of the co-immobilization of PEIs and PGA in Sepabeads derivatives on their stability in organic cosolvent. Inactivation was performed at pH 7.0 and 4°C in 75% dioxane as described in **Subheading 3.10.** PEIs with different MW were tested. Unmodified Sepabeads PGA (rhombus), Sepabeads PGA co-immobilized with low MW PEI (circles), Sepabeads PGA co-immobilized with high MW PEI (squares).

3.7. Multilayer PEI-Aldehyde-Dextran

1. Suspend 7 g of the preparation PEI-PGA-aldehyde-dextran in 100 mL of an aqueous solution of 2 g of PEI (25 kDa) adjusted to pH 8.0 by addition of concentrated HCl.
2. After 12 h, wash soluble PEI with an excess of 0.5 M phosphate buffer, pH 7.0.
3. Add a new layer of aldehyde-dextran following the protocol described in **Subheading 3.6.**
4. Perform successive additions/washings of PEI and aldehyde-dextran to build 3 layers of PEI/aldehyde-dextran on the support surface (*see* **Note 3**).
5. Increase the reaction volume to 500 mL by adding sodium bicarbonate buffer.
6. Add 500 mg of solid sodium borohydride (*see* **Note 4**), and, after 30 min, wash the modified preparation abundantly with distilled water.

This multilayer of hydrophilic polymer promoted an increment on enzyme stability in the presence of organic solvents **(Fig. 6)**.

3.8. Generation of Polymeric Salts

1. Suspend 10 mL of the preparation obtained in **Subheading 3.7.** in 150 mL of a solution of 32 mg/mL sulfate-dextran (6000 Da). pH was previously adjusted to 7.0 by addition of 4 M phosphate buffer, pH 7.0.
2. After 1 h reaction, wash the preparations with an excess of distilled water.

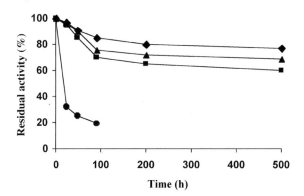

Fig. 6. Stability in organic media of Sepabeads-PGA derivatives after the addition of several layers of PEI-aldehyde-dextran. Inactivation was performed at pH 7.0 and 20°C in 90% dioxane as described in **Subheading 3.10.** Unmodified Sepabeads-PGA (circles), one layer PGA-Sepabeads derivative (squares), two-layer PGA-Sepabeads derivative (triangles), three-layer PGA-Sepabeads derivative (rhombus). The MWs of the PEIs used were 1×10^6 Da in the first layer and 25,000 Da in the second and third.

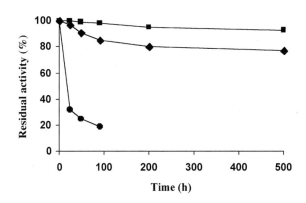

Fig. 7. Increase of stability in organic co-solvent of the multilayer Sepabeads-PGA derivative when dextran sulfate is added. Inactivation was performed at pH 7.0 and 20°C in 90% dioxane. Unmodified Sepabeads-PGA (circles), three-layer PGA-Sepabeads derivative (rhombus), three-layer PGA-dextran sulfate Sepabeads derivative (squares).

The complete treatment of hydrophilization promotes a slight decrease in the enzyme activity (mainly by the first aldehyde-dextran treatment) by around 20%. The treatment promotes a new increment on enzyme stability in the presence of organic solvents **(Fig. 7)**. This immobilized preparation was fully sta-

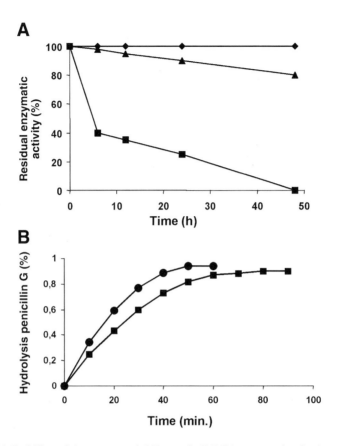

Fig. 8. (**A**) Stability of the commercial Eupergit C-PGA preparation in the presence of water-organic solvent biphasic system. Preparations were incubated in water-MIBK or water-butanone biphasic systems at 32°C. Water medium (rhombus). Water-MIBK biphasic system (triangles). Water-butanone biphasic system (squares). (**B**) Hydrolysis course of penicillin G by using commercial Eupergit C-PGA, 140 mM penicillin G, pH 8.0, and 32°C. Fully aqueous system (circles). Aqueous-MIBK biphasic system (squares).

ble on systems saturated with organic solvents similar to those used to extract penicillin G (**Fig. 8A**), enabling the hydrolysis of this compound in this kind of systems with low inhibition (**Fig. 8B**).

3.9. Thermal Inactivation

1. Resuspend 1 g of the desired immobilized enzyme in 50 mM phosphate buffer, pH 7.0, at the desired temperature.
2. Periodically withdraw samples using tip-cut and measure the activity as described in **Subheading 3.1.**

3.10. Inactivation of PGA Preparations in the Presence of Organic Solvents

1. Wash the enzyme preparation 10 times with 3 volumes of the desired organic solvent/buffer mixture at 4°C.
2. Periodically withdraw samples using tip-cut and measure the activity as described in **Subheading 3.1.**

4. Notes

1. Blocking of epoxy groups on Sepabeads supports is necessary to prevent hydrophobic interactions between support and enzyme. This has been extensively discussed by Mateo et al. *(22)*.
2. Aldehyde-dextran is quite unstable at basic pH. If the solution takes some tanned color it should be discarded.
3. The washing should ensure the full elimination of the aldehyde-dextran or polyethylenimine to prevent reaction between soluble forms that could close the porous structure of the support.
4. Reduction with sodium borohydride requires alkaline pH; for this reason pH should be increased simultaneously to the addition of sodium borohydride.

References

1. Harrison, F. G. and Gibson, E. D. (1984) Approaches for reducing the manufacturing costs of 6-aminopenicillamic acid. *Proc. Biochem.* **19,** 33–36.
2. Van der Wielen, L. A. M., Van Buel, M. J., Straathof, A. J. J., and Luyben, K. Ch. A. M. (1997) Modelling the enzymatic deacylation of penicillin G-Equilibrium and kinetic considerations. *Biocatal. Biotransform.* **15,** 121–146.
3. Arroyo, M., Torres, R., De la Mata, I., Castillón, M. P., and Acebal, C. (2000) Prediction of penicillin V acylase stability in water-organic cosolvent monophasic systems as a function of solvent composition. *Enzyme Microb. Technol.* **27,** 122–126.
4. Klibanov, A. M. (1982) Stabilization of enzymes against thermal inactivation. *Adv. Appl. Microb.* **29,** 1–28.
5. Gianfreda, L. and Scarfi, M. R. (1991) Enzyme stabilization: State of the art. *Mol. Cell. Biochem.* **109,** 97–128.
6. Nanalov, R. J., Kamboure, M. S., and Emanuiloda, E. I. (1993) Immobilization and properties of *Bacillus stearothermophillus* pulanase. *Biotechnol. Appl. Biochem.* **18,** 409–416.
7. Desmukth, S. S., Duta, M., Choudohury, S., and Shanker, V. (1993) Preparation and properties of glucose isomerase immobilized on indon 48-R. *Appl. Biochem. Biotechnol.* **42,** 95–104.
8. Rody, L. G. and Shanke, V. A. (1983) Immobilized nucleases. *Crit. Rev. Biotechnol.* **13,** 255–273.
9. Guisán, J. M. (1988) Aldehyde gels as activated support for immobilization-stabilization of enzymes. *Enzyme Microb. Technol.* **10,** 375–382.

10. Guisán, J. M., Bastida, A., Cuesta, C., Fernández-Lafuente, R., and Rosell, C. M. (1991) Immobilization-stabilization of chymotrypsin by covalent attachment to aldehyde agarose gels. *Biotechnol. Bioeng.* **39,** 75–84.
11. Mozhaev, V., Klibanov, A. M., Goldmacher, V. S., and Berezin, I. V. (1990) Operational stability of copolymerized enzymes at elevated temperatures. *Biotechnol. Bioeng.* **25,** 1937–1945.
12. Bickerstaff, G. F. (1997) Immobilization of enzymes and cells, in *Methods in Biotechnology* (Bickerstaff, G. F., ed.), Humana Press, Totowa, NJ.
13. Chibata, I., Tosa, T., and Sato, T. (1986) Biocatalysis: immobilized cells and enzymes. *J. Mol. Cat.* **37,** 1–24.
14. Gupta, M. N. (1991) Thermostabilization of proteins. *Biotechnol. Appl. Biochem.* **4,** 1–10.
15. Hartmeier, W. (1985) Immobilized biocatalysts—from simple to complex systems. *TIBTECH* **3,** 149–153.
16. Katchalski-Katzir, E. (1993) Immobilized enzymes-learning from past successes and failures. *TIBTECH* **11,** 471–478.
17. Kennedy, J. F., Melo, E. H. M., and Jumel, K. (1990) Immobilized enzymes and cells. *Chem. Eng. Progr.* **45,** 81–89.
18. Klibanov, A. M. (1983) Immobilized enzymes and cells as practical catalysts. *Science* **219,** 722–727.
19. Rosevear, A. (1984) Immobilized biocatalysts- a critical review. *J. Chem. Technol. Bioctechnol.* **34B,** 127–150.
20. Royer, G. P. (1980) Immobilized enzymes as catalyst. *Catal. Rev.* **22,** 29–73.
21. Mateo, C., Abian, O, Fernández-Lafuente, R., and Guisán, J. M. (2000) Increase in conformational stability of enzymes immobilized on epoxy-activated supports by favoring additional multipoint covalent attachment. *Enzyme Microb. Technol.* **26,** 509–515.
22. Mateo, C., Abian, O., Fernández-Lorente, G., Predoche, J., Fernández-Lafuente, R., and Guisán, J. M. (2002) Sepabeads: a novel epoxy-support for stabilization of industrial enzymes via very intense multipoint covalent attachment. *Biotechnol. Progr.* **18,** 629–634.
23. Mateo, C., Fernández-Lorente, G., Cortés, E., García, J. L., Fernández-Lafuente, R., and Guisán, J. M. (2001) One step purification, covalent immobilization and additional satabilization of poly-hys-tagged proteins using a novel heterofunctional chelate-epoxy supports. *Biotechnol. Bioeng.* **76,** 269–277.
24. Melander, W., Corradini, D., and Hoorvath, C. (1984) Salt-mediated retention of proteins in hydrophobic-interaction chromatography. Aplication of solvophobic theory. *J. Chromatogr.* **317,** 67–85.
25. Smalla, K., Turkova, J., Coupek, J., and Herman, P. (1988) Influence of salts on the covalent immobilization of proteins to modified copolymers of 2-hydroxyethyl-metacrylate with ethylene dimetacrylate. *Biotechnol. Appl. Biochem.* **10,** 21–31.
26. Wheatley, J. B. and Schmidt, D. E. (1993) Salt-induced immobilization of proteins on a high-performans liquid chromatographic epoxide affinity support. *J. Chromatogr.* **644,** 11–16.

27. Wheatley, J. B. and Schmidt, D. E. (1999) Salt-induced immobilization of affinity ligands onto epoxide activated supports. *J. Chromatogr. A.* **849,** 1–12.
28. Kramer, D. M., Lehmann, K., Pennewiss, H., and Plainer, H. (1979) Oxirane acrylic beads for protein immobilization: a novel matrix for biocatalysis and biospecific adsorption. 26 International IUPAC Symposium on Macromolecules.
29. Fernández-Lafuente, R., Rosell, C. M., Guisán, J. M., Caanan-Haden, L., and Rodes, L. (1999) Facile synthesis of artificial enzyme nano-environments via solid-phase chemistry of immobilized derivatives dramatic stabilization of penicillin acylase versus organic solvents. *Enzyme Microb. Technol.* **24,** 96–103.
30. Abian, O., Mateo, C., Fernández-Lorente, G., Palomo, J. M,. Fernández-Lafuente, R., and Guisán, J. M. (2001) Stabilization of immobilized enzymes against organic solvents: Generation of hyper-hydrophilic micro-environments fully surrounding enzyme molecules *Biocatal. Biotransfor.* **19,** 489–503.
31. Abian, O., Wilson, L., Mateo, C., et al. (2002) Preparation of artificial hyper-hydrophilic micro-environments (polymeric salts) surrounding immobilized enzyme molecules. New enzyme derivatives to be used in any reaction medium. *J. Mol. Cat. B Enzymatic.* **19–20C,** 295–303.
32. Guisán, J. M., Rodríguez, V., Rosell, C. M., Soler, G., Bastida, A., Fernández-Lafuente, R., and García-Junceda, E. (1997) Stabilization of immobilized enzymes by chemical modification with polyfunctional macromolecules, in *Methods in Biotechnology. Immobilization of Enzymes and Cells* (Bickerstaff, G. F., ed.), Humana Press, Totowa, NJ, pp. 289–298
33. Fernández-Lafuente, R., Rosell, C. M., and Guisán, J. M. (1996) Dynamic reaction design of enzymic biotransfomations in organic media: equilibrium controlled synthesis of antibiotics by penicillin G acylase. *Biotechnol. Appl. Biochem.* **24,** 139–143.

18

Methods for Bioencapsulation of Enzymes and Cells

Thomas M. S. Chang

Summary

Methods to microencapsulate enzymes and cells including recombinant enzymes, stem cells, and genetically engineered cells have been described in this chapter. More specific examples of enzyme encapsulation include the microencapsulation of xanthine oxidase for Lesch Nyhan disease, phenylalanine ammonia lyase for phenylketonuria, and microencapsulation of multienzyme systems with cofactor recycling for multistep enzyme conversions. Methods for cell encapsulation include the details for encapsulating hepatocytes for liver failure and for gene therapy. This also includes the details of a novel two-step method for encapsulation of high concentrations of smaller cells. The recent co-encapsulation of bone marrow stem cells with hepatocytes is also included. Another new approach is the detailed method for encapsulation of genetically engineered *Escherichia coli* DH5 cells and other microorganisms for the removal or conversion metabolites. Another new approach is the nanoencapsulation of hemoglobin and enzymes to form 150-nm diameter nanocapsules; the details are available elsewhere.

Key Words: Artificial cells; bioencapsulation; microencapsulation; nanoencapsulation; enzyme; cells; genetically engineered cells; blood substitutes; liver cells; stem cells; microorganisms.

1. Introduction

Microencapsulation of biologically active material in the form of "artificial cells" was reported as early as 1964 *(1–4)*. However, it is only in the last 10 yr that many centers are developing this process extensively *(5,6)*. More recently we have concentrated on artificial cells for blood substitutes, enzyme therapy, and cell therapy. Space allows only a few examples to be given here.

HIV has stimulated extensive development in the last 10 yr. The early idea of cross-linked hemoglobin *(1,4)* has now been developed as a first-generation blood substitute now in phase III clinical trials by a number of groups *(6–9)*. New generations of hemoglobin-based blood substitutes are also being developed *(9)*. In this research center we are looking at a polyhemoglobin blood substitute with antioxidant enzyme activities *(10–14)*. This is prepared by

From: *Methods in Biotechnology, Vol. 17: Microbial Enzymes and Biotransformations*
Edited by: J. L. Barredo © Humana Press Inc., Totowa, NJ

crosslinking hemoglobin (PolyHb) with superoxide dismutase (SOD) and cata-lase (CAT) into PolyHb-SOD-CAT. This is important in cases with potentials for oxidant damage, as in severe sustained hemorrhagic shock or in stroke and other conditions. We are also developing a further generation of blood substi-tutes based on the use of nanotechnology to prepare 150-nm diameter biodegradable polymeric membrane nanocapsules containing hemoglobin and enzymes. The details for this are available elsewhere (12–15).

Injection of enzyme artificial cells is effective in enzyme therapy for inborn errors of metabolism (16) and for cancer therapy (17). However, accumulation of the implanted artificial cells is a problem. This has been recently by giving the enzyme artificial cells orally as in the following example. Lesch Nyhan is a very rare disease consisting of an inborn error of metabolism with accumula-tion of hypoxanthine. Being lipid-soluble hypoxanthine can diffuse rapidly into the intestine. Therefore the enzyme microcapsules can be given orally to avoid the need for injection. Daily oral administration of microencapsulated xanthine oxidase resulted in the lowering of hypoxanthine in the plasma and cerebral-spinal fluid in a patient with Lesch Nyhan disease (18,19).

Phenylketonuria (PKU) is the most common inborn error of metabolism with enzyme defects resulting in the elevation of the amino acid phenylalanine. However, amino acids from the body do not diffuse rapidly into the intestine. This can now be solved based on recent new findings of extensive enterorecir-culation of amino acids between the body and the intestine (20,21). Large vol-umes of digestive juice containing enzymes and other proteins are secreted into the intestine. These are digested by tryptic enzymes in the intestine into amino acids. The amino acids formed this way in the intestine are reabsorbed into the body. It has been shown for the first time that amino acids formed from this source are hundreds of time higher than those from protein in ingested food (20,21). This means that when given orally, microcapsules containing a specific enzyme can remove the corresponding specific amino acids in the intestine and therefore prevent them from returning to the body. The net result is a depletion of this specific amino acid from the body. For example, microen-capsulated phenylalanine ammonia lyase given orally once a day can selec-tively remove phenylalanine from the enterorecirculating amino acids. This explains earlier observation of the effectiveness of this approach for lowering the elevated systemic phenylalanine in the PKU rats (22) and more recently in the ENU2 phenylketonuric mice (23). This approach can also be used to remove any of the other 26 amino acids in the body with potential for treating other inborn errors of metabolism and for amino-acid-dependent tumors. This is now being developed for clinical trials. This has resulted in renewed inter-est in the microencapsulation of enzymes. For example, our recent studies show that microencapsulated tyrosinase can be giving orally in rats to lower the systemic tyrosine level *(24)*. Microcapsules containing multienzyme systems with cofactor recycling for multistep enzyme conversions have also been studied *(25,26)*.

The third major area of this research center's ongoing interest is based on the microencapsulation of cells and microorganisms. As early as 1965 Chang successfully encapsulated cells and wrote *(27)*: "Microencapsulation of intact cells or tissue fragments ... the enclosed material might be protected from destruction and from participation in immunological processes, while the enclosing membrane would be permeable to small molecules of specific cellular product which could then enter the general extracellular compartment of the recipient. For instance, encapsulated endocrine cells might survive and maintain an effective supply of hormone. ... The situation would then be comparable to that of a graft placed in an immunologically favorable site." "There would be the further advantage that implantation could be accomplished by a simple injection procedure rather than by a surgical operation." "Microencapsulation of intact cells ... erythrocytes were suspended in hemolysate rather than in the diamine solution; and a silicone oil was substituted for the stock organic liquid. Microencapsulation was then carried out ... by the principle of interfacial polymerization for membranes of cross-linked proteins. ... A large number of human erythrocytes suspended in hemolysate within a microcapsule of about 500 µm diameter was prepared by the syringe (drop) method." He also described this in other publications *(2,4)*. However, it is only with the more recent interest in biotechnology that many groups around the world have extended this approach of cell encapsulation *(5,6,29,30)*. Some of the recent research interest in this research center is as follows.

Encapsulated hepatocytes has been used here as a model for cell and gene therapy. Implantation of encapsulated hepatocytes increases the survival of fulminant hepatic failure rats *(31)*. Encapsulated rat hepatocytes are not rejected after being implanted into mice *(32)*. Instead, there is an increase in viability after intraperitoneal implantation *(32)*. This is due to the retention of hepatostimulating factors inside the microcapsules as they are secreted by the hepatocytes *(33)*. The Gunn rat is the model for the Crigler-Najjar syndrome in humans due to defects of the liver enzyme UDP-glucuronosyltransferase (UDPGT). Intraperitoneal implantation of artificial cells containing hepatocytes lowered the high systemic bilirubin levels *(34,35)*. Kinetic analysis shows that the hepatocytic enzyme (UDPGT) in artificial cells conjugated bilirubin to the monoconjugated and diconjugated form for excretion in the urine as in normal animals *(36)*.

For encapsulation of a high concentration of small cells, a two-step method has been developed *(37,38)*. This is because when using the standard method of cell encapsulation, some cells are exposed on the surface of artificial cells *(37)*.

Therefore a new method has been developed to prevent this problem (37,38). This will be particularly useful for the encapsulation of hepatocytes and genetically engineered cells. In order to increase the viability and the function of the encapsulated hepatocytes after implantation even further, we have coencapsulated hepatocytes with bone marrow stem cells (39–42). This resulted in increased viability in vitro and also in vivo (40,41). This also has a significantly longer effect on lowering the high systemic bilirubin levels in Gunn rats (42).

Another area is the study of artificial cells containing nonpathogenic microorganisms including genetic engineered microorganisms. There is much potential in genetically engineered microorganisms. However, they cannot be injected. We are studying the possibility of using oral microcapsules containing these cells (43,44). We started with basic research using Escherichia coli with Klebsiella aerogenes gene for urea removal. This is because for more than 30 yr investigators have been unable to find an oral treatment for uremia, mainly because of the inability to remove the large amount of urea (45). We have found a new approach combining artificial cells with gene-expression genetic engineering technology (43,44). E. coli DH5 cells do not have a large capacity for removing urea. This can be increased very significantly by metabolic induction (43,44). Oral administration of a small amount of artificial cells containing genetic engineered E. coli DH5 cells once a day resulted in the decrease of high urea level in the uremic rats to normal (44). This was maintained during the 21 d of treatment. However, after metabolic induction the resulting increase in the capacity for removing urea is not sufficiently stable. This results in large variations in activity between batches of the induced E. coli DH5 cells. We have recently used metabolic induction to increase the urea-removal activities of Lactobacillus delbrueckii and also to encapsulate this in artificial cells (46). We have also studied the encapsulation of two other microorganisms, one that removes cholesterol (47) and another that converts substrates to L-DOPA (48).

Our other research includes the microencapsulation of erythropoietin (EPO)-secreting renal cells (49).

2. Materials

2.1. Microencapsulation of Enzymes

2.1.1. Cellulose Nitrate Membrane Microcapsules

1. Hemoglobin solution containing enzymes: 15 g of hemoglobin (bovine hemoglobin type 1, 2X crystallized, dialyzed and lyophilized, Sigma, St. Louis, MO) was dissolved in 100 mL of distilled water and filtered through Whatman No. 42 paper (Whatman, Kent, UK) (see **Note 1**).
2. Water-saturated ether: Shake analytical-grade ether with distilled water in a separating funnel. Then leave standing for the two phases to separate so that the water can be discarded.

3. Fisher magnetic stirrer.

4. Cellulose nitrate solution: Spread 100 mL of USP Collodion (USP) in an evaporating disk in a well-ventilated hood overnight. This allowed the complete evaporation of its organic solvents, leaving a dry thin sheet. Cut the thin sheet of polymer into small pieces and dissolved in a 100-mL mixture containing 82.5 mL analytical-grade ether and 17.5 mL analytical-grade absolute alcohol (*see* **Note 2**).

5. Tween-20 solution. The 50% (v/v) concentration solution is prepared by mixing equal volumes of Tween-20 (Atlas Powder, Montreal, Canada) and distilled water, and then adjusting the pH to 7.0. For 1% (v/v) concentration solution, mix 1% of Tween-20 into the buffer solution used as the suspending medium for the final microencapsulated enzyme system.

2.1.2. Preparation of Polyamide Membrane Microcapsules by Interfacial Polymerization

1. Span 85 organic solution: 0.5% (v/v) Span 85 (Atlas Powder) in chloroform:cyclohexane (1:4).

2. Terepthaloyl organic solution: Add 100 mg of terephthaloyl chloride (ICN Pharmaceuticals, Costa Mesa, CA) to a 30 mL organic solution (chloroform:cyclohexane, 1:4) kept in an ice bath. Cover and stir with a magnetic stirrer for 4 h, and then filter with Whatman no. 7 paper (Whatman). Prepare just before use (*see* **Note 3**).

3. Diamine-polyethyleneimine solution: Dissolve 0.378 g $NaHCO_3$ and 0.464 g 1.6-hexadiamine (J. T. Baker Chemical, Phillisburg, NJ, USA) in 5 mL distilled water that contains the material to be encapsulated. Adjust pH to 9.0. Add 2 mL 50% polyethyleneimine (ICN Pharmaceuticals) to the diamine solution, readjust pH to 9.0, and make up the final volume to 10 mL with distilled water. Prepare just before use (*see* **Note 3**).

4. Hemoglobin solution 10 g/100 mL: Prepare as described above (**Subheading 2.1.1.**) for cellulose nitrate microcapsules, but material to be encapsulated is dissolved in 5 mL of hemoglobin solution instead of distilled water. The final pH is adjusted to 9.0.

2.1.3. Lipid-Polymer Membrane Microcapsules that Retain Cofactors

1. Glutamic dehydrogenase, bovine liver, type III, 40 U/mg (Sigma).

2. Alcohol dehydrogenase, yeast, 330 U/mg (Sigma).

3. Urease, 51 U per mg (Millipore, Bedford, MA).

4. Lipid-organic liquid: 1.4 g lecithin and 0.86 g cholesterol were added to 100 mL tetradecane and stirred for 4 h at room temperature. If a more permeable lipid membrane is required to allow urea to diffuse across it, then the lipid compositions should be 0.43 g cholesterol and 0.7 g lecithin.

2.2. Microencapsulation of Cells and Microorganisms

2.2.1. Standard Method

1. Calcium-free perfusion buffer: 142 mM NaCl, 6.7 mM KCl, and 10 mM HEPES, pH 7.4.

2. Collagenase perfusion buffer: 67 mM NaCl, 6.7 mM KCl, 5 mM CaCl$_2$, 0.05% collagenase, and 100 mM HEPES, pH 7.5.

3. William's E medium (Gibco Laboratories, Burlington, ON).

4. Streptomycin and penicillin (Gibco).

5. Nylon monofilament mesh 74 μM (Cistron Corp., Elmford, NY).

6. Buffered saline: 0.85% NaCl, 20 mM D-fructose, and 10 mM HEPES, pH 7.4.

7. Stock solution of sodium alginate: 4% sodium alginate, and 0.45% NaCl.

8. Iscove's modified Dulbecco's medium (IMDM) (GIBCOBRL Life Technologies, Good Island, NY).

9. Nylon filter, 85 μm.

10. Hepatocytes: obtained from Wistar rats as described under the method in **Subheading 3.2.1.1.**

11. Bone marrow stem cells: obtained from the bone marrow of Wistar rats as described in **Subheading 3.2.1.2.**

12. Luria-Bertani (LB) medium: 10 g/L bacto-tryptone, 5 g/L bacto-yeast extract, and 10 g/L sodium chloride. Adjust pH to 7.5 with 1 N NaOH.

13. Genetically engineered *E. coli* DH5. *E. coli* DH5 is a nonpathogenic bacterium.

14. Alginate solution: 2% sodium alginate, and 0.9% sodium chloride (*see* **Note 4**). Sodium alginate is Kelco Gel® low-viscosity alginate, Keltone LV, MW 12,000–80,000 (Merck & Co., Clark, NJ). Sterilize before use, either by filtration or by heat for 5 min.

15. Syringe pump, compact infusion pump model 975 (Harvard Apparatus, Mill, MA).

16. Poly-L-lysine, MW 15,000–30,000: 0.05% poly-L-lysine (Sigma), and 10 mM HEPES buffer saline, pH 7.2.

17. Citrate solution: 3% citrate and 1:1 HEPES buffer saline, pH 7.2.

18. Calcium chloride solution: 1.4% calcium chloride, pH 7.2.

19. Poly-L-lysine, MW 16,100: 0.05 % poly-L-lysine (Sigma), and 10 mM HEPES buffer saline, pH 7.2.

20. CaCl$_2$ solution: 100 mM CaCl$_2$, 20 mM D-fructose, and 10 mM HEPES buffer, pH 7.4.

21. Hank's balanced salt solution.

22. Poly-L-lysine-fructose solution: 0.05% poly-L-lysine, 0.85% NaCl, 20 mM D-fructose, and 10 mM HEPES buffer, pH 7.4.

23. Sodium alginate 0.2%: 0.2% sodium alginate, 0.85% NaCl, 20 mM D-fructose, and 10 mM HEPES buffer, pH 7.4 (*see* **Note 5**).

24. Sodium citrate solution: 50 mM sodium citrate, 0.47% NaCl, and 20 mM D-fructose, pH 7.4.

25. Collagenase: type IV (Sigma).

26. Trypsin: type I-S trypsin inhibitor (Sigma).

27. HEPES: (4-(2-hydroxyethyl)-1-piperazine ethane sulphonic acid) buffer (Boehringer Mannheim, Montreal, PQ) (*see* **Note 6**).

28. Droplet generator 1: contains 2 coaxially arranged jets: (a) the central jet consisted of a 26G stainless steel needle (Perfektum) (Popper & Sons, New Hyde Park, NY), and (b) a 16G surrounding air jet, through which the sample and air, respectively were passed. To prevent the extruding sample from occluding the outlet of the sur-

rounding air jet, the tip of the sample jet was constructed such that the tip project-ed 0.5 mm beyond the end of the air jet.

29. Droplet generator 2: a larger and a slightly modified variant of the droplet genera-tor 1. It was constructed with a 13G sample jet and an 8G surrounding air jet. The ends of the jets were cut flush to each other. A 1.7×1.1 mM PTFE capillary tube (Pharmacia P-L Biochemicals, Montreal, PQ) was inserted into the sample jet until it protruded approximately 15 mM from the outlet of the sample jet. The end of the capillary tubing was tapered to facilitate shearing by the flow of passing air from the air jet. The capillary tubing is approximately 3.2 m in length, and has the capac-ity of be filled with microspheres suspended in 2.5 mL of sodium alginate.

30. Commercial generator: used for the preparation of larger samples.

2.2.2. Two-Step Method for High Concentration of Cells or Microorganisms (see **Note 7**)

1. *Pseudomonas pictorum* ATCC 23328 (ATCC, Manassus, VA) was used because of its ability to degrade cholesterol.
2. Inoculum medium: bovine calf serum (Sigma). This was used in all experiments unless otherwise specified.
3. Nutrient agar plates: 80 g/L nutrient agar (Difco Laboratories, Detroit, MI). Autoclave for 15 min at 121°C, allow cooling at 50°C, and pour into plastic Petri dishes. Store at 4°C for up to 2 mo.
4. Cholesterol medium: 0.1% ammonium nitrate, 0.025% potassium phosphate, 0.025% magnesium sulfate, 0.0001% ferric sulfate, 0.5% yeast extract, and 0.1% cholesterol. Adjust pH to 7.0 and autoclave for 15 min.

3. Methods

3.1. Microencapsulation of Enzymes

3.1.1. Cellulose Nitrate Membrane Microcapsules

Cellulose nitrate membrane microcapsules were prepared using an updated procedure based on earlier publications *(1–4,6)*.

1. Dissolved or suspend enzymes and other materials to be microencapsulated were in 2.5 mL of the hemoglobin solution. Adjust the final pH to pH 8.5 with Tris-HCl, pH 8.5, and hemoglobin concentration to 100 g/L.
2. Add 2.5 mL of this solution to a 150 mL glass beaker, and add 25 mL of water-saturated ether.
3. Stir the mixture was immediately with a Fisher magnetic stirrer at 1200 rpm (set-ting of 5) for 5 s.
4. While stirring was continued, add 25 mL of a cellulose nitrate solution continue stirring for another 60 s.
5. Cover the beaker and allow to stand unstirred at 4°C for 45 min.
6. Decant the supernatant and add 30 mL of n-butyl benzoate. Stir the mixture for 30 s at the same magnetic stirrer setting.

7. Allow the beaker to stand uncovered and unstirred at 4°C for 30 min. Then remove the butyl benzoate completely after centrifugation at 350g for 5 min.
8. Add 25 mL of the 50% (v/v) Tween solution. Start stirring at a setting of 10 for 30 s.
9. Add 25 mL of water and continue stirring at a setting of 5 for 30 s, then add 200 mL of water.
10. Remove the supernatant and wash the microcapsules three more times with 200 mL of the 1% Tween-20. Suspend the microcapsules in a suitable buffer, e.g., phosphate buffer (pH 7.5). In properly prepared microcapsules, there should not be leakage of hemoglobin after the preparation (*see* **Note 8**).

3.1.2. Preparation of Polyamide Membrane Microcapsules by Interfacial Polymerization (see **Note 9**)

Polyamide membrane microcapsules of 100 μM mean diameter were prepared using an updated method based on earlier methods (*1–4,6*).

1. Add enzyme to 2.5 mL of the hemoglobin solution with pH and concentrations adjusted as in **Subheading 3.1.1.**
2. Mix 2.5 mL of the diamine-polyethyleneimine solution for 10 s in a 150-mL beaker placed in an ice bath.
3. Add 25 mL of Span 85 organic solution and stir in the Fisher magnetic stirrer at speed setting of 2.5 for 60 s.
4. Add 25 mL of terephthaloyl organic solution and allow the reaction to proceed for 3 min with the same stirring speed.
5. Discard the supernatant and add another 25 mL of the terephthaloyl organic solution.
6. Carry out the reaction with stirring for another 3 min. Discard the supernatant.
7. Then, add 50 mL of the Span 85 organic solution and stir for 30 s. Discard the supernatant.
8. After this, use the procedure of Tween-20 as described for cellulose nitrate microcapsules (**Subheading 3.1.1.**) for the transfer of the microcapsules into the buffer solution.

3.1.3. Lipid-Polymer Membrane Microcapsules that Retain Cofactors (see **Note 10**)

Cofactors covalently linked to macromolecules like dextran or polyethyleneime can be retained within semipermeable microcapsules to be recycled enzymatically. However, linkage of cofactors to macromolecules increases steric hindrance and reduces their rate of reactions with enzymes. In biological cells like erythrocytes, free cofactors and multienzyme systems are all retained within the cells in free solution. Thus, studies were carried out here to immobilize free cofactors inside microcapsules with membranes impermeable to cofactors but permeable to the initial substrates. In this way, the free cofactor can function without steric hindrance in close proximity to the enzymes. Furthermore, all enzymes and cofactors inside the microcapsules are in free solution. Lipid-polyamide membrane microcapsules have been prepared *(4)*. These are permeable to lipophilic molecules but with little or no permeability to hydrophilic molecules as small as K^+ and Na^+.

Lipid-polyamide microcapsules of 100 μm mean diameter containing multienzyme systems, cofactors, and alpha-ketoglutarate were prepared *(50)*. The first part is similar to the procedure described earlier in this chapter under the basic procedure for the preparation of polyamide microcapsules (**Subheading 3.1.2.**).

1. To 2 mL of the hemoglobin solution add 12.5 mg glutamic dehydrogenase, 6.25 mg alcohol dehydrogenase, 0.5 mg urease, 1.18 mg ADP, and either NAD^+ (0.52 mg, 105 mg, 2.11 mg, or 21.13 mg) or NADH (21.13 mg) dissolved in 0.25 mL of water. Finally, add 56.5 mg alpha-ketoglutarate, 2.5 mg $MgCl_2$, and 0.93 mg KCl in 0.25 mL.
2. Add 2.5 mL of the hemoglobin-enzyme solution so prepared to 2.5 mL of the diamine-polyethyleneime solution.
3. The remaining steps were the same as described above (**Subheading 3.1.2.**) except that the Tween-20 steps were omitted here. Instead, after washing with the Span 85 organic solution, the following steps were carried out to apply the lipids to the polyamide membranes.
4. Rinse the microcapsules twice with 10 mL of the lipid-organic liquid.
5. Then, add another 10 mL of the lipid-organic liquid and slowly rotate the suspension for 1 h at 4°C on a multi-purpose rotator.
6. After this, decant the supernatant and recover the lipid-polyamide membrane microcapsules and leave in this form at 4°C without being suspended in aqueous solution until it was added to the substrate solution just before the reaction.

The procedure takes practice. The microcapsules prepared must be tested for the absence of leakage of enzymes or cofactors before being used in experimental studies.

3.2. Microencapsulation of Cells and Microorganisms

3.2.1. Standard Method

3.2.1.1. Preparation of Rat Hepatocytes

1. Anesthetize each rat with sodium pentobarbital and cannulate via the portal vein.
2. Cut the thoracic vena cava and perfuse the liver with calcium-free perfusion buffer for 10 min at 40 mL/min.
3. Afterward, perfuse the liver with the collagenase perfusion buffer for an additional 15 min at 25 mL/min.

4. Excise the liver, place in William's E medium supplemented with 100 μg/mL streptomycin and penicillin, and shake gently to free loose liver cells from the liver tissue.
5. Collect the cells, filter through a 74-μm nylon monofilament mesh, and centrifuge to remove connective tissue debris, cell clumps, nonparenchymal cells, and damaged cells.
6. Prepare isolated hepatocytes for encapsulation by first washing and suspending the cells with buffered saline.
7. Mix the cells with a 4% stock solution of sodium alginate, to make a cell suspension consisting of 20×10^6 cells/mL of 2% sodium alginate.

3.2.1.2. PREPARATION OF RAT BONE MARROW STEM CELLS

1. Anesthetize each rat with sodium pentobarbital, and isolate both femurs.
2. Use Iscove's modified Dulbecco's medium (IMDM) to flush out bone marrow cells from the femurs using a 5 mL syringe with a 22-gage needle.
3. Filter the cell suspension through a nylon filter (85 μm).
4. Wash the bone marrow cells with IMDM and centrifuge at 50g for 10 min at 4°C; this was repeated three times.
5. After the last wash, keep the bone marrow cells (nucleated cells) on ice until use.

3.2.1.3. GENETICALLY ENGINEERED *E. COLI* DH5 CELLS AND MICROORGANISM (see **Note 11**)

1. Use genetically engineered bacteria *E. coli* DH5, containing the urease gene from *K. aerogenes*. Use LB growth medium for primary cell cultivation. Carry out incubation in 5 mL LB in 16 mL culture tubes at 37°C in an orbital shaker at 120 rpm. For the large-scale production of biomass, for microencapsulation purpose, use a 250-mL Erlenmeyer flask containing 100 mL of the suitable medium.
2. Harvest Log-phase bacterial cells by centrifuging at 10,000g for 20 min at 4°C.
3. Discard the supernatant and wash the cell biomass five times with sterile cold water to remove media components by centrifugation at 10,000g for 10 min at 4°C.
4. Suspend bacterial cells in autoclaved ice-cold sodium alginate solution.
5. Press the viscous alginate-bacterial suspension through a 23-gage needle using a syringe pump.
6. Use compressed air through a 16-gage needle to shear the droplets coming from the tip of the 23-gage needle.
7. Allow the droplets to gel for 15 min in a gently stirred ice-cold solution of calcium chloride (1.4%).
8. After gelation in the calcium chloride, coat alginate gel beads with poly-L-lysine for 10 min.
9. Wash the beads with HEPES and coat with an alginate solution (0.1%) for 4–8 min.
10. Wash the alginate-poly-L-lysine-alginate capsules in a 3% citrate bath to liquefy the gel in the microcapsules.

3.2.1.4. ENCAPSULATION USING THE STANDARD METHOD

1. Suspend hepatocytes, hepatocytes and bone marrow cells, or bacterial cells in an autoclaved ice-cold 0.9% sodium alginate solution.
2. Press the viscous alginate suspension through a 23-G stainless steel needle using a syringe pump. Use sterile compressed air, through a 16-G coaxial stainless steel needle, to shear the droplets coming out of the tip of the 23-G needle.
3. Allow the droplets to gel for 15 min in a gently stirred, heat-sterilized and ice-cold calcium chloride solution. Upon contact with the calcium chloride buffer, alginate gelation is immediate.
4. After gelation in the calcium chloride solution, react alginate gel beads with poly-L-lysine (PLL), MW 16,100, for 10 min. The positively charged PLL forms a complex of semipermeable membrane.
5. Wash the beads with HEPES, pH 7.2, and coat with an alginate solution (0.1%) for 4 min.
6. Wash the alginate-poly-L-lysine-alginate capsules thus formed in a 3% citrate bath to liquefy the gel in the microcapsules.
7. Store the APA microcapsules formed, which contains entrapped hepatocytes or bacterial cells, at 4°C and use for experiments. Keep the conditions sterile during the process of microencapsulation.

3.2.2. Two-Step Method for High Concentration of Cells or Microorganisms

The standard method described above is not optimal for encapsulating high concentrations of cells or microorganisms. Cells or microorganisms may be trapped in the membrane matrix. This can weaken the membrane. If cells are exposed to the surface, this may also result in loss of immunoisolation and rejection. As a result, the two-step method has been developed to prevent this problem *(37,38)*.

1. Entrap the hepatocytes or hepatocytes and bone marrow stem cells suspended in sodium alginate were within solid calcium alginate microspheres. This is done by filling a 5-mL syringe with the cell suspension, and extruding the sample with a syringe infusion pump through the sample jet of the first droplet generator. Allow the droplets formed at the end of the sample jet to fall dropwise into a Pyrex dish (125.65 mm) containing 300 mL $CaCl_2$ solution. Resuspend every 5 min the cells in the syringe by gentle inversion of the syringe to minimize the effect of cells sedimenting in the alginate solution. The air flow and infusion rate through the droplet generator are 2–3 L/min and 0.28–0.39 mL/min, respectively, and the clearance height between the end of the sample jet and the surface of the calcium solution is set at approximately 20 cm. Fit a strainer cup inside the dish to collect the droplets and to facilitate the removal of the formed microspheres.
2. Allow the microspheres to cure for approximately 15 min, after which they are removed and temporarily stored in Hank's balanced salt solution supplemented with 10% 100 m*M* $CaCl_2$.

3. Collect 1.0 mL of formed microspheres and wash three times with buffered saline.
4. Aspirate the final saline washing and add 1 mL of 1.2–1.6% sodium alginate to the 1.0 mL of washed microspheres. Prepare the sodium alginate by diluting the 4% stock solution with buffered saline. With a 5-mL syringe, fill the length of the PTFE capillary tubing with the sodium alginate and suspension of microspheres. Insert the tapered end of the capillary tubing through the top of the sample jet of the second droplet generator until the tip of the tubing extended approximately 15 mm beyond the end of the sample jet. The air flow and extrusion rate through the modified droplet generator are 7–9 L/min and 0.28–0.39 mL/min, respectively. Set the tip of the capillary tubing approximately 20 cm above the surface of the calcium solution. With the 5-mL syringe still attached to the other end of the tubing, extrude the microspheres suspension in the tubing with the Harvard infusion pump. Similarly, allow the drops formed at the end of the sample jet to fall dropwise into a Pyrex dish containing a strainer cup and filled with 300 mL of 100 mM CaCl$_2$.
5. Allow the spheres to cure in the calcium solution for approximately 15 min, after which they are removed and washed with buffered saline.
6. Stabilize the alginic acid matrix on the surface of the sphere with poly-L-lysine by immersing 5 mL (settled volume) of macrospheres in 80 mL of 0.05% poly-L-lysine-fructose solution for 10 min.
7. Strain the spheres, wash with buffered saline, and immerse into 200 mL of 0.2% sodium alginate for 10 min to apply an external layer of alginate.
8. After 10 min collect the spheres and immerse in 200 mL 50 mM sodium citrate solution to solubilize the intracapsular calcium alginate. This may require up to 30 min with frequent changes of the sodium citrate solution.

3.2.3. Macroporous Microcapsules

When using cells or microorganisms to act on macromolecules, the above methods cannot be used. Thus, in using microorganisms to act on cholesterol bound to lipoprotein, the microorganisms have to be encapsulated in macroporous microcapsules *(47)*.

1. Culture *Pseudomonas pictorum* in nutrient broth at 25°C, followed by harvesting and resuspension in a cholesterol medium. After culturing this suspension for 15 d at 25°C, use it as an inoculum for biomass production. Grow the culture in bovine calf serum at 37°C for 36 h, and then harvest. Use this to prepare bacterial suspensions for immobilization. The concentration was about 0.4 mg of dry cell/mL (*see* **Note 12**).
2. Autoclave a solution of 2% agar and 2% sodium alginate for 15 min and cool to 45°C to 50°C.
3. Add *P. pictorum* suspended in 0.4 mL of 0.9% NaCl drop by drop to 3.6 mL of agar alginate solution at 45°C, while stirring vigorously.
4. Keep 3 mL of the mixture obtained at 45°C while it was being extruded through the syringe. Collect the extruded drops into cold (4°C) 2% calcium chloride and allow hardening. These agar-alginate beads are about 2 mm in diameter.

5. After 15 min, discard the supernatant and resuspend the beads in 2% sodium citrate for 15 min.
6. Then they are washed and stored in 0.9% saline at 4°C.

When testing for immobilized bacterial activity, 1 mL of beads/microcapsules are placed in a sterile 50-mL flask. Five mL of serum are added and a foam plug is fitted. Samples are withdrawn at specified intervals. When empty beads or microcapsules are prepared, the bacterial suspension is replaced by saline, and all the other steps are kept the same.

4. Notes

1. Hemoglobin at a concentration of 100 g/L is necessary for the successful preparation of cellulose nitrate membrane microcapsules. Furthermore, this high concentration of protein stabilizes the enzymes during preparation and also during reaction and storage *(51)*. When the material (e.g., NADH) to be encapsulated is sensitive to the enzymes present in hemoglobin, highly purified hemoglobin is used. This requires the use of purification using affinity chromatography on an NAD$^+$ Sepharose column.
2. When using cellulose nitrate microcapsules containing enzymes for oral administration, the permeability of the membrane may need to be decreased. This is to prevent the entry of smaller tryptic enzymes. Permeability can be decreased by decreasing the proportion of alcohol used in dissolving the evaporated cellulose nitrate polymer.
3. Failure to prepare good microcapsules is frequently due to the use of diamine or diacids that have been stored after they have been opened. A new unopened bottle will usually solve the problems. Unlike cellulose nitrate microcapsules, in interfacial polymerization the hemoglobin solution can be replaced by a 10% polyetheleneimine solution adjusted to pH 9.0. However, the microcapsules prepared without hemoglobin may not be as sturdy. Cross-linking the microencapsulated enzymes with glutaraldehyde after the preparation of the enzyme microcapsules could also be carried out to increase the long-term stability of the enclosed enzymes *(51)*, although this decreases the initial enzyme activity.
4. Alginates are heteropolymer carboxylic acids, coupled by 1–4 glycosidic bonds of β-D-mannuronic (M) and α-L-gluronic acid unit (G). Alkali and magnesium alginate are soluble in water, whereas alginic acids and the salts of polyvalent metal cations are insoluble. Thus, when a drop of sodium alginate solution enters a calcium chloride solution, rigid spherical gels are formed by ionotropic gelation.
5. Alginate concentration in the tested range, 1.00–2.25% (w/v), does not affect the bacterial cell viability or cells growth. Quality of microcapsules improves with increasing alginate concentration from 1% to 1.75% (w/v). The use of 2% (w/v) alginate resulted in perfectly spherical shape and sturdy microcapsules with maximum number of encapsulated bacterial cells. An increase in the liquid flow rate of the alginate-cell or bacterial suspension through the syringe pump from 0.00264 to 0.0369 mL/min resulted in an increase in microcapsule diameter. The flow rate in

the range of 0.00724 to 0.278 mL/min resulted in good spherical microcapsules. At an air flow rate of 2 L/min, the microcapsules had an average of 500 ± 45 μm diameter. At air flow rates above 3 L/min, microcapsules were irregular in shape. These results indicate that alginate concentration, air flow rate, and liquid flow rate are critical for obtaining microcapsules of desired characteristics and permselectivity *(44,52)*. We find that the following composition is most suitable for our purpose: 2% (w/v) alginate, 0.0724 mL/min liquid flow rate, and 2 L/min air flow rate. Microcapsules prepared this way are permeable to albumin, but impermeable to molecules with higher molecular weights *(53)*. Thus, hepatostimulating factors *(34)* and globulin *(52,53)* cannot cross the membrane of the standard microcapsules.

6. All the solutions are kept in a cold-ice bath before use and during the process of bioencapsulation. The pH of the solutions is kept at 7.4 by buffering with HEPES. Except for sodium alginate, the solutions are sterilized by filtering through a sterile 0.2-μm Millipore filter.

7. The two-step method prevents the entrapment of small cells in the membrane matrix. Microcapsules prepared this way when implanted are much more stable with a decrease in rejection *(37,38)*.

8. The long-term stability of microencapsulated enzyme activity can be greatly increased by cross-linking with glutaraldehyde *(51)*. This is done at the expense of reduced initial enzyme activity.

9. In multienzyme reaction requiring cofactor recycling, the cofactor can be cross-linked to dextran-70 and then encapsulated together with the enzymes. For example, $NAD^+-N^6-[N-(6-aminohexyl)-acetamide]$ was coupled to dextran T-70, polyethyleneimine, or albumin to form a water soluble NAD^+ derivative and then encapsulated together with the multienzyme systems in the microcapsules *(25,26)*. This way both the cellulose nitrate microcapsules and polyamide microcapsules can be used. This allows for high permeation to substrates and products. However, linking the cofactor to soluble macromolecules resulted in significant increases in steric hindrance and diffusion restrictions of the cofactor.

10. Lipid-polyamide membrane microcapsules containing multienzyme systems, cofactors, and substrates can retain cofactors in the free form. Thus, analogous to the intracellular environments of red blood cells, free NADH or NADPH in solution inside the microcapsules is effectively recycled by the multistep enzyme systems that are also in solution. However, only lipophilic or very small hydrophilic molecules like urea can cross the membrane. For example, ammonia and urea equilibrate into the microcapsules to be converted into amino acids *(25,26)*. However, external alcohol instead of glucose had to be used as substrate for the conversion and recycling of NAD^+ to NADH. Some substrates, e.g., alpha-ketoglutarate, had to be encapsulated because they cannot enter the lipid-complexed microcapsules *(25,26)*.

11. Genetically engineered cells to produce desired biologically active materials are being successfully produced every day because of increasing advancement in molecular biology research. Genetically engineered cells have been prepared that can produce hormones like insulin used in diabetics, dopamine used in Parkinson's disease, and a number of genetically engineered and mammalian cells that can

exhibit therapeutically important nature. The use of these valuable genetically engineered cells, however, is very limited. A method for preparing alginate-poly-L-alginate membrane to prepare biocompatible artificial membrane was developed. This method has potential to facilitate use of genetically engineered cells for various applications.

12. Temperature is a very critical parameter in the immobilization of *P. pictorum (47)*. A low temperature produces gelation of the polymer in the syringe or conduits. A high temperature prevents gelation but increases the mortality rate of *P. pictorum*. Exposing *P. pictorum* to 55°C for 10 min or more can completely inhibit enzymatic activity. However, up to 20 min exposure to 45°C does not significantly inhibit cholesterol activity. Open-pore agar beads stored at 4°C did not show any sign of deterioration. The beads retain their enzymatic activity even after 9 mo of storage.

Acknowledgments

This research has been carried out with support to TMSC from the Canadian Institutes of Health Research, the Quebec MESST Virage Award of Centre of Excellence in Biotechnology and the Quebec ministry of health's MSSS-FRSQ team award on blood substitutes in transfusion medicine.

References

1. Chang, T. M. S. (1964) Semipermeable microcapsules. *Science* **146,** 524–525.
2. Chang, T. M. S., MacIntosh, F. C., and Mason, S. G. (1966) Semipermeable aqueous microcapsules: Preparation and properties. *Can. J. Physiol. Pharmacol.* **44,** 115–128.
3. Chang, T. M. S., MacIntosh, F. C., and Mason, S. G. (1971) Encapsulated hydrophilic compositions and methods of making them. Canadian Patent 873, 815, 1971.
4. Chang, T. M. S. (1972) *Artificial Cells.* Monograph. Charles C. Thomas Publisher, Springfield, Ill. Online at www.artcell.mcgill.ca.
5. Chang, T. M. S. (1995) Artificial cells with emphasis on bioencapsulation in biotechnology. *Biotech. Ann. Rev.* **2,** 267–295.
6. Chang, T. M. S. (2003) Artificial cells for replacement of metabolic organ functions. *Artif. Cell Blood Sub Biotech.* **31,** 151–162.
7. Chang, T. M. S. (1997) *Blood Substitutes: Principles, Methods, Products and Clinical Trials.* Karger/Landes Co. Austin, Texas. Online at www.artcell.mcgill.ca.
8. Chang, T. M. S. (2002) Oxygen carriers. *Curr. Opin. Investig. Drugs* **3,** 1187–1190.
9. Chang, T. M. S. (2003) New generations of red blood cell substitutes. *J. Int. Med. Res.* **253,** 527–535.
10. D'Agnillo, F. and Chang, T. M. S. (1998) Polyhemoglobin-superoxide dismutase-catalase as a blood substitute with antioxidant properties. *Nat. Biotech.* **16,** 667–671.
11. Powanda, D. and Chang, T. M. S. (2002) Cross-linked polyhemoglobin-superoxide dismutase-catalase supplies oxygen without causing blood brain barrier disruption or brain edema in a rat model of transient global brain ischemia-reperfusion. *Artif. Cells Blood Sub. & Immob. Biotech.* **30,** 25–42.

12. Chang, T. M. S. and Yu, W. P. (1996) Biodegradable polymer membrane containing hemoglobin for blood substitutes. US Patent 5,670,173.

13. Yu, W. P. and Chang, T. M. S. (1996) Submicron polymer membrane hemoglobin nanocapsules as potential blood substitutes: preparation and characterization. *Artif. Cells Blood Sub. & Immob. Biotech.* **24,** 169–184.

14. Chang, T. M. S., Powanda, D., and Yu, W. P. (2003) Analysis of polyethylene-glycol-polylactide nano-dimension artificial red blood cells in maintaining systemic hemoglobin levels and prevention of methemoglobin formation. *Artif. Cell Blood Sub Biotech.* **31,** 231–248.

15. Yu, W. P. and Chang, T. M. S. (2003) Nano-dimension red blood cell substitutes based on a novel ultrathin polyethylene-glycol-polylactide copolymer membrane nanocapsules containing hemoglobin and enzymes. Submitted.

16. Chang, T. M. S. and Poznansky, M. J. (1968) Semipermeable microcapsules containing catalase for enzyme replacement in acatalasemic mice. *Nature* **218,** 242–245.

17. Chang, T. M. S. (1971) The in vivo effects of semipermeable microcapsules containing L-asparaginase on 6C3HED lymphosarcoma. *Nature* **229,** 117–118.

18. Chang, T. M. S. (1989) Preparation and characterization of xanthine oxidase immobilized by microencapsulation in artificial cells for the removal of hypoxanthine. *J. Biomat. Artif. Cells Artif. Organs* **17,** 611–616.

19. Palmour, R. M., Goodyer, P., Reade, T., and Chang, T. M. S. (1989) Microencapsulated xanthine oxidase as experimental therapy in Lesch-Nyhan Disease. *Lancet* **8664,** 687–688.

20. Chang, T. M. S., Bourget, L., and Lister, C. (1992) Oral administration of microcapsules for removal of amino acids. US Patent 5,147,641.

21. Chang, T. M. S., Bourget, L., and Lister, C. (1995) A new theory of enterorecirculation of amino acids and its use for depleting unwanted amino acids using oral enzyme-artificial cells, as in removing phenylalanine in phenylketonuria. *Artif. Cell Blood Sub. & Immob. Biotech.* **25,** 1–23.

22. Bourget, L. and Chang, T. M. S. (1986) Phenylalanine ammonia-lyase immobilized in microcapsules for the depletion of phenylalanine in plasma in phenylketonuric rat model. *Biochim. Biophys. Acta* **883,** 432–438.

23. Safos, S. and Chang, T. M. S. (1995) Enzyme replacement therapy in ENU2 phenylketonuric mice using oral microencapsulated phenylalanine ammonia-lyase: a preliminary report. *Artif. Cells Blood Sub. & Immob. Biotech.* **25,** 681–692.

24. Yu, B. L. and Chang, T. M. S. (2002) In-vitro kinetics of encapsulated tyrosinase. *Artif. Cells Blood Sub & Immob Biotech* **30,** 533–546.

25. Chang, T. M. S. (1985) Artificial cells with regenerating multienzyme systems. *Methods in Enzymology* **112,** 195–203.

26. Gu, K. F. and Chang, T. M. S. (1990) Production of essential L-branched-chained amino acids, in bioreactors containing artificial cells immobilized multienzyme systems and dextran-NAD+. *Appl. Biochem. Biotech.* **26,** 263–269.

27. Chang, T. M. S. (1965) *Semipermeable Aqueous Microcapsules.* Ph.D. thesis, McGill University, Montreal, Canada.

28. Lesney, M. S. (2001) Going Cellular. *Modern Drug Discovery.* ACS Publication **4**, 45–46.

29. Orive, G., Hernandex, R. M., Gascon, A. R., Calafiorer, R., Chang, T. M. S., et al. (2003) Cell encapsulation: promise and progress. *Nat. Medicine* **9**, 104–107.

30. Lim, F. and Sun, A. M. (1980) Microencapsulated islets as bioartificial endocrine pancreas. *Science* **210**, 908–909.

31. Wong, H. and Chang, T. M. S. (1986) Bioartificial liver: implanted artificial cells microencapsulated living hepatocytes increases survival of liver failure rats. *Int. J. Artif. Organs* **9**, 335–336.

32. Wong, H. and Chang, T. M. S. (1988) The viability and regeneration of artificial cell microencapsulated rat hepatocyte xenograft transplants in mice. *J. Biomat. Artif. Cells Artif. Organs* **16**, 731–740.

33. Kashani, S. and Chang, T. M. S. (1991) Effects of hepatic stimulatory factor released from free or microencapsulated hepatocytes on galactosamine induced fulminant hepatic failure animal model. *J. Biomat. Artif. Cells Immob. Biotech.* **19**, 579–598.

34. Bruni, S. and Chang, T. M. S. (1989) Hepatocytes immobilized by microencapsulation in artificial cells: Effects on hyperbilirubinemia in Gunn rats. *J. Biomat. Artif. Cells Artif. Organs* **17**, 403–412.

35. Bruni, S. and Chang, T. M. S. (1991) Encapsulated hepatocytes for controlling hyper-bilirubinemia in Gunn rats. *Int. J. Artif. Organs* **14**, 239–241.

36. Bruni, S. and Chang, T. M. S. (1995) Kinetics of UDP-glucuronosyl-transferase in bilirubin conjugation by encapsulated hepatocytes for transplantation into Gunn rats. *J. Artif. Organs* **19**, 449–457.

37. Wong, H. and Chang, T. M. S. (1991) A novel two step procedure for immobilizing living cells in microcapsules for improving xenograft survival. *Biomater. Artif. Cells Immob. Biotech.* **19**, 687–698.

38. Chang, T. M. S. and Wong, H. (1992) A novel method for cell encapsulation in artificial cells. US Patent 5,084,350.

39. Chang, T. M. S. (2001) Bioencapsulated hepatocytes for experimental liver support. *J. Hepatology* **34**, 148–149.

40. Liu, Z. C. and Chang, T. M. S. (2000) Effects of bone marrow cells on hepatocytes: when co-cultured or co-encapsulated together *Artif. Cells Blood Sub. Immob. Biotech.* **28**, 365–374.

41. Liu, Z. C. and Chang, T. M. S. (2002) Transplantation of co-encapsulated hepatocytes and marrow stem cells into rats. *Artif. Cells Blood Sub. Immob. Biotech.* **30**, 99–112.

42. Liu, Z. C. and Chang, T. M. S. (2003) Co-encapsulation of stem sells and hepatocytes: in-vitro conversion of ammonia and in-vivo studies on the lowering of bilirubin in Gunn rats after transplantation. *Int. J. Artif. Organs* **26**, 491–497.

43. Prakash, S. and Chang, T. M. S. (1995) Preparation and in-vitro analysis of genetically engineered *E. coli* DH5 cells, microencapsulated in artificial cells for urea and ammonia removal. *Biotech. Bioeng.* **46**, 621–626.

44. Prakash, S. and Chang, T. M. S. (1996) Microencapsulated genetically engineered live *E. coli* DH5 cells administered orally to maintain normal plasma urea level in uremic rats. *Nat. Medicine* **2**, 883–887.

45. Friedman, E. A. (1999) Predicting nephrology in the 21st century. *ASAIO J.* **45,** 363–366.

46. Chow, K. M., Liu, Z. C., Prakash, S., and Chang, T. M. S. (2003) Metabolic induction of *Lactobacillus delbrueckii. Artif. Cells Blood Sub. Biotech.* **34,** 425–434.

47. Garofalo, F. and Chang, T. M. S. (1991) Effects of mass transfer and reaction kinetics on serum cholesterol depletion rates of free and immobilized *Pseudomonas pictorum. Appl. Biochem. Biotech.* **27,** 75–91.

48. Llyod-George, I. and Chang, T. M. S. (1995) Characterization of free and alginate-polylysine-alginate microencapsulated *Erwinia herbicola* for the conversion of ammonia, pyruvate and phenol into L-tyrosine and L-DOPA. *J. Bioeng. Biotech.* **48,** 706–714.

49. Koo, J. and Chang, T. M. S. (1993) Secretion of erythropoietin from microencapsulated rat kidney cells: preliminary results. *Int. J. Artif. Organs* **16,** 557–560.

50. Yu, Y. T. and Chang, T. M. S. (1981) Lipid-polymer membrane artificial cells containing multienzyme systems, cofactors and substrates for the removal of ammonia and urea. *Trans. Am. Soc. Artif. Intern. Organs* **27,** 535–538.

51. Chang, T. M. S. (1971) Stabilization of enzyme by microencapsulation with a concentrated solution o or by crosslinking with glutaraldehyde. *Biochem. Biophys. Res. Comm.* **44,** 1531–1533.

52. Coromili, V. and Chang, T. M. S. (1993) Polydisperse dextran as a diffusing test solute to study the membrane permeability of alginate polylysine microcapsules. *J. Biomat. Artif. Cells Immob. Biotech.* **21,** 323–335.

53. Chang, T. M. S. and Prakash, S. (1998) Therapeutic uses of microencapsulated genetically engineered cells. *Mol. Med. Today* **4,** 221–227.

Index